Anatomy and Physiology of Diseased Plants

Authors

Dragoljub D. Šutić, Ph.D.
Professor
Faculty of Agriculture
University of Belgrade
Belgrade, Yugoslavia

James B. Sinclair, Ph.D.
Professor
Department of Plant Pathology
University of Illinois at Urbana-Champaign
Urbana, Illinois

CRC Press
Boca Raton Ann Arbor Boston

Library of Congress Cataloging-in-Publication Data

Anatomy and physiology of diseased plants/authors, Šutić, Dragoljub D. and Sinclair, James B.
p. cm.
Includes bibliographical references.
ISBN 0-8493-4806-4
1. Anatomy. 2. Physiology. 3. Plants—disease.
I. Šutić, Dragoljub D. II. Sinclair, James B.
[DNLM: 1. Plants. RS 619 P125]
SB381.M68 1990
635'.11--dc21 90-33272

Direct all inquiries to CRC Press, Inc., 2000 Corporate Blvd., N.W., Boca Raton, Florida, 33431.

International Standard Book Number 0-8493-4806-4

Library of Congress Card Number 90-33272
Printed in the United States

INTRODUCTION

Pathological changes in diseased plants are the result of changes in the structure, organization, and functions of the cells and tissues affected. Such changes in structure and organization are the subject of the science concerned with the anatomy of diseased plants. Pathological changes in the functions of cells and tissues are the study of the physiology of diseased plants.

The science of the anatomy of healthy plants includes the use of biochemical, biophysical, serological, scanning and electron microscopy, and other methods. New information concerning the healthy cell, such as that of the cell organelles and constituent ultrastructure, cell development, intracellular communication, genetic message coding, etc., are continually being added to our knowledge base. The application of these innovative methods and the new information derived from their use has enabled scientists to survey cell differentiation and organization during plant ontogenetic development.

The anatomy of diseased plants is a separate scientific discipline within plant pathology based on the knowledge initially obtained from research on the anatomy of healthy plants. Comparative studies on healthy and diseased plants has provided data showing how causal agents of disease and damage to plants alter plant structure and organization in the course of their activities associated with plants.

Changes in the internal structure and organization of diseased plants are a result of numerous and various disorders in cytogenesis, histogenesis, and organogenesis. These changes are mutually linked and therefore difficult to separate for individual study. However, they can be studied individually when viewed within the framework of plant anatomy. In the study of the relationship between the causal agent and the plant, plant anatomists have discovered new information on the function of healthy plant parts. Also, plant pathologists have discovered how causal agents bring about reduced yield losses in infected and damaged plants. Such knowledge provides basic information for the development of timely and successful control measures.

The anatomy of diseased plants is based on knowledge derived from diseased cells, tissues and organs and has developed into a scientific discipline along with cytopathology, histopathology, and organopathology.

PREFACE

This book represents a translation of the original volume by Prof. Dr. Dragoljub Šutić titled: "Anatomija i Fiziologija Bolesnih Biljaka" and published by Nolit in Belgrade, Yugoslavia. Prof. Šutić has lectured for over 20 years on this subject in the Department of Plant Protection of the Faculty of Agriculture University of Belgrade. He decided it was necessary to write this book based in part on his lectures, since no other book had been published under such a title in Yugoslavia or in any other country, which he was aware.

The primary aim of the book was to present information and interpretation of it by approaching the diseased plant as a "patient" taking into consideration all changes that occur in the structure and function of a diseased plant which were of biological, diagnostic and therapeutic importance. Thus, creating a scientific discipline substantially different from general phytopathology. It is a discipline which deals with plant disease development concentrating on the study of individual plant diseases, using information from botany and plant pathology as well as the anatomy and physiology of diseased plants. This combination of information from various disciplines could then be used as a basis for the development of therapeutic and applied studies in phytomedicine.

The approach taken in the preparation of the book was to present the mutual linkages among all pathological changes particularly those in the affected cell and tissue, which, in turn, affected the structure and function of the plant. This concept was used to organize the information in the major categories of cytopathological and histopathological changes. Organopathological or morphological changes were considered only as an external expression of internal and anatomical changes in the diseased plant.

It was hoped that this book would serve as a reference guide for students of agriculture, botany, forestry, plant pathology, plant physiology, and related disciplines. The book was written to develop concepts and theories using published facts. The survey of the literature was not meant to be exhaustive, but in part selective using data and information to illustrate certain principles.The book should be useful to students and teachers studying principles of plant pathology and as a guide to the literature. It is to serve a wide variety of plant scientists and to stimulate further research in plant pathology with the ultimate goal of developing new and innovated methods for the control of plant diseases.

Prof. James B. Sinclair, the second author, contributed greatly to the preparation of the English version of this book, completing the text with new data and with results from his own research, as well as with original photographs. He also enriched the contents of the book with new information, interpretations and ideas of particular importance for the affirmation and further development of diseased plant anatomy and physiology as a separate discipline within the framework of plant pathology. Dr. Sinclair contributed particularly to the final editing of the text.

The authors with to thank the University of Belgrade (UB) and the University of Illinois at Urbana-Champaign (UIUC) for the use of library facilities in the preparation of the manuscript. Dr. Šutić expresses a special appreciation to Mrs. Vesna Vidas for assistance in the initial translation of the original text into English. Thanks are given to Mrs. Dr. Marija Šutić (UB) and Nancy J. David (UIUC) for typing the initial and subsequent drafts, respectively, of the manuscript.

THE AUTHORS

Dragoljub Šutić, Ph.D., is Professor of Pathohistology and Pathophysiology of Plants and Plant Virology, Faculty of Agriculture, at the University of Belgrade, Belgrade, Yugoslavia.

Dr. Šutić received his diploma in Agricultural Engineering in 1949, his diploma in Biology in 1954, and his Ph.D. in Plant Pathology in 1955 from the University of Belgrade. He was appointed in 1949 as a member of the Faculty of Agriculture, University of Belgrade; became a Professor of Plant Pathology in 1959, Head of the Department of Plant Pathology in 1970, and was a member of the Council of the Faculty of Agriculture from 1972 to 1973 and 1976 to 1978. He studied at the Virology Station Center of Vegetable Pathology, Versailles, France from 1957 to 1958.

He has been the President and Vice President, Plant Protection Association, Belgrade; President, Microbiological Society, Servia; President and Vice President, Microbiological Society, Yugoslavia; a member of the Council of the Mediterranean Phytopathological Union; and a member of the International Committee on the Taxonomy of Viruses.

He has authored three books: *Plant Viruses, Plant Virus Diseases,* and *Vegetable Diseases and Their Control;* three laboratory manuals: *Plant Virology, Phytobacteriology,* and *Plant Pathophysiology;* and published over 150 scientific papers.

He received the Honorary Plaque, Faculty of Agriculture, University of Belgrade in 1970; was elected a member of the French Agricultural Academy, Paris, in 1976; and received the Medal of the Republic with a Silver Wreath in 1979.

He continues to teach and do research in plant pathology in the Faculty of Agriculture, University of Belgrade.

James B. Sinclair, Ph.D., is Professor of Plant Pathology in the Department of Plant Pathology, College of Agriculture, University of Illinois at Urbana-Champaign (UIUC), Urbana, Illinois.

Professor Sinclair received his B.Sc. degree from Lawrence University, Appleton, Wisconsin in 1951 and his Ph.D. in Plant Pathology from the University of Wisconsin, Madison in 1955 under J. C. Walker with whom he continued to work with a postdoctoral appointment until 1956, when he accepted a position in the Department of Plant Pathology, Louisiana State University (LSU), Baton Rouge, Louisiana. At LSU he served as an Assistant Professor, Associate Professor, and then Professor until 1968. Also, he was an Administrative Assistant to the Chancellor from 1966 to 1968. He joined the Department of Plant Pathology, UIUC, in 1968 as a Professor of International Plant Pathology. He was campus, and all-university coordinator for the Illinois-Tehran Research Unit, 1974 to 1978.

Professor Sinclair has taught five graduate courses in Plant Pathology; has planned, participated in and given invitational lectures at numerous national and international conferences and workshops; has worked in over 40 countries professionally; and has directed the research of over 60 graduate students of whom 15 have completed a portion of their thesis research at an overseas institution. He is a member of many national and international professional organizations. Professor Sinclair formed and was Chairman of the Inter-American Seed Pathology Group and initiated the formation of seed pathology committees in the American Phytopathological Society and International Society of Plant Pathology. He served from 1979 to 1983 as the Chairman, Seed Pathology Committee, International Society of Plant Pathology.

Professor Sinclair's research has been primarily on seed- and soilborne pathogens of soybeans and other crops and their control, and on the uptake and translocation of systemic fungicides in various crop plants. He has done pioneer research in the pathology of soybean seeds and a computer-based program using artificial intelligence for the diagnosis of soybean diseases. He is currently doing research on the biological control of plant diseases and serodiagnosis of soybean diseases

He has published over 260 refereed research papers and 200 research abstracts; authored, edited, or co-edited 17 books and over 200 other articles. He has received a number of awards and honors. The most recent recognition for his accomplishments was the presentation of the ICI/American Soybean Association Research Recognition Award in 1983, the UIUC Paul A. Funk Award for outstanding faculty research in 1984, the U.S. Department of Agriculture Distinguished Service Award in 1988, and the American Soybean Association Production Research Award in 1989. He was the first plant pathologist to be elected as an honorary member of the Illinois Crop Improvement Association, Inc., Urbana in 1990.

TABLE OF CONTENTS

Chapter 1

Chapter 2

Chapter 4

Chapter 1

PLANT CYTOPATHOLOGY

I. INTRODUCTION

Plant cytopathology is the study of the organization, structure, and vital activities of diseased or damaged plant cells, while plant cytology is the study of healthy, intact cells. The two disciplines have a common denominator as the subject of research, the cell. The development of more sophisticated methods for research has stimulated a rapid development of plant cytology in general, and more specifically, cytopathology.

Plant cytopathology developed into an exacting science with the development of instruments such as electron microscopes, spectrophotometers, and ultracentrifuges. The development of methods and techniques using these and other instruments has allowed for precise studies on the internal structure and organization of plant cells. The development of analytical and biochemical methods has allowed for the study of biochemical processes within cells, including molecular interrelationships. Use of these methods has allowed for surveying the sequence of pathological changes that take place in infected or damaged plant cells at all stages of pathogenesis.

Many details about pathological changes in diseased or damaged plant cells, membranes, or organelles, and in the internal organization have been described. Biochemical, biophysical, and physiological cell research also has contributed to a better understanding of changes in the vital activities of diseased cells, such as those involved in photosynthesis, respiration, and protein and nucleic acid synthesis. The synthesis of some virus particles and their distribution within the cell protoplast also have been studied.

II. THE DISEASED PLANT CELL

Disease occurs only in living plant cells; therefore, the concept of disease is linked inseparably with living cells and tissues. Infection or damage usually begins with one cell or a group of cells, then develops with the spread of the causal agent(s) from the diseased or damaged cells to tissues, organs, and in some cases throughout the entire plant.

The diseased or damaged cell is characterized by pathological changes or disorders caused by the biotic or abiotic factor(s). Biotic causal agents are found in the following groups of plant pathogens: bacteria, mycoplasmalike organisms, fungi, parasitic higher plants, protozoa, viruses, viroids, and nematodes. Insects and other animals cause various types of damage. Abiotic factors include mechanical, physical and chemical injuries, nutritional deficiencies or excesses, and meteorological factors.

Cytopathological changes occur in the plant disease system: host cell - pathogen - environment (Figure 1). All parts of this system are in constant flux. Individual host cell structure and function(s) differ depending on the tissue(s) in which it is associated. Plant pathogens differ in their structure, nutrition, and pathogenic specialization. Changing environmental conditions can influence the host cell and pathogen simultaneously. Thus, in such a system. pathogenesis is not static, but changes from moment to moment and the cells and tissues involved are subject to constant alterations.

Cytopathological changes in diseased or damaged cells are linked and mutually conditioned since they occur in unique cell systems. It is often difficult to separate these changes or group them systematically. However, for convenience, they are classified, generally, into changes in the structure, organization, or functions of the cell(s).

Changes in cell structure, whether external or internal, can be defined as constitutional ones and pertain to the make up of the living and nonliving parts of the cell. Changes in cell

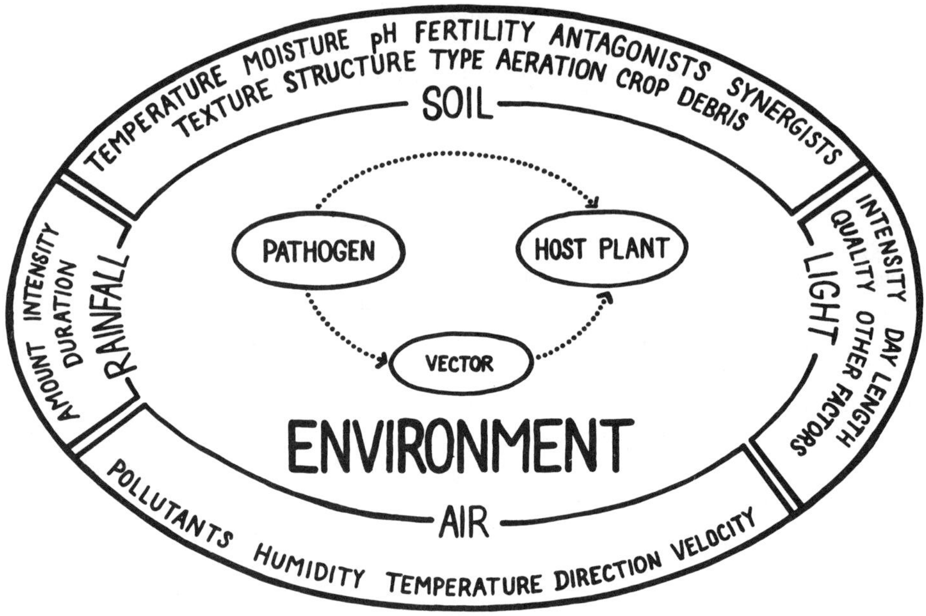

FIGURE 1. Environmental influences on plant diseases caused by infectious agents. (From Sinclair, J. B. and Backman, P. A., APS Press, Inc., St. Paul, MN, 1989. With permission)

organization can be defined as those that affect the relationship between cell components, and changes in cell function as those affecting the biochemical and physiologic processes.

All changes in cell structure are accompanied by changes related in cell function. For example, changes in chloroplast structure will hinder or impair photosynthesis activity within the cell; likewise, any changes in mitochondria structure will hinder or prevent respiratory activities. Damage to the cell membrane causes changes in permeability,which leads to uncontrolled movement of substances into and out of the cell, and eventually death of the cell. Weakened or hindered cell biosynthesis activity prevents the production of new substances indispensable for the recovery of infected or damaged cells. Changes in cell structure and their eventual death disrupt the linkages among cells, disturbing the permanent communication among cells, tissues, and the entire plant.

Pathological changes disturb and jeopardize the unity of the biological system within the cell. Some pathogens cause changes in nuclear division and in chromosomes threatening and, in some cases, even changing the genetics of the cell.

Disorders in certain parts of the cell lead to disruption of metabolic processes, the continuity of which ensures the economic transport of nutrients and energy within the cell. Changes in the biochemical system, nucleic acid, and protein synthesis in the host-pathogen relation, adversely affects the synthesis of substances indispensable for cell formation, development, and function.

Changes in the pathological cell give rise to changes in the structure and function of the cell group or tissues to which they belong. Pathological changes extend from the anatomical and morphological structures of infected and damaged cells to other parts of the plant.

Thus, all stages of pathogenesis take place in cells of tissues and organs, and plant cytopathology focuses on the diseased cell, and therefore, plays a key role in phytopathology.

III. INFECTION IN PLANT CYTOPATHOLOGY

The primary stages of cytogenesis in vascular plants are cell division, cell growth and differentiation, and their linkages into more or less specialized tissues and tissue systems.

Vascular plants are formed by continuous cell division beginning with that of a fertilized egg (embryogenesis) and proceeding throughout the life of the plant with division of meristematic cells. Somatic cells formed from the process of division grow and develop until maturity. During the course of the growth process, cells are formed to meet the needs of the tissue to which they belong. At maturity cells reach full cytohistological differentiation and specialization.

Infection of cells can take place at any time during cytogenesis, i.e. from division to mature cytohistological differentiation. Some example pathogens that attach meristematic tissues are presented.

Downy mildew fungi, such as *Peronospora tabacina* on tobacco, *P. arborescens* on poppy, and *Plasmopora balstedii* on sunflower cause systemic infections affecting the apical meristems of their respective hosts disrupting cell division. Systemically infected plants are dwarfed or stunted.

Fungi causing bunt, such as *Tilletia caries,* and loose smut, such as *Ustilago tritici* and *U. nuda,* on wheat and barley, respectively, partially infect the meristematic tissues of the host's floral parts. When apical meristems of the host initiate floral parts, hyphae of the pathogens penetrate the inflorescences and infect all floral parts. *Claviceps purpurea,* cause of ergot, prevents division of embryonic cells in the ovaries of certain grasses so that no seeds are formed in infected flowers.

By infecting cambium layers, some pathogens, such as *Nectria galligena* on apple branches, or *Plasmodiophora brassicae* on crucifer roots, intensify the activity of noninfected cambium cells, which give rise to hyperplastic tissue formations.

Certain plant viruses, such as potato virus Y and red clover mottle virus, spread systemically in susceptible plants into meristematic cells, which can result in necrosis of meristematic cells and thus, distorted plant growth.

Somatic cells, formed as a result of cell division, are susceptible to various plant pathogens throughout the growth and maturation period. Generally, susceptibility is greatest in the earlier rather than later stages of development. In the process of elongation and maturation of plant cells, changes in their cell wall composition take place altering the degree of susceptibility. In young herbaceous seedlings, the main constituents of middle lamella of cell walls are pectic substances[1] which allows the seedlings to be susceptible pectolytic enzymes produced soil fungi such as *Botrytis cinerea, Pythium debaryanum*, and *Rhizoctonia solani*, which cause seedling diseases of a wide variety of plants. Lignin becomes the main component of the middle lamella as cells mature, accounting for 20-30% of primary and secondary constituents.[2] With the lignification of cell walls, plants become more resistant to pectolytic enzyme activity and more resistant to seedling diseases.

As another example, *Alternaria alternata* can penetrate the underdeveloped cuticle on epidermal cells of young tomato fruits and cause 100% infection.[3] However, when the cuticle becomes mature, hyphae of the pathogen cannot penetrate, which must preclude fruit infection.

Most phytopathogenic bacteria develop in the intercellular spaces of parenchyma cells and by secreting pectolytic enzymes, degrade pectic compounds in the middle lamella. Changes in pectin quantity directly influence the rapidity and intensity of bacterial development and alters cell and tissue susceptibility to bacterial pathogens.

Phytopathogenic viruses also may cause cell destruction in the course of cytogenesis. However, virus particles generally multiply within the cytoplasm of younger cells, and since multiplication is related to the biosynthesis activity of the host cell, cell wall destruction may not be directly related to virus activity.

In the early stages of cell histogenesis, disorders occur in cell differentiation of infected cells. Such changes may be caused by a variety of pathogens, although they are primarily characteristic of biotrophic pathogens living relatively long periods within host cells. Examples of changes in cell lignification were reported in plants systemically infected with the downy mildew and rust fungi.[4] Also, in pear fruits infected with pear stony pit disease, a greater number of lignified parenchyma (stone) cells occurred hindering normal fruit development. In contrast, apple trees infected with apple rubbery wood mycoplasma, which prevents normal lignification, have branches that are flexible and twisted downward.

Certain pathogens may cause the loss of traditional differentiation traits in mature cells and take on nontraditional ones, For example, somatic cells (most often parenchyma) may be transformed into meristematic cells (meristematization), vascular cells (vascularization), or stones cells (sclerenchymization.)

Thus, cells may become infected at any time during cytogenesis, and interaction between the pathogen(s) and host cell(s), determines the type of changes that take place in these infected and damaged cells.

IV. TYPES OF CYTOPATHOLOGICAL CHANGES

Changes in structure, organization, and function of diseased and damaged cells are the result of numerous linked, individual changes which are impossible to consider in detail. Each change may arise from one or several processes with specific and/or interdependent mechanisms. To understand these processes and their mechanisms, it is necessary to understand the life processes in diseased and damaged cells, tissues, and plants. This is the continuing goal of plant pathologists, but at the same time, one which, in spite of extensive research, remains impossible to achieve fully at this time.

In order to more easily understand them, cytopathological changes can be classified into a number of basic types based primarily on the processes and mechanisms which determine the characteristic traits of the changes. Classification of such changes leads to an over simplification of complex processes taking place in diseased and damaged cells, but it is necessary for didactic reasons.

A. CYTOPLASMATIC

Cytoplasmatic (plasmatic) activity generally relates to the cell division of plant cells. Cell division is a trait of primary meristematic cells. However, under certain conditions somatic cells also may be transformed into secondary meristematic cells, which can act as primary meristematic cells taking part in the structure and physiological activities of plants. Cytoplasmatic cell activity is the basic process on which the structure and organization of tissues, organs, and plants is based. It ensures both spatial and continual growth of plants and simultaneously influences growth rate.[5]

The process of cell division may be changed by various abiotic or biotic causal agents. Depending on the causal agent, cytoplasmatic changes may be classified as physiological, traumatic, or infectious.

Physiological factors include plant nutrition and environmental conditions necessary for cell division and other cellular activities which are normally subject to change in ontogenic development and may affect cell division. Under unfavorable physiological or environmental conditions, normal cell division is slowed and individual tissues are formed with a reduced number of cells, resulting in dwarfed plants.

Traumatic changes following mechanical injuries of plant parts may cause an intensification of cell division (meristematic or somatic) in the cells surrounding the damaged area, ensuring partial restitution of damaged parts and protection against other unfavorable external factors.

Cytoplasmatic changes caused by pathogens develop in different ways in the course of pathogenesis. Infections may cause changes in primary meristematic cell activity and inhibit or stimulate formation of secondary meristematic and somatic (parenchyma) cells. Consequently, these disorders may become apparent in one of two ways: hypoplastic (reduced) or hyperplastic (increased) tissues.

1. Hypoplasia

In the process of hypoplasia, a lesser number of cells is formed than in nonaffected tissues (Figure 2). Reduced cell division results from infected meristematic cells or indirectly from metabolic byproducts from other cell or tissue infections.

In infected meristematic cells, metabolic processes are disturbed or inhibited. Similar effects may be caused by changes in the composition and activity of hormones that affect cell division. Infection of meristematic cells, particularly those of apical meristems, may cause necrosis which results in cessation of activity.

Hypoplasmatic changes may be caused indirectly by pathogens infecting roots and vascular bundles. Root pathogens such as *Armellaria Mellea, Agrobacterium tumefaciens* and beet necrotic yellow vein virus, disturb or completely inhibit the uptake of nutrients from the soil, while vascular pathogens, such as *Verticillium* spp. and *Pseudomonas solanacearum* disrupt their transport through the plant, leading to disruptions in cell division and other cellular activities.

One of the consequences of hypoplasmatic changes is an insufficient number of cells to form the structure of individual tissues. Such tissue development lags more or less in growth, and as a rule have an altered appearance. Processes of differentiation of parenchyma and conductive or vascular tissues, forming the basic structure of plant organisms, are particularly susceptible to such changes. Slower growth rates of some plant parts are more pronounced when hypoplasia occurs simultaneously with hypotrophic processes.

Hypoplasmatic changes are most frequent in partially or completely systemically infected plants. Causal agents of such infections usually are biotrophic pathogenic fungi, viruses, or mycoplasmas.

Among biotrophic fungi that caused hypoplasia are: *Peronospora arborescens, P. schachtii, Plasmopora halstedii, Taphrina pruni,* and *Ustilago tritici.*

P. arborescens infects apical meristematic cells of poppy with the result that all organs developing from this infected meristem are colonized by the mycelia of the pathogen. Such infected organs are smaller than normal, the stems are thicker and shorter, leaves are smaller and deformed, and flowers either do not develop, or if they do develop, do not open.

The apical meristem of sugar beet infected with *P. schachtii* gives rise to shortened stems with dense leaves in the form of rosettes and reduced root development.

Sunflower seedlings infected at germination with *P.halstedii* become completely colonized by mycelia resulting in infected plants that are four to five times smaller than nonifected ones. Inflorescences do not form.

As a result of infection of plum flower buds by *T. pruni*, infected fruits lack seeds or have abnormal seeds. Wheat plants originating from seeds infected with *U. tritici* are infected systemically, which results in stunted plants producing spikes with no grains.

Most viruses are systemic in plants and result in hypoplasmatic changes. Maize dwarf mosaic virus, for example, was reported to reduce plant growth in some cultivars by 4.5 to 19% and grain yield by 4 to 21.5%.[6] Maize plants infected with this virus in Romania were 10 to 17% shorter and their cobs 8 to 15% lighter.[7]

Wheat streak mosaic virus causes dwarfing ranging from 16 to 25% and lower yields by 53 to 59% depending upon wheat cultivar.[8] Sugar beet yellow vein virus causes root stunting and reduces root yields by 35 to 40%.[9] Hypoplasmatic changes in tomato plants caused by the tomato aspermy virus result in fruits that are smaller than normal and seedless.

Prune dwarf virus inhibits cell division and somatic cell differentiation resulting in dwarfing of infected stem tissues. In the susceptible Italian plum, the virus causes leaf narrowing and dwarfing with shoots shorter than normal because of shortened internodes and flowers with stunted pistils. The fanleaf virus causes deformation of grape leaves on which irregular lobes or no lobes are formed.

The stolbur mycoplasma disrupts growth and flower formation in pepper plants resulting in fruits which are smaller than normal and practically unusable. In apple stems, infected with the apple flat limb mycoplasma, normal cell division and somatic cell and tissue formation are inhibited resulting in flatness and asymetrical stem twisting giving the disease its name.[10]

2. Hyperplasia

Hyperplasmatic growth is characterized by excessive meristematic division or proliferation of cells in infected plant tissues resulting in the formation of conspicuous overgrowths. Both primary and secondary (induced) meristematic cells can be involved. In the case of induced division, somatic cells increase in size (hypertrophy) and volume, then divide similar to meristematic cells. Hyperplasia is stimulated by increased hormonal activity.

Usually, the cells formed during hyperplasia develop into somatic cells, which are similar to the surrounding somatic cells. Hyperplasmatic somatic cells form abnormal secondary parenchyma tissue in a process referred to as pathological parenchymization.

Excessive cell division outside the control of the host gives rise to pathological formations called "neoplasmatic formations", which are more or less localized near the point of infection. Neoplastic tissue do not take part in the function of the tissue(s) from which they have originated and, in a sense, represent independent formations.

Characteristically, neoplasmatic formations appear on stems, branches, or on underground organs, such as roots and tubers, and seldom on leaves. Neoplasias may be simple or complex. All neoplastic formations, whether simple or complex, formed by cell division in plant tissues have been called histoid galls.[10] A simple neoplasium is made up of a cluster of undifferentiated cells of varying size and amorphous in appearance. Examples of simple neoplasias are cankers, enations, nodules, overgrowths, and tumors.

A variety of plant pathogens induce simple neoplasmatic formations. *Synchytrium endobioticum,* cause of potato wart, is a typical representative from among the fungi. During its life cycle, the fungus forms sporangia in the epidermal cells of potato tubers. In the course of pathogenesis infected cells hypertrophy, which is followed by hyperplasia of contiguous cells, induced hyperplasia. These activities result in the formation of the typical amorphous tumorlike growths.

Nectria galligena and *N. ditissima* cause hyperplasmatic changes in apple and pear stems and branches. As wound parasites, these fungi initially colonize bark tissues causing cell necrosis and decay resulting in depressions of various sizes around the infection points. Adjacent cambium cells are stimulated to divide rapidly (hyperplasia) which results in irregular folds of tissue surrounding the depression followed by the formation of a canker. This type of canker, formed by necrotic and hyperplasmatic cell growth, is referred to as a necroplasmatic type.

Ustilago maydis, cause of maize smut, can infect all parts of maize plants at any time during growth. Young infected tissue cells divide intensely around points of infection resulting in hyperplastic growths of varying size and shape (Figure 3). The mycelium of the fungus colonizes the neoplastic cells and using portions of their constituents, transforms then into a multitude of chlamydospores.[11] The macrocyclic rust fungus, *Gymnosporangium sabinae*, in its aecial stage, causes pear leaf rust, induces swelling around points of infection on its alternative host, *Juniperus sabina*, which bears the teleutospore stage.[3]

Bacterial and viral pathogens cause hyperplasmic changes in infected plants. Bacterial cells of *Pseudomonas syringae* pv. *savastanoi,* cause of olive knot, penetrate indirectly through bark injuries on stems and branches, and stimulate meristematic cell division which results in wartlike formations around infection points. The white clover tumor virus stimulates hyperplasmatic formations on clover stems and roots. Certain plant viruses stimulate formation of leaf enations on infected plants. Enations usually form on lower leaf surfaces as fringed or winged excrescences, usually consisting of epidermal and mesophyll cells permeated with vascular bundles. They appear most frequently between two pronounced lateral leaf veins, spreading in the direction of the veins. Smaller enations may appear between smaller veins of infected plants. Butler and Jones[12] described enations on tomato leaves as composed of two or three layers of hypertrophied epidermal cells covering hyperplastic mesophyll cells. Among some of the plant viruses that cause enations are pea enation mosaic virus (Figure 4), grapevine fanleaf virus (Figure 5), and cherry rasp leaf virus.

In contrast, complex neoplasias are characterized by differentiation of somatic cells with specific functions. For example, the fungus *Plasmodiophora brassicae,* cause of crucifer clubroot, induces tumors composed of hyperplastic and hypertrophic cells in which secondary conductive tissues occur. Also, tumor cells of crown gall, caused by the bacterium *Agrobacterium tumefaciens,* which contain a genome of the bacterium in the host cell, divide independently of host cell control.[13] Conductive tissue also is induced in the tumor cell which may establish links with that of the plant conductive tissue.[14]

The most complex neoplastic growths involve the enlargement of whole plant organs, which have been called organoid galls.[12] This type of gall usually develops in plant organs systemically infected in the initial stages of development. For example, cherry shoots systemically infected with the fungus, *Taphrina cerasi*, are formed from buds colonized by the fungus. During growth, rapid cell division takes place in the shoots resulting in a bushy appearance and plums caused by *T. insititiae.*

The bacterium, *Rhodococcus fascians,* cause of witches' broom in sweet peas, is characterized by numerous thickened and fleshy shoots 2 to 5 cm long at the nodes of cotyledons or first true leaf.[15]

Cucumber mosaic virus causes the bushy disease in pepper by stimulating prolific branching and leaf deformation on susceptible cultivars.

Certain plant mycoplasmas, through systemic infection, stimulate shoot development through hyperplasia and abnormal branching, namely the apple proliferation mycoplasma, stolbur bindweed (*Convolvulus arvensis*) mycoplasma (Figure 6), and white clover (*Trifolium repens*) proliferation mycoplasma.

B. CYTOTROPHIC CHANGES

Cytotrophic changes refer to change in cell size and volume. Generally, the changes are accompanied by a corresponding change in cell form. The size and volume of cells are determined and characteristic for a given plant species. Unfavorable environmental conditions frequently cause cytotrophic changes in plant cells. Cytotrophic changes may be caused by a variety of abiotic or biotic factors. Cytotrophic changes usually are associated with nutritional disorders, but the term is broader and should include all other factors, abiotic or biotic, with similar effects. Cytotrophic changes can be caused by a variety of mechanical, physiological, traumatic, or infectious agents. Most frequent among the abiotic factors are deficiencies and excesses in light, nutrients, temperature, and water.

Mechanical cytotrophic changes are caused by deficiencies or excesses of cell constituents, such as water, which influences cell turgor. When in excessive quantities, water can cause excessive cell elongation and an increase in cell volume; when deficient, cell volume and size can be reduced.

Physiological cytotrophic changes result from disorders in metabolism and nutritional imbalances, which may inhibit or stimulate cell growth and size. Stimulated cell enlargement usually is accompanied by hyperplasia.

Traumatic cytotrophic changes occur in cells associated with wounds or injuries in mechanically-damaged plant parts. These cells become hypertrophied and the tissues become hyperplastic forming a mass of irregular, undifferentiated cells.

Infectious cytotrophic changes are caused by a variety of plant pathogens including, among the fungi-*Plasmodiophora brassicae,* cause of crucifer clubroot; *Synchytrium endobioticum,*

PLATE I. FIGURE 2. Maize ear with galls filled with chlamydospores caused by *Ustilago maydis*. FIGURE 3. Tumors on an olive branch as a result of hyperplastic meristematic cell division caused by *Pseudomonas syringae* pv. *savastanoi*. FIGURE 4. Mosaic (left) and enations (right) on garden pea leaves caused by the pea enation mosaic virus. (From Miličić et al., *Acta Bot. Croat.*, 35, 25, 1976.) FIGURE 5. Enations on a deformed grape (cv. Muscat-Hamburg) (left) caused by the grape fanleaf virus compared with a uninfected leaf (right). FIGURE 6. Dwarfing and proliferation of branches of bindweed (*Convolvulus arvensis*) caused by the stolbur mycoplasma (right) compared with a uninfected plant (left).

cause of potato black wart; *Taphrina deformans,* cause of peach leaf curl; and *Verticillium albo-atrum,* cause of Verticillium wilt; among the bacteria-*Agrobacterium tumefaciens,* cause of crown gall; among the viruses-prune and wheat dwarf viruses.

Cytotrophic changes, regardless of the cause, may be irreversible or reversible. Irreversible cytotrophic changes usually are caused by pathogens or animals. Following infection or damage, cell constituents are changed or destroyed, cell cytoplasm and organelles decompose with degradation ending in cell death. Reversible cytotrophic changes do not destroy any portion of the cells and remain as long as the unfavorable factor persists. However, with the cessation of the factor, normal cell size and volume is resumed. Such changes are temporary or may be considered as slight modifications. The most frequent cause of reversible cytotrophic changes is abiotic factors.

All living cells of almost all plant tissues are susceptible to cytotrophic changes, however the susceptibility of individual tissues may differ. For example, parenchymal tissue cells are the least differentiated and considered more susceptible to cytotrophic changes than are epidermal tissue cells, which is also true for bark cells compared to pith cells.[5] Also, young cells are more susceptible to cytotrophic changes than mature cells. Susceptibility decreases in somatic cells of tissues involved with perennial dormancy.

Cytotrophic disorders are manifested in two ways; cell hypertrophy and hypotrophy-an increase or decrease in cell size and volume.

1. Cell Hypertrophy

Cell hypertrophy is characterized by an increase in cell size and volume and can be a natural phenomenon as well as a pathological one. For example, in the natural development and survival of plants, groups of hypertrophied parenchyma cells serve as water-bearing tissue. Also the protoplasmic membrane of hypertrophied parenchymatous cells spreads through the pores and into the lumen of conductive tissues to form tyloses, providing reserve nutrients to the system. However, several pathogens, such as *Endothia parasitica,* cause of chestnut blight, and the grapevine fanleaf virus (Figure 5), cause of grapevine degeneration, stimulate an over production of tyloses which interfere with transport in conductive tissues.

Certain cell hypertrophies occur in response to mechanical wounding. Individual cells surrounding such wounds become hypertrophic, followed by rapid cell division, resulting in the formation of a mass of undifferentiated cells called callus parenchyma tissue. Callus formation promotes wound healing by preventing desiccation of internal tissues and providing a barrier against colonization by microorganisms.[16]

Infected cell or plant part hypertrophy is caused primarily by various bacterial and fungal pathogens. In general, hypertrophic changes that occur in diseased plants result in neoplastic growth in the form of galls, warts, tumors and similar types of formations. The process of cell hypertrophy usually begins with the separation of some somatic cells, most frequently parenchmatous cells, from internal tissues, followed by the hypertrophy and commencement of independent development. Later the cells divide rapidly to form a neoplastic growth. Two intracellular fungal pathogens, and intercellular fungal pathogen and a bacterial pathogen can serve as examples.

Synchytrium endobioticum, cause of potato black wart, is primarily an intracellular fungal pathogen. The fungus causes epidermal cell hypertrophy in potato tubers and stolons. The fungus forms its prosorus and sorus stages with in the hypertrophied cells destroying the protoplast and then the cell. *Plasmodiophora brassicae,* cause of curcifer clubroot, invades parenchyma cells, intracellularly which become hyperplastic enlarging eight to ten times normal size, within which it forms plasmodia.[17] These enlarged cells are referred to as "giant cells". Contiguous noninfected parenchyma cells, induced by the pathogen, become hyperplastic, forming a mass of new, poorly differentiated cells. The neoplastic tissue is referred to as "clubs".

The intercellular fungal pathogen, *Taphrina deformans,* cause of peach leaf curl, develops between the palisade and spongy tissue layers of peach leaves causing cell hypertrophy (Figure 8).[3]

In roots of plants infected by the crown gall bacterium, *Agrobacterium tumefaciens,* hypertrophied cells frequently separated into irregular groups of hyperplastic cells.

2. Cell Hypotrophy

Cells which do not reach normal size in the course of their growth process are defined as hypotrophic, and the phenomenon as hypotrophy. If the influence of the unfavorable condition results in the reduction in size of a mature cell, then such cells are atrophic (plant cell atrophy) and become atrophied.

Abiotic cell hypotrophy may result from nutrient deficiencies or unfavorable environmental conditions. Biotic cell hypotrophy is wide spread in diseased plants and may be caused by a variety of plant pathogens.

Pathogenic fungi cause cell hypotrophy directly by restricting nutrient transfer to host cells or indirectly by preventing water and nutrient transfer and absorption in diseased plants. Pathogenic fungi which develop haustoria in host cells, absorb the various cell constituents, impoverish the protoplast and thus, prevent normal plant cell development without killing the cell, are usually obligate parasites. A large number of epiphytic and intercellular endophytic pathogenic fungi belongs to this group. Representatives of epiphytic fungal pathogens include the powder mildew fungi (Erysiphaceae), which produce haustoria in the protoplasts of epidermal cells. Endophytic pathogenic fungi causing similar effects include biotrophs and necrotrophs which cause leaf spots and blights, such as *Plasmopora viticola,* cause of grape downy mildew, *Puccinia* spp., cause of leaf rusts, *Mycosphaerella* spp., cause of leaf blights, *Septoria* spp., cause of leaf spots, and *Venturia* spp., cause of leaf spots; and *Colletotrichum* spp., cause of various anthracnoses.

Other characteristic pathogenic fungi are those that cause root, crown, basal stem and conductive tissue damage which prevents the uptake of water and nutrients and transport in diseased plants. This blockage deprives plant cells of essential water and materials resulting in cell hypotrophy.

Those root pathogens that partially or completely destroy the absorptive root tissues, thus preventing water and nutrient uptake, include many of the soilborne fungi, such as *Armillaria mellea* and *Rosellinia necatrix,* causal agents of fruit tree and grapevine root rots, and *Fusarium* spp., cause of brown root rot of wheat. Basal stem and root-rotting fungal pathogens prevent the movement of absorbed water and nutrients and include *Leptosphaeria* spp. and *Ophiobolus graminis,* cause of take-all of wheat; and *Botrytis* spp., *Fusarium* spp., *Phythium debaryanum,* and *Rhizoctonia solani,* causal agents of postemergence damping-off of seedlings.

Vascular fungal pathogens interrupt transport of water and nutrients in the vascular system and include *Fusarium oxysporum* and *Verticillium albo-atrum,* cause of wilts in a large number of crops and plants.

Cell hypotrophy also occurs in plants infected with plant bacterial pathogens that cause leaf spots and blights, such as *Pseudomonas syringae* pv. *tabaci,* cause of tobacco wildfire, *Xanthomonas campestris* pv. *vesicatoria,* cause of tomato leaf spot, and *Xanthomonas campestris* pv. *junglandis,* cause of walnut blight. Vascular invading pathogenic bacteria including *Corynebacterium michiganense* pv. *michiganense, Curtobacterium flaccumfaciens,* and *Pseudomonas solancearum* prevent water and nutrient movement in vascular tissues, which, similar to pathogenic fungi, prevent normal cell growth and development in infected plants.

All living plant cells are susceptible to hypotrophic and atrophic influences, the consequences of which are detrimental to the development of the whole plant, if such changes effect the conductive tissue cells in the course of their formation and differentiation. Hypotrophy changes due to plant pathogens are by their nature irreversible and, therefore, measures to counteract these effects can only be short-term.

C. CYTOMORPHOLOGICAL CHANGES

Changes in the form of cells during their formation are classified as morphological ones. The form of mature cells varies among uni- and multicellular plants as does structure and function and is a result of long-term specialization in the course of plant evolution. For example, in the spermatophytes, cells in the course of specialization have differentiated to such an extent as to form highly specialized tissues, which form specialized organs and whole-plant organisms with a variety of tissue systems.

Generally, all cells of a tissue maintain the form obtained at maturity. However, the form may change within certain limitations under the influence of external conditions. However, some cells seem to be exempt from this static state. Undivided lactescent cells, theoretically, have unlimited growth. Regardless of this principle of relative stability, cell form may be altered under the influence of abiotic and biotic factors. Such morphological changes usually involve isodiametric and prosenchymatous (elongated) cells, which represent most of the cell forms in the structure of multicellular plants. Any alteration of form is related to plasmatic and trophic changes; together which represent related and mutually-conditioned disorders in cell division and differentiation.

Changes in cell form can be classified as either quantitative or qualitative depending on the degree of change. Quantitative changes occur in the process of cell formation without the disapperance of its basic form. These changes result in the altered dimensions of cells, which either are unstable and may, in the course of further development, be reversed.

Smaller than normal cells may occur under two major conditions; first, when mother cells divide prior to reaching their nature dimensions; and second, after normal cell division, when daughter cells fail to enlarge to mature dimensions.

Quantitative changes in cell formation may occur under the influence of unfavorable physiological, mechanical, or pathological conditions causing plasmatic and/or trophic disorders.

Physiological factors, such as changes in plant nutrition, can lead to pronounced cell changes or malformations. Under the stress of nutrient deficiences, cell size may be reduced by one-tenth or one-fifth compared to normal cells, and thus a change in cell form. Such diminished cells represent true dwarfed forms.

Mechanical factors, which prevent normal growth ring formation in woody plants, also inhibit normal cell function in parenchyma, medullary ray, tracheid, and fiber tissues within the affected areas. Similar effects can be caused by wood-decaying fungi, for example, which destroy normal growth ring element formation. Quantitative morphological changes also can occur under the influence of other plant pathogens. For example, the fungal pathogen *Plasmodiophora brassicae* causes hypertrophy and hyperplasia in the parenchyma cells of crucifer roots resulting in a multitude of hypotrophic, undifferentiated cells.

Qualitative morphological changes are characterized by the transition of cells from their typical to an atypical form. Isodiametric cells for example, may become prosenchymatous and vice versa. As a general rule, formative changes of this type usually are irreversible for individual cells. Qualitative changes in form are similar to quantitative ones, resulting from hyperplasia and hypertrophy. Therefore, any factors, whether infectious, physiological, or traumatic, that can influence these changes, also may cause corresponding changes in cell form.

Under conditions of abundant moisture or of the distribution of moisture within a plant resulting in the release of surplus water from plant tissues, the excess water accumulates in the cells of other tissues resulting in their becoming hyperhydrous (water-bearing). Parenchyma cells which are most susceptible to such processes, enlarge and become transformed into prosenchymatous forms.

Hypertrophic cells form swellings in and around points of hydration. For example, under high moisture conditions, swellings appear around lenticels (lenticelosis) in potato tuber tissue. Such affected mesophyll cells elongate and by exerting pressure on surrounding tissues, prevent normal formation of palisade cells.

The healing of wounds caused by traumatic processes is accompanied by hyperplasia and hypertrophy, results in quantitative changes in cell form the points of mechanical injury, which may grow in the same direction, enlarging several times. These enlarged cells within the framework of other isodiametric parenchyma cells, have a flattened appearance. Some hypertrophic cells may divide rapidly and form a mass of parenchyma cells called "callus tissue". This type of tissue represents another example of qualitative morphological change.

Qualitative changes in cell form also have been studied in plant parts infected with various fungal pathogens, such as "giant cells" originating from isodiametric parenchymal cells in crucifer roots infected with *Plasmodiophora brassicae;* epidermal potato tuber cells in which *Synchytrium endobioticum* forms it is prosori and sori, resulting in an elongated apperance of the cells; sugar beet cells colonized by *Urophlyctis leproides* enlarge, ramify, and squeeze between normal cells;[5] and palisade cells in peach leaves infected with *Taphrina deformans* are compressed by the hyperplasmatic cells, and thus are unable to attain their normal, elongated form.

Cytomorphological changes also may result from fusion of contiguous cells. In the course of fusion, complete or partial fermentative degradation of separating walls may take place, giving rise to enlarged fused cells called "symplasts", which are changed in form. Cytoplasma and nuclei move freely between these fused cells. Such fusions occur in plants infected by such pathogens as the fungi of *Synchytrium* spp. and the fungus *Urophlyctis pulposa,* which causes perforative cell fusion in plants of *Chenopodium rubrum.*[5]

Finally, it should be stressed that changes of cell form represent a general reaction of all living cells, and particularly of the least differentiated parenchyma cells. Morphological changes also occur as nonspecific reactions when they result from either abiotic or biotic factors.

D. CYTOTRANSFORMATIVE CHANGES

Cytotransformative changes refer to the change of one kind of cell into another, accompanied by changes in form and structure as well as function(s). Transformation is a trait of both animal and plant cells. Alterations in cell function and structure represent an ability acquired by the organisms in the process of the evolution as part of the struggle for survival. There are a variety of transformative plant cell changes, and although all living cells are capable of them, changeability is most pronounced in parenchyma cells.

Cell transformations take place in the course of normal growth and development, particularly in perennial plants, and after plant trauma caused by unfavorable environmental conditions, by mechanical injuries, by animal damage, and those caused by plant pathogens.

Cells do not react differently because of causal agent, but in approximately identical ways by nonspecific reactions. All transformations may be classified into two groups based on their primary characteristics: reversions and metaplasia.

1. Reversion

Reversion is a transformation of one kind of cell back to the function(s) of the parent cells. This transformation causes a differentiation of already differentiated cells and, therefore, the process is defined as "recurrent differentiation" or "de-differentiation".

One of the most frequent forms of reversion is the transformation of somatic cells into secondary meristematic cells (meristematic reversion). This form of transformation also may be defined as parenchyma cell meristemization.[18] For example, this is the process by which meristematic cells form lateral secondary meristematic tissues known as "phellogen" or "cork cambium"

Phellogen formation in plants may be autonomous or the result of external abiotic or biotic factors. Autonomous phellogen formation occurs in perennial woody plants giving rise to peridermal protective tissues (peridermis). In stems, phellogen results from cell reversion in subepidermal and occasionally in epidermal tissues; but in the roots, it is most often formed from pericycle cells. External phellogen cells are transformed into peridermal cells by cell division forming a peridermal layer. Mature internal cells form a special tissue known as the phelloderm. The peridermal layer and phelloderm together with the phellogen, comprise the peridermis, which is a complex secondary protective tissue.

Reversion, based on meristemization, plays an important role in the defense and protection of plants against mechanical injuries, insect damage, or pathogen invasion, as well as in natural plant development. For example, phellogen activity results in the formation of callus tissue from a mass of hyperplasmatic cells at points of injury or invasion enabling wound healing and thus, protection of healthy from damaged plant parts. Similar processes also take place at grafting points where the formation of callus tissue contributes to the merger of rootstock and scion tissues.

Phellogen activity, beneath an infection point, is an important mechanism for defense, isolation, and/or rejection of infected plant parts. In the reversion processes, the phellogen is formed beneath the points of pathogen penetration forming an external peridermal layer surrounding the affected area and blocking further pathogen movement. The peridermal tissue becomes impervious, preventing nutrient and water movement, which causes rapid desiccation of infected cells. Such desiccation restricts the development of infected plant parts, which die and may drop out. This process enables plants to free themselves of pathogens in the diseased parts (Figures 7 and 9). Such defensive reactions based on specific cellular activity also has been defined as "histogenic reactions".[3] Such histogenic reactions were observed in a number of diseases caused by bacteria, fungi, and viruses (Figure 8).

The most numerous group of pathogens stimulating histogenic reactions is represented by those causing lesions, spots, or cankers on herbaceous plant parts, such as *Cercospora beticola* on beet, *Clasterosporium carpophilum* on peach and *Phyllosticta* spp. on soybeans (Figure 10). Necrotic areas with the spots become surrounded with peridermal tissues, then desiccate, contract, and drop out so that infected leaves have a shot-hole appearance.

Histogenic reactions occur in some potato cultivars whose tubers are infected with the fungi *Spongospora subterranea* and *Synchytrium endobioticum*. According to Josifović[11], a corky periderm layer is formed beneath the points of infection by these pathogens blocking them and isolating the diseased cells. Also, the same author showed that peridermal tissue was formed around cankers caused by the fungus *Nectria galligena* in different tree species, which prevented fungal development and canker formation in resistant plants.

Histogenic defense reactions also occur in roots of some infected plants. For example, such reactions have been described in tobacco plants infected with the fungus *Thielavia basicola* and in flax roots infected with the fungi *Fusarium lini*.[3]

Formation of secondary peridermal tissue also has been observed as a defensive mechanism in plants infected with certain plant viruses. For example, in potato tubers infected by potato mosaic and other viruses that cause local cell necrosis, peridermal tissue was formed in the vicinity of necrotic cells isolating the internal necrotic areas.[18] The formation of peridermal tissue around stone cells isolates these cells within pear fruits infected with the pear stony pit disease.

Meristematic reversion on which histogenic defense reactions are based is a type of cell transformation having biological significance in plant development and survival regardless of whether the reversion occurs naturally or because external factors.

2. Metaplasia

Metaplasia is a type of transformation in which cells of one type transform into another of

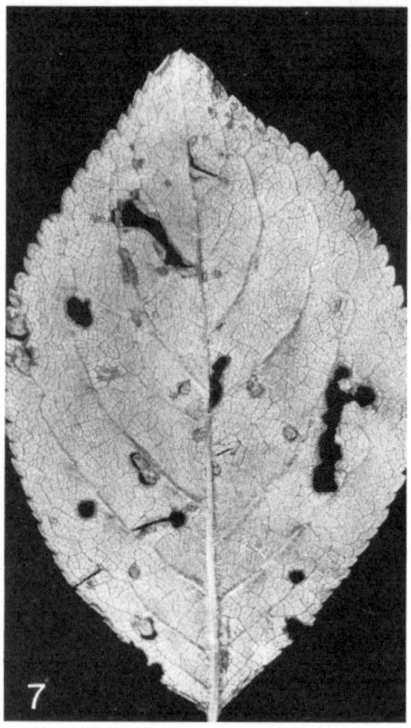

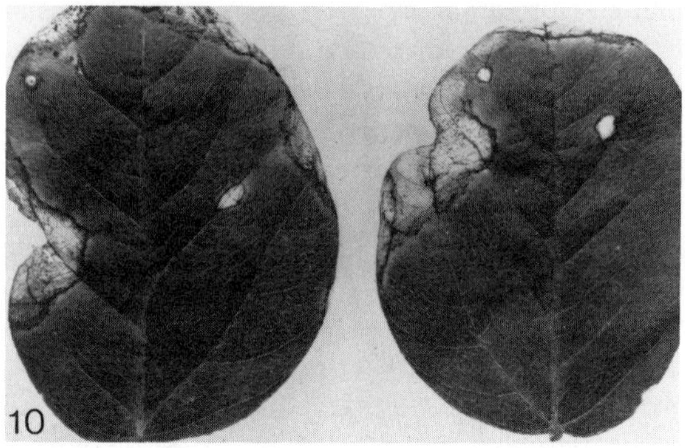

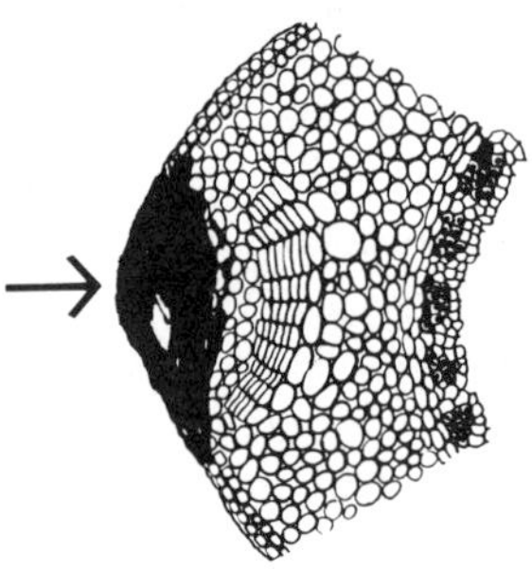

8

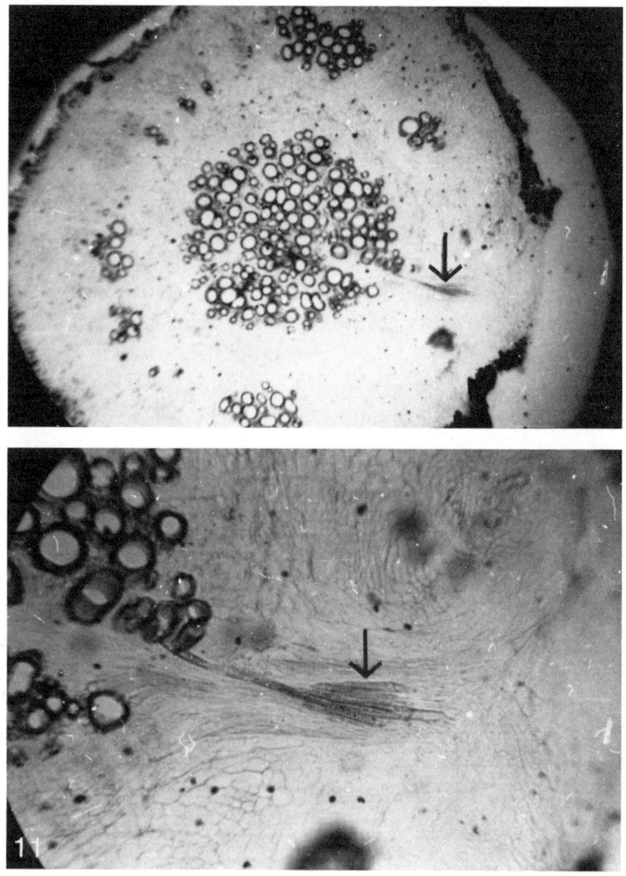

PLATE II. FIGURE 7. Necrotic spotting and shot-hole on a plum (cv. Crvena Ranka) caused by *Pseudomonas syrinae* pv. *morsprunorum*. FIGURE 8. Schmematic presentation of hypertrophied tissue with destroyed chloroplasts (arrow) in peach leaf tissues caused by *Taphrina deformans* (peach leaf curl). (From Wallace, E. and Whetzel, H.H., New York *Agr. Exp. Sta.*, Bul 276, 1910.) FIGURE 9. Necrotic spotting and shot-hole on a plum (cv. Italian) caused by the Prunus necrotic ringspot virus. FIGURE 10. Necrotic areas associated with margins of unifoliolate soybean leaves caused by *Phyllosticta sojicola*. (From Sinclair, J. B. and Backman, P. A., APS Press, Inc., St. Paul, MN, 1989. With permission.) FIGURE 11. Formation of secondary peridermal tissue (arrows) in cotton leaves beneath points of infection by *Xanthomonas campestris* pv. *malvacearum* (above and below). (From Simova, D., *Zăstita bilja,* 73, 287, 1953.)

a different form and function. Metaplasmic changes arise from cells of permanent living tissues which are not unchangeable, but may be susceptible to changes and transformation under certain conditions. Metaplasmic changes in such cells may involve both the nonliving (cell wall and inclusions) and living (protoplasm) components. All structural metaplasmic changes result in corresponding changes in cell activity. These changes are considered nonspecific since they may occur in the same or similar manner regardless of the causal agent involved. In a biological sense, they are reactions of adaption to environmental changes or defensive reactions to unfavorable external factors. Metaplasmic cell transformation, because of specific initiation processes, occur as sclerenchymization, vascularization, and/or virescence.

Sclerenchymization is a neoplasmic change in which parenchymal cells become lignified (Figure 11).[18] This transformation occurs with lignin accumulating in the primary and partially in the secondary parenchymal cell walls. In their mature or final state, the cells are known as stone cells, sclerenchyma cells, or scleroids.

Natural scleroid formation is important in the development of mechanical or support tissues and in the preservation of structure and function of plant parts, such as on pear and quince fruits, plum and sour cherry pits, and in fir pine tree bark.[16]

According to Küster,[5] transformation of parenchyma cells into stone cells is not a rare phenomenon in normal histogenesis and the process is part of parenchyma cell senescence (sclerosis). For example, the tissue of bark medullary rays may appear sclerotic due to the presence of stone cells. In addition to their natural occurrence, stone cells may develop in plant tissues in response to injuries caused by animals or to diseases. Stone cells have been found in insect galls and in pear fruits attacked by pear stony pit disease. This disease induces an excess of stone cells in pear fruits of susceptible cultivars and because of this transformation the disease was named pear stony pit. Infectious stony pit disease of a similar nature also has been described in quince fruits.[19]

Vascularization is a process in which parenchyma or similar cells are transformed into secondary vascular tissues, generally into tracheids similar to primary xylem conductive elements. Differentiated procambium cells of primary meristem or cambial ring cells give rise to tracheid elements. However, under certain conditions, tracheids are formed from somatic cells, such as parenchyma, medullary ray, or mesophyll leaf cells, which ar most prone to this process. In the transformation process, isodiametric parenchyma cells become elongated tracheid cells.

Induced vascular bundle element formation, under certain conditions, may be induced by external abiotic or biotic factors. Therefore, the process can be considered a natural reaction to establish normal functions within such tissues. In the vascularization of callus tissue formed as a result of mechanical injury, or grafting, or other such abiotic factor, for example, new vascular bundles are formed to promote functioning of the new tissue and to facilitate wound healing.

Similarly, secondary vascular bundle element formation occurs in hyperplasatic tissues induced by certain fungal plant pathogens such as *P. brassicae*. The fungus stimulates secondary vascular bundle formation in the clubs on crucifer roots. Such induced vascular bundle derived from hyperplasmatic cells also are characteristic of those induced by the crown gall bacterium *Agrobacterium tumefaciens*.

Virescence is the pathological greening of flower parts, which normally are not green. Virescence may affect all flower parts including petals, sepals, anthers, and ovules. The greening process occurs simultaneously with complex histological and morphological changes in the affected flower flower parts. For example, the transformation of corolla petals into green, leaflike organs represents an irreversibly qualitative alteration.

Virescence has been recorded as part of a disease syndrome caused by biotrophic pathogens. The pathogens involved are partially or wholly systemic in the plant and may explain, in part, the profound changes which occur in flower formation and development. Causal agents capable of inducing greening of flower parts include fungi belonging to the genus *Albugo* (Figure 12). For example, *A. candida,* which causes white rust of crucifers, induces the enlargement of infected parts in general and of inflorescence stems, peduncles, and flower parts in particular. Chloroplasts are formed in affected flower parts, resulting in their greening. Affected cells bearing chromoplasts are transformed into those bearing chloroplasts. Infected flowers are sterile, which is detrimental to commercial crucifer seed production. A similar disease develops in flowers of the case-weed, *Capsella bursa-pastoris,* which has been used for experimental studies.

Another group of pathogens that cause greening in plant parts is plant mycoplasmas. For example, tomatoes infected with stolbur tomato mycoplasma produce green flowers with elongated, fused sepals and a calyx which is several times larger than normal. The enlarged calyx encloses the inner flower parts, which develop abnormally. The anthers and pistil also become green and are malformed and smaller than normal and in cases of early infection, remain sterile. A similar floral virescence is caused by mycoplasmas affecting canola and strawberries.

Thus, virescence is a specific type of metaplasmic change involving complex cytological and histological changes in flowers and is a phenomenon of particular interest to plant biologists and pathologists.

E. CYTOCHROMATIC CHANGES

Many indispensible activities of animals and plants are associated with plant coloration. These fundamental functions of natural coloration are aided by specific plant structures which have been developed over the course of plant evolution. Any alteration in the natural or characteristic coloration of a plant or plant part is referred to generally as a cytochromatic change.

Plants present a great diversity of color and because of their significance in nature, coloration has been and remains a subject of interest in many applied and theoretical studies. Theoretical studies have been concerned with color biosynthesis, function of coloration in the cell, and in genetic studies. Changes in coloration have been used in plant disease detection and diagnosis and, in a sense, indirectly for plant and pest control. Color changes rank as the most common and conspicuous sign or symptom of disease in plants. These manifestations and their specific color changes are of particular interest to plant pathologists, who study and describe them.

All chromatic changes can be divided, according to color carriers and mechanisms of formation, into three types; permanent pigment change, formation of dark pigments, and formation of white or light-colored pigments.

1. Chromatic Changes of Permanent Pigments

Plant pigments are specialized complex organic compounds which are synthesized in plant cells and take part in plant activities as well as coloration. The number of plant pigments is large, approaching several hundred and are located either in specialized cell organelles called plastides or in the cell sap of vacuoles. Plastides, as carriers of plant pigments, have a particular adapted structure for pigment functions.

2. Chromatic Changes of Chlorophyll

Chlorophylls a and b are located in the chloroplasts of all green plants, which are found in cells of herbaceous rather than woody tissues. The chlorophylls absorb the blue-green and infrared wave lengths of the solar spectrum and are the only pigments that impart a green color in plants. The chlorophylls are the most important of all plant pigments, transforming light energy into chemical energy.

Chloroplasts are part of a specific tissue known as either chlorenchyma or photosynthetic parenchyma and are usually found in leaf or leaflike structures. Cells of such tissues usually differentiate into photosynthetic palisade and spongy tissues. However, chloroplasts are found in cells of other tissues such as epidermal, stem parenchyma, or those of unripen fruits. The entire system of chlorophyll-chloroplasts-photosynthetic cells and tissues, in general, represents the highest level of pigment organization and function in plant life.

The chlorophylls are the most abundant of all plant pigments, three to four and even ten times that of carotenoids, the second most numerous pigment group found in plants. The overall quantity of the chlorophylls in leaves is estimated at 0.13 to 0.35% of dry matter. Chlorophyll molecules are unstable and easily degradable and must be constantly resynthesized in chlorenchyma cells. The dynamics of these changes depend on many factors, although about 10% of the chlorophylls are regenerated daily under average conditions in most plant species; 6 of 7% of chlorophyll a and 2 to 3% of chlorophyll b.

Changes in the quantity of the chlorophylls, directly or indirectly, are the most common of plant color alterations. Direct changes in green color are the most wide spread. Indirect changes result in reducing the quantity of chlorophyll with an accompanying of revealing various and vivid carotenoid pigments.

Among the specific mechanisms resulting in a change in green color are: physical, physiological, genetic, or infectious.

Physical factors that cause a reduction in green color usually are extremes in environmental conditions, such as extremes in light, moisture, and temperature, which cause reductions in the quantity and synthesis of the chlorophylls. Extremes in temperature are most influencial on the synthesis of the chlorophylls, the optimum of which is between 26 and 30° C but varies between and within plant species. Temperatures below the optimum have a more profound affect than those above the optimum. The effect of low temperature on chlorophyll production can be observed in crops sowed in the spring. When temperatures are below normal, spring crops such as wheat and onions, will have chloratic leaves. Maize seedlings actually cease to synthesize chlorophyll below 6°C.[20] Plants of a variegated cultivar of *Brassica oleracea acephala* overwintered under cold conditions developed pale green or colorless leaves, which regained their normal green color at 12 to 15°C or above.[5]

The influence of low temperature on chlorophyll formation depends on the susceptibility of the plant to such temperatures. Winter hardy plants, for example, produce chlorophyll at lower temperatures than plants susceptible to low temperatures. Reduction of chlorophyll formation or its absence in plants at lower temperatures is a consequence of the slowing down or prevention of chlorophyll biosynthesis.

Light intensity also affects chlorophyll synthesis with the optimum occurring at 2500 lux in the red spectrum at 20°C.[21] Plant developing in optimum conditions but deficient in light or in the dark, in addition to other changes, become increasingly chloratic and etiolated. Normal chlorophyll formation is absent in etiolated plant cells and newly formed chloroplasts in such cells are smaller than normal. The normal growth and differentiation of etiolated cells in plants are disturbed resulting in abnormal histogenesis or organogenesis and thus, result in the development of deformed plants.

The abundance or deficiency of water has a detrimental effect on the formation of chlorophyll. Under moisture stress, plant cells loose turgor pressure and, depending on the intensity and duration of the deficiency, results in reduced chlorophyll regeneration and thus, chlorosis. Soil water deficiencies also cause plant dehydration resulting in decreased biosynthesis of the chlorophylls as well as reduced plant growth. During periods of drought, plants tend to loose their green color before they become permanently wilted.

Surplus water results in increased cell and tissue turgor pressure and the accumulation of moisture in the intercellular spaces. If such conditions persist hyperhydrous (water-bearing) tissues are formed and, in addition to a series of other changes, chloroplasts and other plastids disappear from these cells, leading to chlorosis of those affected plant parts, organs, or the entire plant.

Long-term waterlogging impairs or prevents the uptake of adequate quantities of oxygen by the root systems of terrestrial plants leading to, in the most serious situations, plant asphyxia.

Physiological changes also affect the biosynthesis of the chlorophylls. Chlorophyll synthesis requires a variety of chemical elements such as magnesium and nitrogen, which are part of the molecular structure, and copper, iron, manganese, and zinc, which act as catalysts in chlorophyll synthesis. To ensure normal chlorophyll formation in sufficient quantities, calcium, phosphorus and other elements must be available to the plant.[22] A deficiency of any of these elements prevents chlorophyll synthesis resulting in green color deterioration.

Nitrogen deficiency, for example, becomes evident when normally green leaves become pale or yellow and later, depending on plant species, red, yellow-orange or violet.

Magnesium deficiency also results in a decrease or cessation in chlorophyll synthesis with young leaves becoming chloratic and older ones orange or red. In addition to chlorosis, necrosis of tissues between leaf veins develops in pome and stone fruit crops.

Copper, iron, manganese, and zinc are not part of the chlorophyll molecular, but are indispensible for its synthesis by taking part in all oxidoreduction and other reactions in chlorophyll synthesis. Disturbed or hindered chlorophyll synthesis caused by a deficiency of one of these elements is evident by plant chlorosis. In addition, iron and nitrogen deficiency can bring about plant chlorosis because of their role in the structure of chloroplast protein.

Deficiencies of individual elements is difficult, if not impossible, to determine based solely on chlorosis or other chromatic changes, since the symptoms are similar for the deficiency of one or more than one element.[23]

Normal or induced changes in plant genotype may result in loss of chlorophyll coloration and the occurrence of mutants colored with other pigments. Neuffer et al.[24] described a number of such mutants in maize and classified them as green, colored, or white. Green mutants were characterized by various green leaf discolorations with interveinal chlorotic spots, green-striped with chlorotic stripes on leaves, zebra-striped mutants with transverse yellow stripes, and others. They described colored mutants as those with yellow leaves and classified them as yellow-striped leaves with yellow tissues between veins, as gold yellow mutants, oily yellow seedlings, and others. Their white mutants were characterized by the lack of any dominant pigment and thus appeared white and were classified as white-striped with long white stripes on the leaves; albino mutants with the greater part of the plant parts being white, and others.

Generally, bacterial and mycoplasmalike, fungal and viral pathogens cause a reduction in green color by inhibiting chlorophyll production, and in rare cases, stimulate greening of plant parts which normally are not green. This reduction in chlorophyll may result from inhibition of chlorophyll synthesis or chlorophyll degradation by the action by enzymes or toxins during the infection process. The quantitative reduction of chlorophyll may be caused by direct or indirect means. Direct reduction usually is associated with the activities of bacteria, mycoplasmas, and fungi; while indirect reduction is associated with viral activity. Reduction or inhibition of chlorophyll may be localized or systemic in infected plants.

Localized decoloration is characterized by the development chloratic lesions or spots of various shapes and sizes. The main characteristic of this type of decoloration is the rapid and complete disapperance of chlorophyll from infected cells and tissues. The most rapid decoloration often is associated with pathogens that produce toxins as part of pathogenesis. Because of this reduction in chlorophyll, the color of other plant pigments, such as anthocyanins, carotenoids, and falvonoids, appear within or around the affected areas. Symptoms of localized decoloration are associated primarily with bacterial and fungal pathogens; for example, bacterial pathogens that cause leaf spots such as *Pseudomonas syringae* pv. *syringae, P. syringae* pv. *tabaci* (Figure 13) and *P. syringae* pv. *tomato;* and fungal pathogens, such as *Phytophthora infestans, Plasmopara viticola, Polystigma rubrum, Phoma* spp., *Septoria* spp., *Uromyces phaseoli* (Figure 14) and *Venturia* spp.

Systemic chlorophyll decoloation usually involves a large portion or an entire plant and, in contrast to localized decoloration, the process develops slowly. In some cases, decoloration is so characteristic of infected plants that the disease is named after such symptoms. Systemic decoloration is characteristic of disease caused by mycoplasmas and viruses, but may be associated with diseases caused by certain bacteria and fungi.

Depending on the mode of colonization and spread within an infected plant, certain fungi can cause either partial or complete decoloration of that plant. Partial decoloration occurs in plant parts which, for example, have been infected with one of the following fungi; *Peronospora arborescens, P. schachtii* or *P. tabacina* (early infection of young plants).

Complete systemic decoloration usually is associated with soilborne pathogens which infect the root systems of plants, thus preventing the absorption of nutrients and water which are essential for chlorophyll synthesis. Members of this group include *Armillaria mellea* (root rot), *Fusarium* spp. (wheat yellowing), and *Rosellina necatrix* (general chlorosis and fruit tree leaf roll). Infection by *Pseudomonas syringae* pv. *syringae* is characterized by severe toxic schock which results in a general leaf necrosis and stem die-back in apricot and peach trees.

Complete or general decoloration also occurs in plants systemically infected by viruses. Such decoloration is characterized by a general chlorosis, or mosaic or mottled leaves. General chlorosis is caused, for example, by onion yellow dwarf virus on onion, potato leaf roll virus on potato, and beet yellows virus on sugarbeet.

Mosaic symptoms are the most common type of systemic plant decoloration and appear as alternating chloratic, light or dark green patches of various shapes and sizes on infected leaves. A large number of virus-caused diseases are named based on this symptom, such as alfalfa mosaic (Figure 15), cucumber mosaic, potato mosaic, sugarbeet mosaic, tobacco mosaic, and plum and peach mosaic (Figure 16).

Mosaic diseases also are manifested in some specifically diagnostic ways, such as grapevine asteroid mosaic, where chloratic plant parts have an asteroid shape; and potato streak mosaic, caused by potato virus Y, which induces necrosis along leaf veins. Plants infected with plum line pattern develop chlorosis in the tissues surrounding veins. Tobacco mosaic virus causes an oakleaf mosaic pattern in pepper around the veins at the base of infected leaves. Prunus necrotic ringspot virus causes green patches surrounded by chlorotic rings on peach leaves. Similar symptoms are caused by tobacco ringspot virus on tobacco leaves. Due to the parallel venation, mosaic symptoms usually spread between veins on monocoytledonous plants to form chlorotic streaks or elongated spots between the veins such as those caused by barley stripe mosaic virus, leek yellow stripe virus, and wheat streak mosaic virus (Figure 17).

It is often difficult to distinguish nottling as a symptom of systemic viral decoloration form various forms of mosaic. Mottling frequently appears as uneven chlorotic and pale zones on leaves resembling large irregular mosaic patterns. Mottling typically is caused by many viruses including the broad bean mottle, carnation mottle, cherry leaf mottle, and strawberry mottle viruses.

Mycoplasmas like other pathogens reduce chlorophyll quantities in infected plants and, like viruses, cause systemic disease accompanied by general plant chlorosis. As obligate parasites, mycoplasmas develop in sieve cells causing structural and functional changes, interfering with normal transport of synthates from leaves. Excessive accumulation fo these compounds in leaves interferes with various metabolic processes including chlorophyll synthesis resulting in general green decoloration. As the chlorophyll fades, leaves turn yellow followed by a decline in vitality and eventual death. Such decoloration is typical of mycoplasma infections that cause stolbur disease on eggplant, pepper, tobacco, and tomato (Figure 18). Stolbur disease is particularly detrimental on pepper because it causes wilting and sterility at the yellowing stage giving the disease its name of pepper yellow wilt.[25]

The reduction of chlorophyll represents a general trait of infected plants, the dynamics of which is not fully understood, and once begun is irreversible. The greatest decrease of chlorophyll can be expected when the processes of synthesis in infected cells or tissues ceases due to physiological changes caused by the pathogen.

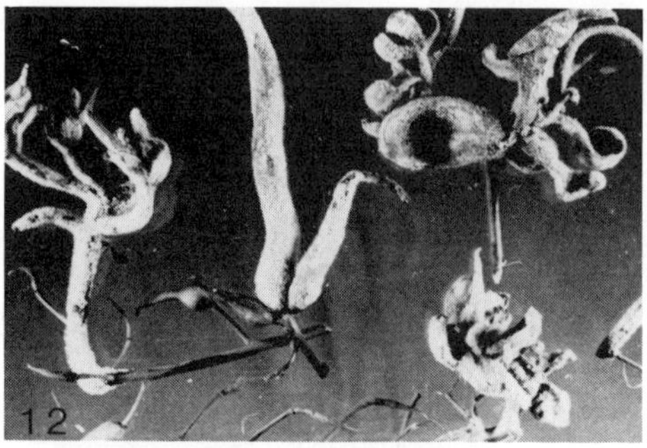

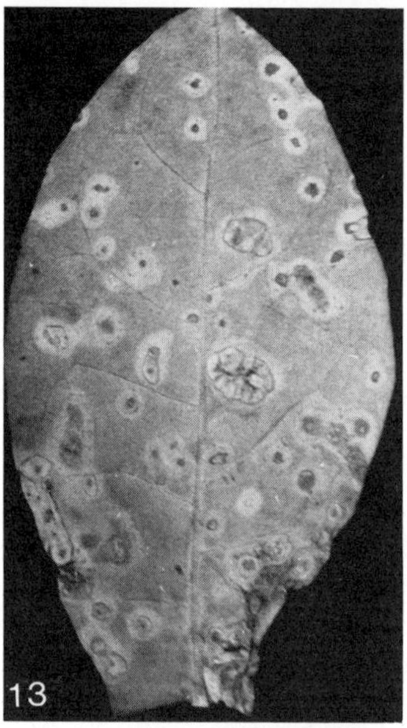

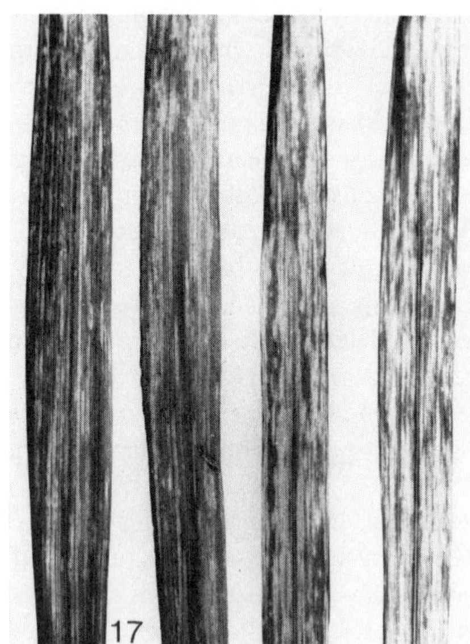

PLATE III FIGURE 12. Enlarged and deformed radish branches bearing virescent flower parts caused by *Albugo candida* (white rust). FIGURE 13. Chloratic rings around points of infection on a tobacco leaf caused by *Pseudomonas syringae* pv. *tabaci*. FIGURE 14. Small green "islands" in the chloratic zones surrounding pustules of teleutospores on a common bean leaflet caused by *Uromyces phaseoli*. FIGURE 15. Chloratic mottling and spots on alfalfa leaflets caused by the alfalfa mosaic virus. FIGURE16. Line mosaic pattern on a plum leaf caused by the plum line pattern virus. FIGURE 17. Chloratic streaks and stripes between veins of wheat leaves caused by the wheat streak mosaic virus.

Gassner and Goeze[26] reported a 13, 28, and 30% reduction in chlorophyll content in bean leaves infected with *Mycosphaerella pinoides* at 9, 13, and 16 days after infection, respectively, compared to control plants. Kokin and Tumarinson[27] and Tumarinson[28] showed that the greatest decrease of chlorophyll in oat leaves infected with *Puccinia coronifera* occurred at teleutospore formation. Reduction of chlorophyll in leaves of cotton plants infected with either *Verticillium albo-atrum* or *V. dahliae* depended upon plant nutrition; plants receiving fertilizer had a reduction of 14.7% while those that did not had a 81.7% reduction.[29] Dvoretskaya and Skurtul[30] found that *Cladosporium fulvum,* cause ot tomato leaf mold, reduced chlorophyll by 50% or more. It was reported that the reduction in chlorophyll in oak, *Quercus pedunculata,* leaves, caused by *Microsphaera alphitoides,* depended on the degree of infection and ranged from 10 to 48%, while in poplar tree leaves infected by *Melampsora tremulae* the reduction amounted to 36%.[31]

Phytopathogenic bacteria, like fungi, can influence chlorophyll content. Kuprevich[31] found that potato (cv. Rana Rosa) leaves infected by *Xanthomonas campestris* pv. *vesicatoria* had reduced chlorophyll content ranging from 32 to 64% depending on symptom development.

Reduction in chlorophyll content has been recorded in virus-infected plants. McKinney[32] investigating the influence of tobacco mosaic virus on chlorophyll content in tobacco plants found that the common strain reduced chlorophyll by 25.3% the yellow strain by 56.7%, and the mild dark green strain by 12.8%.

According to Tošić[8] wheat streak mosaic virus decreased chlorophyll content at the heading stage in the wheat cultivars San Pastore, Bancuta 1205, and Bezostaya 1 by 35, 37, and 25%, respectively, and brome mosaic virus in the same cultivars reduced it 36, 33, and 19%, respectively compared to noninoculated plants. Maize plants infected with the maize dwarf mosaic virus had a total reduction in chlorophyll ranging from 15 to 30% at 6 to 20 days after inoculation.[33]

Members of all groups of plant pathogens cause various amounts of chlorophyll reduction. However, certain pathogens may cause an increase in chlorophyll content resulting in either localized or generalized intensified greening of infected plant parts.

Localized intensive greening occurs in cells surrounding points of entry on the host is termed "green islands". The increase in chlorophyll content in these situations is localized and temporary usually without significant influence on the total chlorophyll content of the infected plant. Some of the pathogens that cause greeen islands include obligate parasites such as *Uromyces phaseoli,* cause of bean rust (Figure 14), and *Erysiphe graminis,* cause of powdery mildew of cereals; and the facultative parasite, *Rhytisma acerinum,* cause of sycamore leaf blotch.[3,15,17]

Little information is available on the development of green islands. Wang[34] showed that green islands caused by *Uromyces phaseoli* on bean plants were photosynthetically active with starch accumulation at a higher concentration than in surrounding tissues. Green islands caused by the wheat powdery mildew fungus developed after an initial degradation of chlorophyll.

Virescence or greening of plant parts occurs in response to certain pathogens. Plant parts which are usually not green, such as flowers and flower parts, are transformced into green organs which generally are sterile. Only a few pathogens are known to cause this type of symtoms and the pathogens involved usually cause systemic infection. For example, the fungus *Albugo candida* causes flower greening on plants of *Capsela bursa-pastoris.* General greening of plant parts often develops with infection by mycoplasmas, such as stolbur, the causal agent of flower greening on strawberry (green petal), raspberry (stunt), sunflower, tobacco, tomato, and other plants. šutić[35] reported flower greening possibly caused by mycoplasma infection on the following plants; *Cineraria hybrida, Gladiolus* sp., *Petunia hybirda, Phlox* sp., *Primula obconica, Tropaeolum majus,* and other plants.

No data are available on the nature of the greening process. However, since greening occurs from systemic pathogen infections, it is assumed that greening is caused by abnormal hormone activity induced by the pathogen in susceptible plants.

3. Chromatic Carotenoid Changes

Carotenoids are a group of orange, red, and yellow pigments located in chloroplasts and plastides. The most important among them are the *alpha-*, *beta-*, and *gamma*-carotenes (orange-red), lutein (yellow), and xanthophyll (yellow). The chloroplasts and plastids carrying these pigments are found primarily in leaves, but occur in other above- and below-ground organs.

The role of the carotenoids, which are widespread in plants, is many fold. It is assumed that, due to their fast oxidation that they protect chlorophyll from excessive light. They also have a determinative role in plant development and growth. For example, *beta*-carotene is an important source of vitamin A. As "auxillary" pigments, carotenoids absorb light in the blue and partially in the green range of the visible spectrum and transmit energy from absorbed sunlight to chlorophyll a.[21]

The amount of carotenoids in plants varies but averages between 0.1 and 0.2% of dry matter depending on many factors. Carotenoids are synthesized throughtout the entire growth period, but changes throughout the season with greater amounts produced during the period from first leaf apperance to full leaf development. These pigments become evident after chlorophyll destruction in the surface tissues of various maturing fruits, such as those of tomato where fruits which are at first green, then turn orange or yellow, and finally red.[36]

Carotenoid pigments may be degraded and regenerated depending on external factors. The dynamics of their synthesis, which is similar to that of chlorophyll, may be influenced by various genetic, physical, physiological, or disease factors.

Genetic factors have a significant effect on carotenoid production. For example, when the number of chromosomes in yellow maize plants was doubled, the amount of these pigments in grains increases by 40%.[20] Also, Neuffer et al.[24] described maize mutants with brown, orange, and red pigments in the fruit pericarp and ear axises, as well as mutants with lemon- or yellow-white endosperm.

Low temperatures inhibit and high temperatures increase carotenoid production. Carotenoid formation as well as other biochemical processes requires adequate plant nutrition.

Research has shown that certain pathogens may unfavorably influence carotenoid production. Sullivan and Chilton[37] showed that *Uromyces trifoliirepentis* caused a 20% reduction of carotenoids in infected leaves of *Trifolium repens*; and Kuprevich[31] proved that *Microshpaera alfitoids*, cause of oak powdery mildew, caused a reduction in xanthophyll ranging from 4 to 18% simultaneously with a reduction of chlorophyll by 13 to 21%. In contrast, *Puccinia suaveolens*, causal agent of rust on thistle (*Cirsium arvense*), stimulated an increase in xanthophyll content by 19%, and a carotenoid content of 52% with a simultaneous decrease in chlorophyll to 33% in infected leaves.

Results of research on the effect of plant virus infection on these plant pigments are contradictory. For example, Elmer[38] reported a reduction of chlorophyll and xanthophyll was accompanied by a doubling of carotene in the yellow- and dark green parts of tobacco leaf tissues infected by the tobacco mosaic virus. However, Guthrie[39] could not confirm these findings and concluded that chlorophyll was degraded more rapidly than the carotenoids. McKinney[32] showed that the common, mild dark green, and yellow stains of the tobacco mosaic virus significantly reduced the quantities of the most important plastid pigments below that of noninfected tissues. He reported that the common strain reduced carotene and xanthophyll by 27.2 and 25..9%, the mild dark green strain by 13.3 and 14.2%; and the yellow strain by 60.7 and 53.1%, respectively. Kuprevich[31] reported that potato leaf roll virus significantly reduced chlorophyll content in young potato leaves by 5% and in moderately developed leaves by 24 to 36%, while carotene and xanthophyll content remained similar to those in noninfected leaves.

Carotenoids play and important role in natural plant coloration. During fruit maturation and just before defoliation of deciduous plants, for example, carotenoids become plastid pigments and serve as color carriers in these organs. As chlorophyll degrades and disappears, carotenoids become predominant. Green fruits take on red, yellow or similar colors during maturation and near the end of the vegetation season, autumn leaf coloration take place as the red and yellow pigments and several shades of their combinations become predominant.

Alterations in carotenoid coloration also may be caused by biotic factors. A common trait of all these induce changes is the reduction of chlorophyll quantities revealing the other pigments. Local and systemic green decoloration is accompanied by local or systemic carotenoid coloration.

Among biotic factors, pathogens play a major role in bringing about these changes by altering the quantities of carotenoids and chlorophyll in infected plants. The extent of color changes in infected plants depends on the degree of alteration in these pigments. Pathogen toxins affecting protoplasts can cause rapid cell extinction and thus, play an active role in chlorophyll degradation, resulting in the unmasking of other pigments. In such processes, dissolved pigments, such as anthocyanins and flavonoids, are released from the cell sap and spread into infected tissues, thus influencing color changes. Also, other biochemical reactions lead to the formation of colored degradation products in cells and tissues as part of the necrosis process.

Color changes in infected plant cells and tissues usually have a distinctive appearance and result from a series of complex changes and reactions occurring in cells during the infection process. However, insufficient data are available regarding all the individual reactions that take place in an infected plant and that bring about the various color changes.

A large and perhaps the greatest number of plant pathogens cause color changes in the carotenoid pigments. The involvement of color change in carotenoids is so characteristic of many diseases that they obtain their popular names from then; for example, orange leaf spot of plum, caused by *Polystigma rubrum;* red leaf spot of almond, caused by *Polystigma ochraceum;* yellow rust of barley and wheat, caused by *Puccinia glumarum;* and leaf stripe of barley, caused by *Helminthosporium gramineum.*

Similarly, virus diseases distinguished by changes from green leaf color to reddish and yellow colors could be classified in a similar manner, such as bean yellow mosaic, plum pox, and potato mosaic. In some viral diseases the green color is completely replaced with yellow on most or all of the leaf areas, such as rice yellows and strawberry yellow edge. General leaf yellowing also can result from infection by a mycoplasma such as stolbur on pepper plants.

4. Chromatic Changes in Anthocynins and Flavonoids

Anthocyanins and flavonoids plant pigments are glucosides composed of phenolic compounds (aglucones) and corresponding sugars (see section on associated phenols), and are found primarily in the cell sap of vacuoles. Flavonoids are wide-spread among plant species and, similar to anthocyanins, play and important role in plant biology. These pigments are found in green plant parts as well as flowers and fruits and impart orange and yellow coloration. The role of anthocyanins and flavonoids in plant biology is not fully understood, but it is assumed that by absorbing ultraviolet rays they protect chlorophyll from exposure to these damaging rays.

Anthocyanins carry blue, red, and violet pigments in different plant parts, such as leaves of red cabbage, roots of red beets, flowers of many species, and fruits of cherry and plum. Anthocyanins, in addition to carotenoids, may accumulate in maturing fruits resulting in blue or red coloration.[36] They may also form under unfavorable enviromental conditions, suggesting a protective reaction under certain conditions.

The synthesis of anthocyanins and other pigments belonging to this group, is influenced by various factors depending on plant species. For example, anthocyanin formation may be reduced or absent in plants grown in the dark while for others, anthocyanin is increased or unaltered.

Seedlings of some plants develop a reddish color only under illumination. Light is necessary for blue and other pigments to develop in flowers of several plants, while in others, pigment formation proceeds normally or almost normally in the dark.

Anthocyanin synthesis also is influenced by plant nutrition. Depending on plant species, lack of sufficient nutrition may reduce or stimulate the synthesis of this pigment. Anthocyanin formation has been observed frequently in plants following mechanical damage or animal injury, suggesting that anthocyanin production is related to a defense mechanism in such plants. Anthocyanin formation in response to injury or damage has been related to an increase in sugar levels in the affected tissues.

Studies on damaged and normal tissues have shown that tissues with little or no chlorophyll, such as nongreen portions of variegated plants or mite galls, are more prone to anthocyanin formation. Küster[5] found an abundance of anthocyanin production in roots exposed to light with little or no chlorophyll production.

The quantity of anthocyanin and related pigments in a plant can be genetically controlled. For example, some maize mutants lack anthocyanin while others have a red pericarp due to anthocyanins in the grain aleurone layer.[24]

Similar to other plant pigments, anthocyanins and flavonoids are susceptible to color changes in infected plant tissue or parts. They appear in green plant parts following green decoloration. Color changes in infected tissues are complex processes in which several other colors appear separately or jointly, and are therefore difficult to distinguish individually. On leaves of some plants, for example, spots occur in and around points of infection. The color of these spots changes in the course of development and the marginal zone becomes violet, which is a characteristic of anthocyanin coloration. A number of diseases are known by these characteristic symptoms: shot-hole of cherry, caused by *Clasteropsorium carpophilum;* spur blight of raspberry, caused by *Didymella applanata*; and violet leaf spot of strawberry, caused by *Mycosphaerella fragariae*. Red leaf spot of grape, caused by *Pseudopeziza tracheiphila,* results in the disappearance of chlorophyll and an increase in the appearance of anthocyanins.[11] In peach leaves infected by *Taphrina deformans*, chlorophyll degradation is followed by the appearance of anthochyanin coloration.[40]

Changes in anthocyanin coloration in fleshy fruits also is characteristic for some virus diseases.[35] For example, peach mosaic virus prevents normal fruit coloration resulting in the reddish appearance of infected fruits. Apple fruits infected with apple ringspot disease develop dark reddish spots during maturation. The cherry leaf roll virus prevents anthocyanin formation and infected fruits remain pale light red at maturity.

Disorders in anthocyanin coloration are particularly pronounced in flower petals containing these pigments. Viruses are the most numerous plant pathogens that cause this time of disorder because of systemic infection. Matthews[41] stated that virus infection of flower petals usually affects the anthocyanins in the vacuoles of the tissues without affecting the chromoplast pigments. Thus, this results in a wide range of anthocyanin color intensity due to the substitution with other plastide pigments. Changes in anthocyanin pigmentation are particularly conspicuous in unicolored petals of ornamental plants. Some viruses that cause these characteristic color changes include the canna flower yellow stripe virus (canna mosaic virus), gladiolus flower mottle virus, petunia ringspot virus, stock flower striped mottle virus (cabbage black ringspot virus), and tulip color-breaking virus.[35]

5. Formation of Dark Pigments

Dark and melanoid pigments are complex organic compounds usually formed in the process of quinone binding with other cell constituents. Quinones are oxidized forms of phenolic compounds which readily bind with cell amino acids and proteins (see section on oxidized phenols). Dark pigment formation occurs simultaneously with cell necrosis, resulting in the dark color or black appearance of dead cells and tissues. These processes can play an important role in limiting the ingress of obligate parasites that are incapable of developing in necrotic cells.

Dark pigments are formed only in association with plant cell and tissue damage regardless of the causal agent involved. Their formation differs substantially from that of carotenoids and chlorophylls, which are synthesized in plastids, and anthocyanins and flavonoids, which are synthesized in vacuoles.

The role of phenolic compounds in dark pigment formation has been the subject of numerous investigations, with the primary focus on the role of chlorogenic acid in dark pigment formation in tobacco leaves associated with downy mildew and in necrotic spots caused by physiological factors.[42] According to these authors, chlorogenic acid begins to increase at the time of lesion formation and then decreases as necrosis progresses. Intense oxidation and chlorogenic acid degradation occurs during these processes under the influence of oxidase enzymes. Products of this degradation are quinones characterized by a high reactive ability and easy binding with amino acids, peptides, and proteins. The binding of quinones with cytoplasmic constituents leads to coagulation or precipitation of proteins, which results in necrosis and the formation of dark pigment.

Goodman et al.[17] stated that the two most important means by which plant ward off infection were first, the hypersensitive reaction, and second, tissue darkening. Tissue darkening, according to these authors, was the result of enzymatic phenol oxidation with quinones as oxidation products condensing into polyquinones or which react with amino acids and proteins to form dark-colored substances. In other studies, Ladigina and Edreeva[42] showed that in the final stages of necrosis, nicotenes (piridine alkaloids) and ruthenium (flavonol) were reduced in quantity suggesting participation in dark pigment formation. Nicotenes and products containing nicotene, form complexes with several organic compounds of different solubilities, which contribute to cell necrosis and dark coloration.

The dark coloration of vascular bundles infected with vascular pathogens, such as species of *Fusarium* and *Verticillium*, has been attributed to the oxidation of phenolic compounds. Vascular discoloration does not take place if the oxidation of these compounds is inhibited. Bossi[43], for example, established that fusaric acid, produced by *Fusarium oxysporum*, may impair the activity of some oxidases taking part in phenol oxidation and thus, inhibit the dark coloration processes.

Melanins, which have a role in the natural coloration of plant tissues, are products of complex chemical reactions in cells in which L-tyrosine is involved with the formation of derivatives of indole, resulting in melanin formation, which is followed by polymerization.[44] Stević[45] showed that tyrosinase, one of the polyphenol oxidases, catalyzes tyrosine oxidation to form melanoid compounds. The formation of these compounds in damaged or infected plant tissues can be defined as "plant melanosis".

Dark pigment formation can be a nonspecific reaction caused by either abiotic or biotic factors. Abiotic factors include mechanical injuries, such as the darkening at the junction of a rootstock and scion after a grafting procedure. Protection of the graft is required to reduce the influence of external factors that could induce dark coloration of the juncture.

Darkening of damaged or injured plant parts caused by various biotic factors, such as pathogens, is wide spread, although the mechanisms inducing these discolorations have not been completely illucidated. Tissue darkening is a primary characteristic of local necrosis or hypersenistive reaction caused by some obligate parasites. For example, *Plasmopora viticola* causes local necrosis on leaves of some American grapevine cultivars including *Vitis reparia, V. rupestris,* and *V. rotundifolia*. Plant viruses cause local lesions on leaves of some plants, such as tobacco mosaic virus on *Nicotiana glutinosa,* turnip mosaic virus on *Chenopodium amaranticolor,* apple chlorotic leaf spot virus on *Chenopodium quinoa,* and plum pox virus on *Chenopodium foetidum.* Such necrotic cells and tissues provide a barrier against the spread of obligate parasitic fungi and viruses.

Several necrotrophic (necrogenic) pathogens, which cause various types of spots on fruits, leaves, and other plant parts, can be classified as agents of tissue darkening or discoloration (Figure 19). The diseases caused by them have been named based on the color of the lesion or spot induced. Among numerous mycoses caused by necrotrophites, the following disease can be used as examples: leaf spot on alfalfa, caused by *Pseudopeziza medicaginis* (Figure 20); maize ashy leaf stripe, caused by *Exserohilum turcicum*; potato and tomato brown leaf spot, caused by *Alernaria solani*; walnut ring spot, caused by *Gnomoia juglandis*; and wheat brown spot, caused by *Septoria tritici*. Several bacterioses also are well known by ther characteristic dark spots produced on leaves, including: cotton angular leaf spot, caused by *Xanthomonas campestris* pv. *malvacearum* (Figure 21), tan spot on sesame, caused by *Pseudomonas syringae* pv. *sesame* (Figure 22), hemp black leaf spot, caused by *Pseudomonas syringe* pv. *cannabina* (Figure 23); and pepper and tomato scab spot, caused by *X. campestris* pv. *vesicatoria.*

Pronounced black or melanoid discoloration occur within leaf spots of some infected plants, but no details have been published on the nature of the process involved in forming them. Diseases which have this type of symptom include: apple black canker, caused by *Physalospora cydoniae*, poppy black spot caused by either *Pleospora papaveracea* or *Xanthomonas campestris* pv. *papavericola;* quince brown leaf spot, caused by *Stigmatea mespili*; sycamore leaf blotch, caused by *Rhitisma acernum;* and tomato bacterial speck, caused by *Pseudomonas syringae* pv. *tomato.*

Quince sooty ringspot causes black pigment formation in the interveinal cuticle tissues given the leaf surface a brown-black appearance. Another disease, caused by some strains of the cherry leaf roll virus, appear as black stipes as a consequence of cambium necrosis in grafting sites of walnut (*Juglans regia*) scions and rootstocks of *J. hindii* x *J. regia.*[35]

6. White Coloration

White coloration is another example of chromatic changes in diseased plants. It may occur locally in the form of irregular areas or spots, or affect whole plant organs and tissues. The coloration ranges from white to creamy-white, milk-white, silvery, or bronze.

White coloration in plants can occur either because of the complete degradation of any plant pigments in plastides and/or vacuoles; or as a result of reflected sunlight from air layers formed in damaged or infected tissues. In the case of the latter, air layers prevent sunlight absorption, thus restricting chlorophyll formation and green coloration. The normal white color of flower petals, however, is the result of a combination of the lack of color and the reflection of light rays.[36]

White coloration can be caused by abiotic factors and genetic changes in plants, as well as by pathogens. Among the abiotic factors are low temeperatures and frost, which initiate a decline of mesophyll cells between epidermal tissues resulting in, at first, chlorotic areas followed by whitish zones along leaf margins. Such changes usually are irreversible and occur in the fall and early spring on summer crops (Figures 24, 25).

In some plants, white coloration appears in the form of varigation on leaves and sometimes on fruits and shoots. In such areas, the cells of white and green tissues are allied and probably originate from a single cell or contiguous allied cells.[5] When such cells divide, they give rise to white and green cells forming the varigated tissues. White sector varigation developing in organs usually originate from the base of the organ from an initial cell. The changes of normal cells into white cells is irreversible, indicating the genetic nature of the change. Some white mutants have appeared with chlorophyll-containing cells resulting in the development of irregular green spots on leaves with a major portion of the plant tissues remaining white.[24] White mottle or varigation attributed to genetic changes sometimes occur on cherry and grape leaves. Such white or light-colored areas of different shapes and sizes are delineated and can be distinguished from similar changes caused by some viruses.

White plant or tissue coloration may be caused by a variety of pathogens and may either be limited or dispersed from the points of infection. Limited white coloration usually is associated with pathogen toxins which cause cell destruction. Air penetrates the spaces left in the

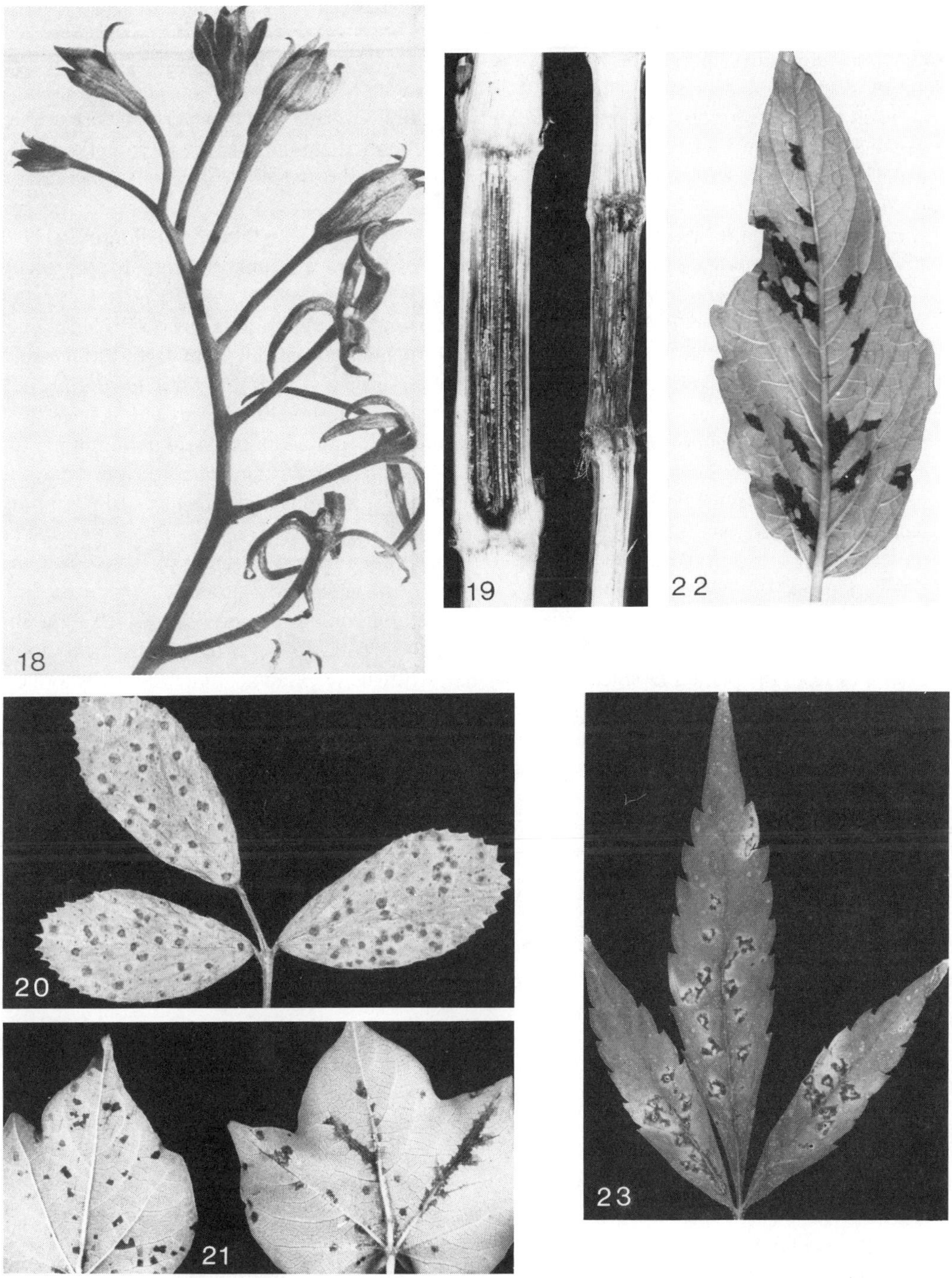

PLATE IV. FIGURE 18. Greening and malformation of blossoms on a tomato plant caused by the stolbur mycoplasma. (From Brčak, J. 1979.) FIGURE.19. Dark discoloration progressing vertically on a maize stem caused by *Gibberella zeae*. FIGURE 20. Sooty colored spots on alfalfa leaflets caused by *Pseudopeziza medicaginis*. FIGURE 21. Coalescing black necrotic spots on the underside of cotton leaves caused by *Xanthomonas campestris* pv. *malvacearum*. FIGURE 22. Spreading tar-like black spots on the underside of a sesame leaf caused by *Pseudomonas syringae* pv. *sesami*. FIGURE 23. Dark sooty spots with chloroatic halos on the underside of hemp leaflets caused by *Pseudomonas syrinae* pv. *cannabina*.

disorganized tissues resulting in a whitish appearance. Such changes usually occur around points of infection forming halos or rings of different shapes and sizes. As tissue disintegration continues, the halos enlarge. Then as desiccation and necrosis develops, the affected areas fall out resulting in the shot-hole symptom.

Such symptoms frequently are caused by a certain phytopathogenic bacteria, whose pathogenicity is a result of enzymatic degradation of pectic elements in the middle lamella of cell walls. Bacterial diseases with halos as symptoms include; halo blight of oats, caused by *Pseudomonas syringae* pv. *coronafaciens*; shot-hole of plum, caused by *P. syringae* pv. *morsprunorum*, halo blight of tomato, caused by *P. syringae* pv. *tomato* (Figure 26) and *Xanthomonas campestris* pv. *gardneri* (Figure 27), and bacterial canker of tomato, caused by *Clavobacter michiganese* pv. *michiganense*.

Toxins produced by certain fungi, which develop long-term infections in woody plants, may spread throughout an infected plant resulting in whitish colorations in plant parts remote from the site of infection. A well known representative of this group is the fungus *Stereum purpureum*, causal agent of silver leaf disease of almond, apple, peach, plum and quince, and of forest tree species of *Acer, Carpinus, Sorbus*, and others. Following root and local stem infection by this fungus, the necrogenic and toxic compounds produced by the pathogen are transported to the crown and leaves of the best. These toxic agents cause epidermal leaf cells to swell beginning near the leaf veins, resulting in their separation from the palisade layer.[3] This separation creates air spaces in the leaf tissue which give a silvery appearance when sunlight is reflected from the affected areas. The pathogen cannot be recovered from leaves showing such symptoms.

In a similar manner, *Phomopsis viticola* causes a whitening of grape shoots. This fungus colonizes the shoots and woody plant parts. In the autumn and winter, the bark of diseased shoots separates from the plant and cracks resulting in a whitish appearance along infected shoots.

Pepper fruits infected with *Phytophthoro capsici* exhibit a whitish or ashy coloration (Figure 28). This symptom occurs when a period of dry, warm weather follows infection. Infected fruits crinkle, desiccate and is followed by decay of the pericarp tissues leaving only a transparent cuticle. Air layers under the cuticle result in whitish coloration giving the disease the name of pepper fruit parchment.[25]

Viruses also can cause white coloration in diseased plants (Figure 29). Systemic infection by a virus can result in such symptoms which are caused primarily by a reduction in chlorophyll. Several viruses are known by such symptoms: apple mosaic (white type), narcissus white streak, and white leaf of rice. Alfalfa mosaic virus causes narrow whitish stripes on pepper fruits and a whitish to ashy irregular mosaic on leaves of some cultivars.

F. CYTOLYTIC CHANGES

Cytolysis implies the degradation or disintegration of whole cells or parts of cells. Such changes usually are considered only as pathological ones. However, they occur spontaneously in the process of plant tissue formation and have general biological significance. Cytolytic changes also may be induced by external factors resulting in different symptoms. Cytolytic changes are classified as either spontaneous (normal) or induced (pathological) changes.

1. Spontaneous Cytolytic Changes

Spontaneous cytolytic changes usually occur during cell differentiation and can be a component part of ontogenetic plant development. These changes occur during cell and tissue specialization and may participate in the formation of conductive and secretory tissues, secondary plant root development, and other growth processes. Good examples of this type of change are the formation of perforations on sieve tube plates in phloem tissues and the partial or complete degradation of cell walls between fused cells forming the tracheae in xylem tissues.

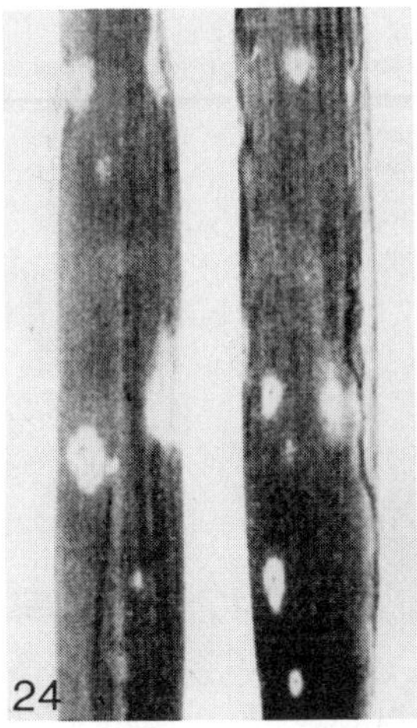

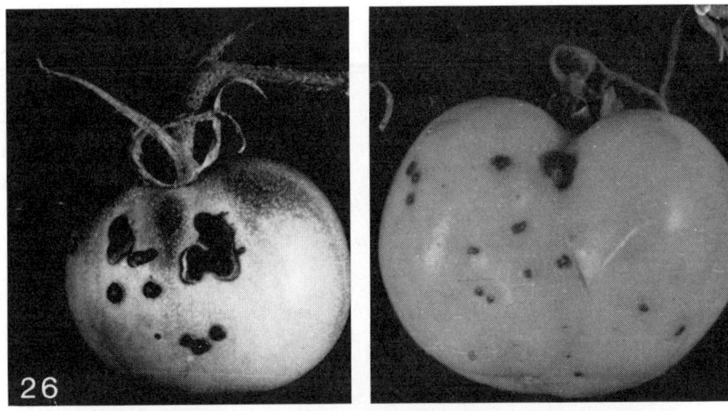

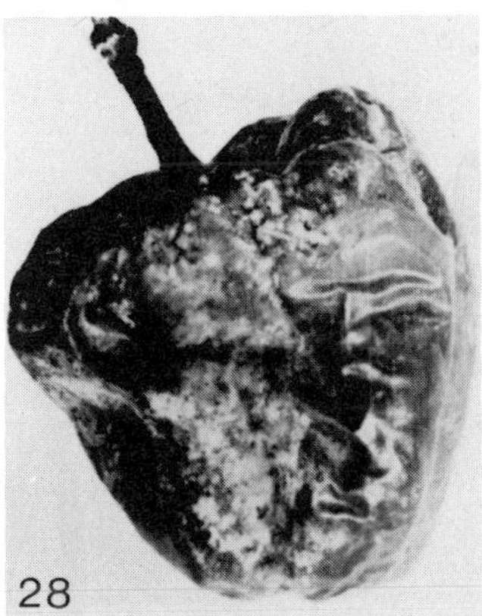

PLATE V. FIGURE 24. Localized chloratic (white) spots on oat leaves caused by *Pseudomonas syringae pv. coronafaciens*. FIGURE 25. Completely rotted young tomato fruit (lower right) caused by *Botrytis cinerea* compared to two uninfected fruits. FIGURE 26. Comparison of localized dark fruit spots on tomatoes caused by *Septoria lycopersici* (left) and *Pseudomonas syrinae* pv. *tomato* (right). FIGURE 27. Complete decoloration (white) around lesions on tomato fruits caused by *Xanthomonas campestris* pv. *gardneri*. FIGURE 28. Complete chlorosis (white parchment-like) on a pepper fruit caused by *Phytophthora capsici*. FIGURE 29. White coloration along the veins of peach leaves (left and right) caused by the plum pox virus compared to uninfected leaf (center).

In some plants cytolytic degradation of transverse walls give rise to the fusion of several secretory cells. Such fused cells form long, undivided lactescent cells, which secrete a milky juice rich in organic matter.

Enzymatic cell degradation occurs during lateral root formation. Lateral roots develop from root primorida, which are formed by division of pericycle or procambium cells. An envelope is formed around the lateral root primordia, which enhances the penetration of the young roots through the cortical tissues. During the penetration process, this envelope exerts both mechanical and chemical influences by secreting enzymes which degrade root cell walls.[36]

2. Induced Cytolytic Changes

Induced cytolytic changes in cells can be caused by a variety of external factors. Even though the factors may vary, the consequences of such changes frequently are similar.

In the final stages of cytolysis, whether caused by an abiotic or biotic factor, the formation of gummosis (gummous, gum flux) and semisolid resinous (resinosis, resin flux) substances, as well as reabsorption of degraded cell parts take place. Gum and resin flux processes usually involve cell constituents as well as whole cells. These processes substantially alter the structure, stability, and strength of affected plant parts. Altered plant tissue and organ structure seemingly are preserved during the course of degraded cell reabsorption.

According to the type of causal agents, induced cytolytic changes may be classified as either mechanical or pathogenic.

a. Mechanically Induced Cytolytic Changes

Mechanical cytolytic changes occur in injured plant parts or by penetration of foreign materials into plant tissues. Mechanical injuries, which are the most frequent cause of cytolysis, include abiotic factors, such as cuts and wounds on plant parts as a result of hail, frost or similar factors; and those caused by pests and other animals.

All living cells, but primarily parenchyma and meduallary ray cells, are subject to cytolytic changes. Gums formed in cell lumina caused by cytolytic changes differ in chemical composition from other cell constituents and are the result of oxidation and other processes within the cell. Gummosis spreads through tissues filling open spaces in conductive tissues. As with transformed periderm cells, they also represent a protective reaction in damaged plant parts.

Cytolysis resins are formed from the degradation of cell walls as well as whole cells. Resin flux occurs following mechanical damages in several plant species and particularly in above ground woody organs of stone fruit trees, such as apricot, cherry, peach, and sour cherry. Resin flux (resinosis) takes place in a variety of plants cells and tissues including parenchyma and vascular cells, wood and fibers, as well as bark, leaf, fruit, and seed tissues. The mechanism for resin formation is not known. Some authors believe that these processes involve enzyme activity associated with dead or dying cells and the oxidation of carbohydrates in cells walls.

Mechanically-induced resin production is used for commercial production. Cuts are made in trunks and branches and the resins collected directly or after hardening. Some well-known commercially-produced resins are; the soft resins or balsams, such as Canada balsam; the hard resins, such as shellac; and the gummous resins, such as amber.

In some plants, these products do not accumulate, but are reabsorbed by continguous cells and eliminated with the result damaged or injured plant parts or organs retain their general appearance.

Reabsorption often occurs in mechanically-damaged plant tissues, such as in grafting, where the process eliminates the dead cells from the adjoining grafted surfaces. As a result, new cells are formed by induced cell division, particularly those of callus tissue, which are responsible for the degradation and restoration processes. Callus tissue, by replacing the dead or degraded portions of a stem, add strength in addition to protection to the new formation. The processes of cytolysis and reabsorption can eliminate thick-walled lignified and suberized cells.

Some authors have shown that hail injury may result in the complete reabsorption of damaged mechanical tissue cells, such as thick-walled collenchyma fibers.[5] In such cases, meristematic cells formed during tissue regeneration and wound healing, play an important role in degraded cell cytolysis and reabsorption.

Cytolysis and reabsorption also occur in the course of root or branch veneer grafting (ablactation). This process involves damaged sieve cells, which have become flattened and disfunctional and finally disappear. By eliminating dead sieve cells, reabsorption contributes to cell regeneration in this tissue.[36]

It should be stressed that gummosis, resinosis, and reabsorption represent protective activities induced by mechanical injury. These cytolysis reactions enable the elimination of needless foreign or unwanted elements, providing favorable conditions for the healing of injuries and wounds. By eliminating degraded plant parts, the plant performs its own environmental cleaning, thus protecting itself from the unfavorable influence of undesirable constituents.

b. Pathogen Cytolytic Changes

All changes in cell degradation caused by pathogens are defined as pathogenic cytolytic changes. Cytolysis in a broader sense refers to cell wall and protoplast degradation, and in a limited sense to cell wall degradation. This degradation is a chemical process based mainly on enzymatic activity. However, other mechanisms of pathogensis by pathogens, such as mechanical, toxin, or hormone production also influence cytolysis and depend on them.

The processes of parasitism by plant pathogens is based on the degradation of organic compounds in plant cell structure, although the exact processes of this degradation are not fully understood. In the course of degradation insoluble organic compounds are transformed into soluble ones, which are absorbed and used by the pathogen. Since enzymatic activity depends on the chemical composition of cells, a brief survey of plant cell composition seems appropriate.

The plant cell wall consists of polysaccharide complexes, including cellulose, hemicellulose, pentosans, and pectic substances. Substantial qualities of structural proteins also are present in walls of rapidly growing cells. Cell walls of older tissues contain lignin, and those of protective tissue layers, cutin and suberin.

The chemical composition of protoplasts is characterized primarily by lipo- and nucleoproteins. Lipoproteins, composed of different ratios of lipids and proteins, are characteristic of plasma membranes such as plasmalemmas and tonopolasts, as well as of various organelles, such as chromatophores, endoplasma reticula, mitochondria, and nuclei. Nucleoproteins consisting of proteins and ribonucleic acids are present in ribosomes and in the deoxyribounucleic acids of cell nuclei.

Plant pathogens produce a large number of enzymes which act consecutively in the degradation of plant cell constituents. Synthesis of several enzymes takes place only when the pathogen comes in contact with the substance to be degraded (i.e. induced enzymes). In addition to enzymes of the pathogen, enzymes from dead host cells also play an active role in cytolysis, although this process has not been thoroughly investigated.

Most pathogens produce enzymes for degradation of polysaccharide complexes in cell walls and membrances, such as pectinmethylesterase, polygalacturonase, and others, which take part in the gradual hydrolytic degradation of pectic substances, and may include liases.[46] Hydrolytic degradation of cellulose and hemicellulose is catalyzed by the enzymes cellulase and hemicellulase, the former of particular importance in cell wall degradation.

Some plant pathogens produce enzymes capable of degrading cell membrane constituents.[47] These include several specific phospholipases which hydrolyze phospholipid ester bonds.[48,49] Proteolytic enzymes (proteinases) degrade proteins by gradual hydrolysis of their peptide bonds. With cell membrance degradation, the main functions of nutrient transport, osmosis regulation, and maintenance of biochemical cell unity cease.

Phospholipases and proteolytic enzymes also play a role in membrane lipoprotein degradation in cell organelles, such as chromatophores, mitochondria, and ribosomes, which results in damage to organelle structure, thus a reduction or hindrance of their functions.

Pathogens vary widely in their cytolytic ability, which is usually characteristic of those which penetrate host cells directly and includes those pathogenic bacteria and fungi whose enzyme system attacks cells directly to establish pathogenesis. Phytopathogenic fungi have the broadest range of enzymatic activity, since they produce almost all the enzymes known to be involved in cytolysis.

Pectolytic enzyme formation by phytopathogenic fungi has been described both *in vivo* and in culture. Polygalacturonase has been shown to be associated with over 30 pathogenic fungi and to occur in tissues infected by *Botrytis allii* and *B. squamosa*[50], *Ceratocystis fimbriata*[51], *Cladosporium cucumerinum*[52], *Penicillium expansum*[53], *Pythium aphanidermatum*[54], *Rhizoctonia solani*[55], and *Colletotrichum trifolii*[56], as well as others.

Pectinesterase production by pathogens is wide spread and is dependent on the media used. It has been shown to be produced experimentally by *Botrytis* sp., *Colletotrichum trifolii, Physalopora* sp., and *Rizoctonia solani.*[15]

Cellulase synthesis has been reported on onion infected by *Botrytis allii, B. cinerea* (Figure 25), and *B. squamosa*[50], on alfalfa infected with *Colletrotrichum trifolii*[56], on maize stems infected by *Fusarium moniliforme*[57], on tomato fruits infected by *Sclerotium rolfsii*[58], and others.

Limited information is available regarding the ability of phytopathogenic fungi to form other enzymes. Xylanase has been observed in alfalfa stems infected by *Colletotrichum trifolii.*[56] Sumere et al.[59] showed that *Puccinia graminis tritici* produced cellulase, polygalacturonase, and hemicellulase, and that germinated uredospores of the fungus synthesized polygalacturonase. Linskens et al.[60] showed that *Penicillium spinulosum* produced cutinesterase. Other reports on cutin-degrading phytopathogenic fungi have stimulated further research.[61]

Gäumann[3] established that wood-destroying fungi, such as *Stereum purpureum,* degraded lignin instead of cellulose in beech trees. Such degradation of lignin and other cell wall constituents in woody stems results in the formation of cavities which are characteristic of the pathogen and host involved.

Proteolytic enzymes were shown to be formed by *Cladosporium cucumerinum* on cucumber seedlings[62] and by *Colletotrichum trifolii* and *Stemphylium botryosum* in culture and in alfalfa stems.[56] The role of proteolytic fungi in cytolysis needs further investigation.

Parasitism of phytopathogenic bacteria, like the fungi, is based on host cell enzymatic degradation, Bacterial cytolytic activities, which are distinct from other plant pathogens, generally is based on the degradation of pectic substances and cellulose in cells walls. Only pathogenesis of some species, such as *Agrobacterium tumefaciens* and a few vascular bacteria, differ from these general characteristics.

Data from numerous studies on phytobacteria indicated that bacterial infections are the result of enzymatic degradation of pectin. Bacterial pectolytic activity, as a general trait, has been established in many trials. For example, polyglacturonase formation have been shown in some bacteria of the following genera; *Erwinia,*[63-65] *Pseudomonas,*[64,66] and *Xanthomonas.*[67] Pectinesterase formation has been reported in *X. campestris* pv. *malvacearum.*[68]

The ability of bacteria to produce cellulase has been shown in a number of species; *Erwinia,*[69] and several species of *Pseudomonas* and *Xanthomonas.*[70] Friedman[71] showed that avirulent strains of *E. carotovora* produced proteinase and polygalacturonase in lesser quantities than an avirulent strain, but that they did not differ from the parent strain in the production of pectinesterase. The enzymatic activity of phytobacteria needs further research.

In contrast to fungi and bacteria, phytopathogenic viruses, lacking the necessary enzymes and metabolic activities, are incapable of performing catolysis, and, therefore, penetrate plant tissues indirectly or passively. Once inside cells, virus particles pass from one cell to another through plasmodesmatas without leaving the host cells.

Phytopathogenic mycoplasmas are similar to the viruses regarding cytolytic activity even though they are metabolitic entities. Mycoplasmas pass from one sieve cell to another through sieve pores in sieve plates, colonizing sieve cells, where they complete their development.

Symptoms resulting from cytolytic changes are interlinked and intertwined with other changes and, therefore, difficult to identify individually. However, most cytolytic changes can be recognized by some common sign or symptom, which results from a specfic cytolytic process. Various kinds of wet, dry, and wood rots have been characterized based on different cytolytic processes and cell wall degradation.

Wet rot occurs primarily as a consequence of the enzymatic degradation of pectic substances in which cells separate along middle lamellae. Such cells, separated from tissues, turn into amorphous semi-fluid masses. Such loss of linkage between cells and tissues is also referred to as maceration. The cells are more susceptible to plasmolysis and the toxic action of pathogen metabolites and die quickly.

Parenchyma tissues are particularly susceptible to wet rot, which usually occurs in plant organs abundant in this type of tissue, including fleshy roots and tubers, leafy vegetables, and others. Enzymes that cause wet rot include primarily the pectolytic enzymes and hemicellulase, but not cellulase.[15] Causal agents of wet rot include some facultative parasitic bacteria and fungi, such as *Sclerotinia libertiana,* a polyphagous pathogen; and *S. fuckeliana* (anamorph *Botrytis cinerea*), both of which cause wet rot in a large number of hosts. Tissue macerating enzymes are produced by other fungi such as *Cladosporium cucumerinum, Pythium* spp., *Phytophthora* spp., *Rhizoctonia solani,* and others, as well as phytopathogenic bacteria such as *Pseudomonas* spp., and *X. campestris* pv. *malvecearum.*[15]

Well-known causal agents of bacterial wet rot are *Erwinia carotovora* and *E. aroideae,* which cause this type of disease in fleshy roots and tubers, and in leafy vegetables (Figure 32).

Some facultative fungal pathogens cause dry rot, which is a result of partial cell degradation, where the degraded cells and tissues remained united and maintain, in part, their external shape well into the vegetative period, are sometimes, in the case of woody plants, for several years. Distinct from wet rot, dry rot develops slower and infected tissues and organs contain less water than normal, resulting in symptoms after which this general class of disease has been named (Figure 30).

Examples of causal agents of dry rot include *Fusarium coeruleum* and *F. solani,* which cause dry white rot of potato tubers. This disease usually occurs on stored tubers infected with the fungus *Phytophthora infestans.*[11] Such infected tubers are unmarketable. Other well-known causal agents of dry rot are the fungi belonging to *Sclerotinia* spp., which cause dry rot of pome and stone fruits. In the initial development of the disease, the fruits become desiccated and grey. The fungus mycelia eventually completely colonizes the fruits and absorb degraded cell constituents resulting in mummification. The mycelium of the pathogen may persist in mummified fruits for many years.

Wood rot of many forest and fruit tree species is the result of enzymatic degradation of cellulose, hemicellulose, and lignin in cell walls, caused primarily by fungi belonging to the class of Basidiomycetes. Some cause a wood rot accompanied by various discolorations which depends on the extent of degradation, such as the fungi *Stereum purpureum,* which causes a white rot of deciduous and fruit trees, and *Merulius lacrymans,* which causes a brown primastic rot of coniferous and deciduous trees. Wood rot may be scattered in trees resulting in individual pockets of decay varying in size and shape, and often referred to as corrosive or alveolar rot. Causal agents of this latter type of rot include the facultative fungi *Stereum frustulosum, Xanthochrous reades,* and others.

Wood decay may develop in a number of ways resulting in specific symptoms. For example, if wood fibers are decayed and degraded in a single direction, it is referred to as a fibrous rot, which usually occurs in white rot. Fibrous white rot frequently is caused by *Armillaria mellea, Phellinus dryadeus,* and *Unginula fomentaria.* However, tissue degradation in two directions results in a lamellar rot, usually associated with fungi belonging to the genus *Poria;* and in three directions, results in prismatic patches, after which the syndrome has been named, prismatic rot. Examples of fungi that cause prismatic brown rot are: *Lenzites quercina, Merulius lacrymans, Polyporus sulphureus,* and *Unginula marginata;* and those that cause prismatic white rot, *Melanopus squamosus* and *Polyporus borealis.*[72]

A common symptom of cytolysis is gummosis caused by a variety of pathogens, such as the facultative fungal parasite *Clasterosporium carphophilum,* which causes gummosis of branches of stone fruit trees, particularly those of sweet and sour cherry. Following infection, the pathogen degrades infected cells. The products thus formed are subject to oxidation and then transformed into gummy materials, which exude from the wounds on infected twigs. Such wounds prevent twig development and result in a rapid decay. Similar substances were observed on apricot and peach twigs infected with fungi of the genus *Monilia.*

The phytopathogenic bacterium *Pseudomonas syringae* pv. *morsprunorum* causes gummosis in infected twigs of apricot, cherry, plum and other stone fruits. In autumn this bacterium penetrates leaf scars and develops cankers on branches and stems. Gummy substances form in infected tissues and are secreted onto the surface. Toxins released by the bacterium following cell degradation and death also take part in these pathological processes.[73]

Arsenijević[74] showed that gum exudation also occurred in cankers on apricot twigs infected by *Pseudomonas syringae* pv. *syringae,* and similar to the aforementioned bacteria, penetrated in the autumn and winter months through crown and stem injuries, resulting in cankers of different shapes and sizes. Host cell necrosis as well as canker and gum formation connected with the toxic effects of this pathogen was thought to be important factors in apricot apoplexy.

Gum formation has been described in the citrus psorosis disease.[35] Psorosis "A" begins its development on the external bark of susceptible citrus species. In the course of disease development, the inner bark layers grown irregularly followed by gum accumulation. Gum formation follows disintegration of meristamatic cells and chemical changes in the starch present in infected parts.

Gummy materials have been found in plum trees infected with plum pox virus, particularly in the cultivar Požegača. Necrotic tissues and irregularly distributed gummy granules appear in the mesocarp or fleshy part of infected fruits (Figure 33). There is no verified explanation for the formation of these gums, but they probably result from chemical transformation of dead cell constituents.

Finally, it should be stressed that one of the most detrimental consequences of cytolysis is the partial or complete degradation of cell walls that protect the protoplasts. Toxins of many phytopathogens penetrate degraded cell walls causing protoplast and cell necrosis, which is of particular importance and interest to plant pathologists.

G. CYTONECROTIC CHANGES

Necrosis in plant diseases pertains directly to the death of cell protoplasts. This cell necrosis is the final result of various pathological processes, but primarily cytolytic ones. However, the same necrotic process cannot be assumed with all cases of cell necrosis, since it differs among the various types of rots and decays related directly to enzymatic cell wall degradation. It is often difficult to define and distinguish between different necrotic changes due to their relationship with other changes that occur in the dying cell. Wood[15] stated that necrotic processes of cells take place both in infected areas and in the contiguous noninfected cell protoplasts, with infected cell groups being clearly differentiated from the healthy contiguous tissues. In some cases, he pointed out that these two zones may be separated further by chlorotic tissues.

Cell necrosis occurs in various plant parts, being most characteristic on leaves where it is manifested in the form of lesions differing in shape and size (Figure 31). Affected tissues with the lesions usually have a dry texture (except under high moisture conditions), a cutaneous structure, and are brittle. It is within these necrotic tissues that pathogens often form their reproductive structures, which are, as a rule, dark colored. The periphery of the necrotic areas often are rimmed with a narrow or broad band of dark tissue, which represents a zone of active parasitism.

Characteristically, necrotic symptoms usually become evident late in the course of pathogenesis due to the lengthly processes resulting in this type of cell extinction, differing from wet rot where symptoms may be visible at the beginning of pathogenesis. In contrast, a necrotic hypersensitive reaction, caused by obligate parasites, does not spread since these pathogens are restricted from further development by the necrotic zone.

In hypersensitive plants, obligate parasites, such as certain fungi and viruses, cause local lesions around the focal points of infection. By restricting these pathogens within the necrotic area, hypersensitive plants protect themselves from initial infection. Hypersensitive reactions represent a form of plant resistance and a unique trait compared to other pathological changes in plants (Figure 34).

In contrast to obligate parasites, necrosis caused by facultative parasites spreads into the contiguous healthy tissues, since dead or dying cells offer a favorable environment for their development. This more or less rapid spread may take the form of a uninterrupted necrotic area or concentric zones surrounding points of infection. Concentric necrotic zone formation occurs in cases where the pathogen develops in successive intervals. Facultative parasites usually are associated with sensitive, but not hypersensitive hosts and therefore, hypersensitive reactions do not occur.

Necrotic changes include a wide range of activity and symptoms. Necrotic processes are wide-spread and detrimental to cells of all plant tissues, including conductive, parenchyma, protective, and meristematic. For example, necrosis of conductive tissues results in the disturbance of the functions of the entire plant, and necrosis of meristematic tissues, particulary the apical meristematic cells, prevents normal plant development and growth. Necrosis of edible portions of crop plants reduce yields and the shelf-life of the harvested product.

Causal agents of plant necrosis generally can be classified as noninfectious and infectious.

1. Noninfectious Necrosis

Noninfectious necrosis is caused by a variety of abiotic factors, can occur at any time, and will continue as long as the causal agent persists. Based on the primary characteristics of the causal agent involved, they can be classified as either mechanical or chemical factors.

Agents of a mechanical nature can cause death of cells or cell groups through injury during crop cultivation or other crop management practices, by insects and other animals, and/or meteorological factors. Mechanical necrosis due to meteorological factors, for example, may occur as a consequence of beating rain, frost, or hail. Sun scald, cause by the insolation of herbaceous plant parts, results in a specific type of mechanical necrosis. Frequently, all cells of sun-scalded tissues die, leaving a whitish or parchment-like epidermal layer.

Chemical causal agents of noninfectious necrosis frequently include toxic gases, and the misuse of crop nutrients, fungicides, or pesticides. Among the more common toxic gases that induce necrosis in plant tissues are sulphur dioxide and sulphur trioxide, which are released into the atmosphere during the burning of coal or the smelting of ores containing sulphur compounds. These and other toxic gases frequently cause damage to perennial plants, such as fruit trees, and to woody ornamental in cities. Such gases, and particulary sulphur dioxide, penetrate leaves through stomatas resulting in chlorophyll and chloroplast degradation in the mesophyll followed by necrosis of the entire protoplast. Sulphur dioxide can have a detrimental effect on plant roots when sufficient moisture forms sulphurous acid with sulphur dioxide in the soil to a level that can cause root necrosis.

Nutrient deficiencies can cause necrosis on all plant parts, which resembles that causes by other agents and therefore, are frequently misinterpreted etiologically. For example, potassium deficiency commonly results in the formation of brown necrotic spots along leaf margins, while manganese deficiency results in brown necrotic spots between leaf veins. A calcium deficieny appears as dark red and brown necrotic spots on apple fruits and flax stems. Boron and copper deficiencies can cause necrosis of apical meristems resulting in death of terminal shoots and thus, irregular plant development and growth. Boron deficiency causes desiccation of apical leaf parts and phloem necrosis of olive tree leaf veins.[75]

Necrosis of plant parts can be caused by various crop protection chemicals through incorrect selection and/or application. For example, spraying with copper-based preparations, such as Bordeaux mixture, during rainy periods, results in scald on apple, pear, and plum leaves. This occurs because of above normal accumulation of atmospheric carbon dioxide on leaf surfaces in the presence of water and contribute to an increase in the release of copper ions. The copper ions penetrate herbaceous plant parts causing cell death, which appears as a necrotic scald.

High concentrations of certain pesticides can cause exosmosis and plasmolysis of herbaceous cells. For example, sulphur pesticides applied during a dry period and under high temperatures, rapidly release sulphur ions, which, in high concentrations, act as a noxious gas causing death.

Misapplied herbicides can result in partial or complete necrosis of plant tissues; contact herbicides by degrading external tissues at points of contact, and systemic herbicides upon reaching lethal levels within plants.

In summary, mechanisms of noninfectious necrosis vary from chemical to mechanical, but all result in protoplast destruction leading to cell, tissue and whole plant death.

2. Infectious Necrosis

Causal agents of infectious necrosis include pathogenic bacteria, fungi, mycoplasmas, rickettsias, and severe viruses. All are capable of being transmitted from infected tissue to healthy plants or tissues. The causal agents may persist for long periods of time in or around points of initial infection. Thus, necrotic tissues can represent a natural source for the spread of pathogens.

Mechanisms of infectious necrosis differ among the pathogen groups, hosts, and the host-parasite relationship. Pathogen toxins and enzymes as well as toxic compounds from infected, damaged cells can affect the occurrence and nature of necrosis.

3.Toxins of Pathogens

Toxins are metabolic byproducts of pathogenic organisms which may cause cell protoplast destruction. The process by which toxins damage or destroy cells is referred to as intoxication. Investigators have shown that toxins produced by pathogenic organisms in culture may cause symptoms similar to those in nature.[76-78]

Wood[15] presumed that the majority of the bacterioses and mycoses in which necrosis was a characteristic symptoms, could be caused by a toxin(s). He also suggested the possibility of the existence of a large number of toxins, but stressed that it would not be surprising if only one toxin may be involved in most diseases.

Many pathogenic fungi produce specific toxins affecting specific hosts.[79-82] According to these investigators, resistance to pathogenic fungi should be an indication of tolerance to a fungal toxin. Since these toxins usually are of low molecular weight, they spread easily from the focal points of infection causing necrosis of contiguous cells. This type of toxin produces characteristic symptoms such as chlorotic zones around infected necrotic tissue.

Protoplast intoxication processes are accompanied by a series of other pathological disorders in the cell. For example, *Helminthosporium victoriae* produces HV toxin or victorin, which causes increased protoplast permeability, increased respiration immediately after infection, loss of electrolytes and other solutes from the cell, stomatal closure, and a series of other physiological changes in oats.[80,81,83]

Toxins cause greater biochemical and physiological changes in cell activities than in cell structure. Depending upon the mode of action, many toxins fix the structural unity of an infected cell, which does not differ substantially from that of healthy cells. Scheffer[84] pointed out that toxins have "receptive" or "sensitive" points in protoplasts of susceptible plants and that resistance to toxins represented a constitutive and not an induced plant characteristic. Specific toxin synthesis has been established in several pathogenic fungi in addition to *H. victoriae; Bipolaris zeicola,* which produces HC toxin in maize; *Bipolaris maydis* race T, which produces HM-T toxin in maize; *Piriconia circinata,* which produces PC toxin in *Sorghum vulgare* var. *subglobrescens;* and others. Other fungal pathogens produce nonspecific toxins and include: *Fusarium oxysporum* f. sp. *lycopersici,* which forms lycomarasmin, vasofuscarin, and fusaric acid in tomato plants; *Piricularia oryzae,* which produces piricularin to which individual rice cultivars react differently; and *Alternaria solani,* which produces alternaric acid. Other pathogenic fungi that produce nonspecific toxins are: *Alternaria tenuis, Armillaria mellea, Endothia parasitica, Dothichiza populea, Fusarium culmorum, Fusicladium* spp., *Phytophthora infestans, Puccinia graminis,* and *Rosellinia necatrix*..

Nonspecific toxins are produced by certain phytopathogenic bacteria including:[15] *Pseudomonas syringae* pv. *tabaci,* which produces "halo" toxin in tobacco, and *P. syringae* pv. *phaseolicola,* which forms a similar halo toxin in beans. Others that produce toxicogenic substances include: *P. syringae* pv. *coronafaciens, P. solanacearum, P. syringae* pv. *tomato* (Figure 26), and *P. syringae* pv. *glycinea,* among others. Polysaccharides, which act as toxins, are produced by the following phytobacteria; *Erwinia amylovora, X. campestris* pv. *phaseoli, X. campestris* pv. *campestris, X. campestris* pv. *hyacinthi, X. campestris* pv. *oryzae,* and *Curtobacterium insidiosum.* The toxic complex polysaccharide-polypeptide-lipid, defined as as "endotoxin", is released from cells of some phytopathogenic bacteria, such as: *Agrobacterium tumefaciens, Erwinia carotovora, X. campestris* pv. *phaseoli, X. campestris* pv. *pruni, P. syringae* pv. *syringae,* and *P. syringae* pv. *morsprunorum* (Figure 7).

The role of enzymes in the necrotic process is less known or understood. According to some authors, it differs substantially from that of toxins. Regardless of these differences, enzymes can have either a direct or indirect role in the necrotic process.

Enzymes which degrade cell walls, and continue their activity, can lead to protoplast death and, thus, take a direct role in necrosis. For example, macerating enzymes, such as hemicellulases, pectrolytases, and proteolytases could take part in such activity. Based on the consequence of their activity, enzymes may act in a similar way to toxins and therefore, can be defined, conditionally, as toxic enzymes or enzyme toxins. Although contributing to cell necrosis, this group of enzymes does not have characteristics typical of toxic enzymes.

The intermediary role of enzymes or toxic enzyme in the necrotic process could be characterized by their participation in cell wall degradation, which enables toxins to reach the protoplasts of damaged cells and perform their toxic action. The participation of such enzymes enables less active toxins or active toxins at lower concentrations to be damaging. Pathogenic fungi that produce toxic enzymes include members of the following genera: *Botrytis, Fusarium, Pythium, Verticillium,* and *Sclerotinia.*[17]

4. Host Cell Toxic Compounds

Many investigators have shown that various biochemical reactions in infected cells result in the necrosis of their own protoplasts and the resulting reaction products are referred to as host cell toxic compounds. The reaction can occur in host cells without particular participation of either pathogenic organisms, enzymes, or toxins. This mechanism involved may be considered as "suicidal", since the cell contributes to its own death.

Toxic compounds of this type can be classified into two groups; the induced toxic compounds that are formed as intermediary products of infected cells in the course of pathogenesis and have a detrimental influence at the moment of formation of cells of both the host and the pathogen;

and conditionally toxic compounds which become toxic only upon the increase in concentration in the cell.

The group of directly toxic compounds includes phenolic oxidation products. For example, it was shown that an increase in the activity of polyphenol oxidase was accompanied by necrotic spot formation,[85] as well as an increase in the activity of mitochondria cell respiration organelles.[86] The comsumption of oxygen also increased prior to necrosis.[87-89]

Oxidized phenol products, such as quinones, react with cell constituents, forming complex colored compounds which are responsible for the dark color of necrotic cells and tissues (see section of dark pigment formation). The binding with quinones inactivates cell constituents and induces precipitaion, resulting in protoplast death. Farkas et al.[85] have shown that the accumulation of quinones around infection points on tobacco leaves infected with tobacco mosaic virus results in local necrotic lesions.

Cell and tissue necrosis caused by the action of oxidized phenolic compounds, is characteristic of hypersensitive plants infected with obligate parasitic fungi and viruses. Necrosis on potato plants caused by *Phytophthora infestans* was one of the first host-parasite necrotic interactions studied.[90,91] Other examples include necrotic spots on *Nicotiana glutinosa* caused by TMV.[85]

The direct role of altered phenolic compound metabolism in the necrotic process in plants infected with phytopathogenic bacteria has not been investigated fully. Some differences from that associated with phytopathogenic fungi may be expected since a majority of these bacteria develop in intercellular spaces and degrade pectic substances. Maine[92] for example, established that phenol oxidase activity increased in tobacco stems infected with *Pseudomonas solanacearum*. Sequeira and Kelman[93] also showed that the same bacterium caused an accumulation of chlorogenic acid and scopoletin, as well as scopoletin glucosides in tobacco plants. These and similar investigations have shown that phenoloic compounds also may have a special importance in bacterial leaf spot formation caused by species within *Pseudomonas* and *Xanthomonas*. Goodman et al.[17] stated that the toxic activity of phenolic compounds was common in diseases caused by pathogenic bacterial, fungi, and viruses.

Regardless of the evidence showing the significance of oxidized phenols in host cell necrosis, the possibility of other compounds participating in the necrotic processes should not be overlooked. The possible role of nicotine or nicotine complexes with different organic compounds can lead to cell necrosis and dark pigment formation in necrotic cells should be mentioned. Also, in the course of amino acid tyrosine oxidation, the activity of tyrosinase gives rise to cell necrosis and melanoid coloration.

Thus, protoplast and whole plant cell necrosis may be caused by several toxic compounds each with a specific range of activity and each needs careful study.

In contrast, conditionally toxic compounds, normally present in plant cells, can become toxic and cause necrosis due to, primarily, an increase in concentration. These types of compounds usually are associated with virus infected plants. Since plant viruses are not capable of producing toxins, they must be formed as a result of an altered host metabolism. Among compounds in the cell with possible toxic effects are some amino acids, amides, and hormones. Increased quantities of the amino acid prolin was found in potato leaves infected with the potato leaf roll virus[94]; of serin in tobacco leaves infected with potato mosaic virus X[96]; and in plants of *Prunus* sp infected with western X-disease mycoplasma.[95] Among the semiamides, for example, asparagine and glutamine was found at higher than normal levels in tobacco plants infected with the potato streak mosaic virus (potato virus Y)[96] and in tomato plants infected with the tomato spotted wilt virus.[97]

Another amino acid, pipecolic acid, was considered to be a possible cause of necrosis.[96] This amino acid normally is not found in fully developed leaves of healthy plants, but accumulates in tissues infected with viruses and under other conditions (see section on amino acid synthesis).[17]

Among the possible hormones that cause necrosis is ethylene, which can accumulate in the form of a gas in some infected plants. Ross and Williamson[98] found greater amounts of ethylene in plants of *Physalis floridana* infected with potato streak virus (PVY) than in noninfected plants. Ethylene accumulation was recorded in plants inoculated with the virus which resulted in the formation of local lesions. Williamson[99] thought that ethylene originated from infected plant tissues and was formed as a consequence of damage to the cells.

Regardless of the examples cited, data concerned with compounds which are conditionally toxic are still lacking. Their possible role in both hypersensitive reactions and in host resistance remain a subject for further studies.

5. Symptoms of Necrotic Changes

The major symptoms of necrosis may be classified into four categories: various types of spots, anthracnoses, necrotic cankers, and vascular necrosis.

6. Spot Diseases

Spot diseases is a general term used for all diseases which result in the formation of spots on herbaceous plant parts. They are the most frequent and wide spread of all plant disease symptoms. Spot formation usually is a consequence of the necrotic processes discussed previously. Their appearance and characteristics are used to identify the necrotic processes and to describe plant necrosis in general. Spotting usually is localized. Localized spots or lesions can be caused by numerous phytopathogenic bacteria, fungi, and viruses on susceptible hosts. Some examples of fungal pathogens causing spotting are *Alternaria solani, Fusicladium dendriticum, Exserohilum turcicum* (Figure 31), *Mycosphaerella sentina, Phytophthora infestans, Pseudopeziza trifollii* and *Septoria tritici* (Figure 34). In some cases, necrotic tissues within the leaf spots fall out, giving a shot-holing appearance. For example, *Clasterosporium carpophilum* and *Phyllosticta prunicola* cause shot-holing in leaves of many stone fruit trees.

Phytopathogenic bacteria that cause spotting include many species of *Pseudomonas*, such as *P. syringae* pv. *cannabina*, *P. syringae* pv. *morsprunorum* (Figure 7), and *P. syringae* pv. *tabaci*; and of *Xanthomonas,* such as *X. campestris* pv. *juglandis*, X. *campestris* pv. *malvacearum,* and X. *campestris* pv. *vesicatoria.*

In plants susceptible to plant viruses, local lesions of different shapes and sizes may be formed. However, these spots may occur as a result of localized infection or systemic infection and differ from those caused by bacteria and fungi. Many virus diseases are named after the necrotic symptoms they cause, such as cucumber necrosis, lettuce necrosis, prunus necrotic ringspot, and tobacco necrosis.

Distinct localized spots also are caused by obligate parasitic fungi and viruses in hypersensitive or incompatible hosts. In these situations, death of uninfected cells surrounding the infection loci takes place preventing further spread of the pathogen. Examples of fungi that cause such localized spots are: *Phytophthora infestans* on potato leaves and tubers,[100,101] *Puccinia dispersa* on *Bromus* spp.,[102] *P. graminis* var. *tritici* and *P. triticina* on wheat,[103,104] and *Plasmopara viticola* on resistant grapevine species, *Vitis riparia* and *V. rupestris.*

Local lesion formation in hypersensitive plants is a general characteristic of viral infection (Figure 38). In some cases, the localized spots are characteristic and can be used as a diagnostic character between different viruses and virus strains. Examples of some of the viruses that cause localized necrotic tissue are alfalfa mosaic spot on bean leaves, plum pox virus on leaves of *Chenopodium foetidum,* prune dwarf virus on cucumber cotyledons, tobacco mosaic virus on leaves of *Nictotiana glutinosa* (Figure 35), and grapevine fanleaf virus on plants of *C. amaranticolor.*[35]

In resistant plants, some phytopathogenic bacteria cause hypersensitive reactions similar to those caused by fungi and viruses, such as has been reported on cotton leaves infected with *Xanthomonas campestris* pv. *malvacearum.*[105]

PLATE VI. FIGURE 30. Dry rot on the upper end of two maize ears caused by *Gibberella zeae*. FIGURE 31. Elongated necrotic spots on a maize leaf caused by *Exserohilum turcicum*. FIGURE 32. Decay and discoloration along the midribs of tobacco (left) and poppy (right) caused by *Erwinia carotovora*. FIGURE 33. Vertical section through plum (cv. Pozegaca) fruit showing the dark discoloration in the mesocarp caused by the plum pox virus. FIGURE 34. Necrotic spots on leaflets of red clover plants caused by *Pseudopeziza trifolii*.

7. Plant Anthracnose

The characteristic symptom of anthracnose is a spreading, roughly circular spot on leaves and sunken circular spots on fruits and stems. Infection usually occurs during the herbaceous growth stages of the host and anthracnose-like necrosis spreads during the growing season. Once infection is established, lesion development is continuous with periodic interruptions which results in anthracnose spots developing concentric rings, stripes, or zones of light and dark depending upon the tissue involved. Anthracnose necrosis on fruits and stems are more or less sunken and involves the inner tissue cells. By developing on pericarp tissues, the mycelium of the fungus can penetrate host seeds resulting in the development of anthracnose spots on seed coats.

The main characteristic of anthracnose necrosis is its continuous development on infected herbaceous plant parts, resulting in open wounds, particularly on fleshy tissue. Hyperplasmatic changes do not occur, which is used to distinguish anthracnose cankers from those caused by bacteria and other fungi. Well-known causal agents of anthracnose-like necrosis are *Ascochyta pisi* on pea and other legumes, *Colletotrichum lageneraium* on a variety of cucurbits (Figures 36, 40), *C. lindemuthianum* on beans, *C. truncatum* on soybeans, and *Gloesporium ampelophagum* on grapevine.

8. Necrotic Type of Canker

In some situations, cell necrosis represents the initial stage of the development of some types of plant cankers. Although the necrotic processes affect living cells of several bark tissues, the parenchyma cells are most susceptible. Depending upon the necrotic processes involved in their development, cankers may be classified as either necrotic or necroplasmatic.

Necrotic cankers are characterized by the spreading of the necrotic process within the bark tissues of branches, stems, and twigs. When the necrotic cell zones girdle these organs, desiccation and death results. This type also may be defined as dry canker. Gum exudation is a frequent symptom accompanying necrotic canker (see section on cytolytic changes). Necrotic canker may be caused by any of several fungi, including *Clasterosporium carpophilum*, cause of canker and gummosis on almond, apricot, cherry, peach, plum, and sour cherry twig; *Fusicladium pirinum* on pear shoots; and *Sphaeropsis malorum*, cause of black canker on apple, pear, and quince stems and branches.

Necrotic canker may be caused by some pathogenic bacteria including *Pseudomonas syringae* pv. *morsprunorum*, cause of canker on stone fruit trees (Figure 7); *P. syringae* pv. *syringae*, cause of canker on apricot and other fruit trees branches; and *Xanthomonas campestris* pv. *citri*, cause of citrus canker.

Local and phloem necrosis caused by pear blister canker (virus-like disease) in systemically infected plants sometimes develop similar necrotic cankers. For example, in blister canker, blasted tissues occur in the form of cankers on young pear tree stems, which leads to tissue desiccation.

The neoplasmatic canker is caused by necrobionts, such as *Nectria ditissima* and *N. galligena*, cause of pear and apple branch canker, respectively. Following tissue penetration, these fungi first cause cell necrosis and later induce cambium cell division. This results in the occurrence of tumor-like formations of different shapes around the points of infection. Both necrotic and hyperplasmatic processes take place in the pathogenesis of this canker.

9. Vascular Necrosis in Plants.

Conductive tissue necrosis is charcterized by some specific traits by which it can be identified. This type of necrosis is caused by pathogenic organisms which colonize conductive tissues or live within them. Xylem and phloem cell necrosis occurs in different ways, depending upon the structure and function of these tissues.

Xylem pathogens generally cause cell necrosis either with toxic metabolic byproducts, or by enzymatic degradation. Toxic byproducts usually are of low molecular weight, therefore, spread easily throughout the vascular system of the infected plant. These compounds penetrate cells and cause precipitation of the cell protoplasts and death. Only toxins with such characteristics can cause vascular cell necrosis, in its true sense.

Another group of noxious compounds found in the xylem of diseased plants is composed of polysaccharides, which may be byproducts of pathogen metabolism or of emzymatic vascular cell degradation. These compounds of high molecular weight are less mobile than toxins, and therefore, remain in the lumen of the vascular tissues. Polysaccharides cause functional disorders in the xylem, and therefore, are considered noxious rather than toxic compounds. Among the polysaccharides belonging to this group are fructosans, glucosans, pentosans, and pectin. Since they are not toxic in the true sense, they will not be considered in this chapter.

Well-known causal agents of toxin formation in xylem elements are facultative pathogenic fungi which may be classified into two groups based on their ecological characteristics.

The first group comprises soil inhabitors which infect plant roots and continue to grow after they reach the xylem. Examples are the vascular Fusaria, *Verticillium albo-atrum* and *V. dahliae*. These pathogens generally are wide spread in soil and cause xylem necrosis. They are difficult to control in the soil.

The second group includes a small number of fungi that develop in shoots and xylem branches, and are transmitted by insect vectors. Examples are *Ceratocystis fagacearum*, cause of oak wilt and vectored by members of the family Nitidulidae; and *C. ulmi*, cause of elm wilt and vectored by *Scolytus multistriatus*, *S. scolytus*, and others.

Lycomarasmin, a metabolic byproduct of *Fusarium oxysproum* f. *lycopersici* in tomato plants, is one of the most investigated toxins of vascular fungi. It spreads through the plant during acute infection causing leaf cell protoplast necrosis.[15] As a result, necrotic spots develop between veins and leaf desiccation occurs spreading from the apex and margin to the leaf base. *F. oxysporum* and *F. oxysporum* f. *lycopersici* also form fusaric acid, a toxin which causes similar symptoms on tomato leaves, but differs from lycomarasmin in that it does not affect bark parenchyma. Other *Fusarium* spp. produce toxins in other plants where similar modes of action are expected.

Limit data are available on low-molecular weight toxin(s) formed by *Verticillium* ssp. The presence of a toxin has been extablished in culture by an isolate of this fungus from hops[106] and alfalfa.[107]

General symptoms of toxin activity are xylem cell necrosis as well as plant stunting and wilting. The extent to which these symptoms become evident depends upon the stage of plant development in which infection occurs. During infection of young plants, toxins cause necrosis of the living conductive cells, i.e. tracheae and tracheids, resulting in a variety of severe symptoms. As the conductive cells mature through further differentiation and lignification, they become more resistant to toxin action. After complete lignification, the resulting dead cells are no longer susceptible to necrosis.

Aleksić et al.[25] established that *V. albo-atrum* caused a variety of symptoms during early infection of pepper plants, which wilted rapidly at the flowering, a disease described as green pepper wilt. The same authors found resistance to the disease among pepper cultivars using indeces to measure plant stunting and xylem necrosis.

Phytopathogenic rickettsias developing in xylem of older plants may result in necrotic spots on leaves of infected plants suggesting that a toxin may be involved (Figure 37). For example, plum leaf scard is characterized by the appearance of necrotic leaf spots.[108] Almond leaf scard, also caused by a rickettsia, results in leaf spotting. The presence of rickettsia with a similar effect on the xylem has been described for Pierce's disease of grapevine,[109] and phony peach disease.[110]

Cell necrosis of phloem tissue develops in a manner different from that in xylem tissue because of the function and structure of phloem tissue. Phloem tissue is made up of sieve cells and companion cells, which are living cells. Thus, phloem tissue provides a favorable environment for the development of obligate parasites, such as viruses and mycoplasmas.

Phloem viruses can cause either a general or local phloem necrosis. One of the best known examples is that caused by the beet curly top virus, which multiplies only in phloem tissues. A diagnostic symptom of the disease caused by this virus is the presence of dark concentric rings on injected beet roots cut in cross sections. Citrus tristeza virus also colonizes phloem tissues, but not parenchyma cells. This virus causes phloem necrosis in grafted areas, particularly on susceptible rootstocks, thus preventing the transport of nutrients to the roots resulting a comparatively rapid wilt and decay. Other examples are the potato leaf roll virus, which causes phloem necrosis in potato stems and tubers; and the potato streak mosaic virus (potato virus Y), which causes local phloem necrosis along potato leaf veins. Localized phloem necrosis has been described for other viruses in other hosts, such as the lettuce necrosis, tobacco necrosis, and tomato spotted wilt viruses.

In addition to viruses, some mycoplasmas can cause phloem necrosis.[111] For example, phloem necrosis develops on trees infected with X disease of wild cherry (*Prunus virginiana*) or pear decline. However, the necrotic processes in phloem tissues infected with mycoplasmas, progress at a slower rate than those caused by viruses. As a result of this adaptation, mycoplasma-infected plants generally remain alive longer than plants infected with viruses that cause phloem necrosis (Figure 39).

H. CYTOHORMONAL CHANGES

Hormones are biologically active substances formed in plant cells which direct their activities, including cell growth and development. Because cell growth and development may be the result of numerous cellular activities, hormones may be defined, in a broad sense, as biologically active substances which regulate and stimulate plant growth and development.

The regulation of metabolic and morphogenetic processes in the course of plant ontogentic development depends upon these hormonal functions. It may be assumed that the highest level of regulation among these processes due to hormones occurs at each stage of plant development. In order to perform their functions, hormones influence the activity of genes and other cell constituents. Glišić[21] stated that their primary influence is on the genetic apparatus and cell membranes. The influence of hormones on genes is significant since genes determine the performance of the individual processes in the cell, such as multiplication, growth, and differentiation.

Hormones are products of cells, which are formed, in contrast to animals, in unspecialized plant cells, although they may be formed in specific plant parts. However, it is difficult to distinguish clearly between the locations of their formation from that of their action. The functions of hormones differ substantially from other naturally-produced, biologically active substances in plants, such as enzymes and antibiotics.

Pathogens alter the metabolic processes as well as the structure of infected plants and therefore, inevitably cause changes in hormone content and thus, functions in such plants. The influence of pathogens on the hormonal activity of infected cells is complex. Sufficient data are not available to make a more precise statement. Nevertheless, it can be assumed that this activity can be both direct or indirect.

A direct influence may be characterized by pathogens which produce hormones in infected cells and thus, bring about changes in quality and quantity of host cell hormones. However, limited data are available concerning the role of pathogens in hormonal changes in infected plants, most of which pertains to the possibility of auxin degradation.

35

38

36

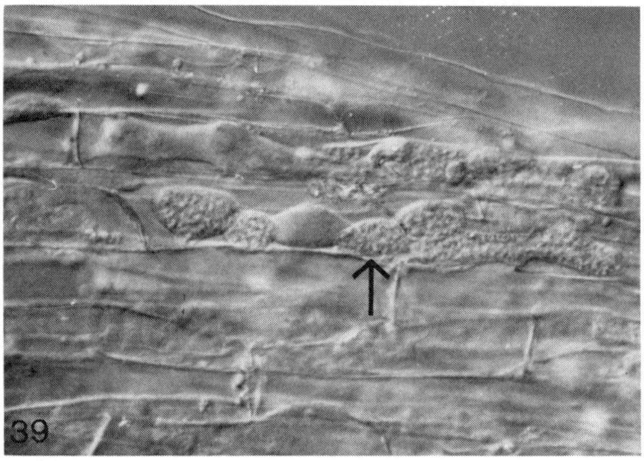
39

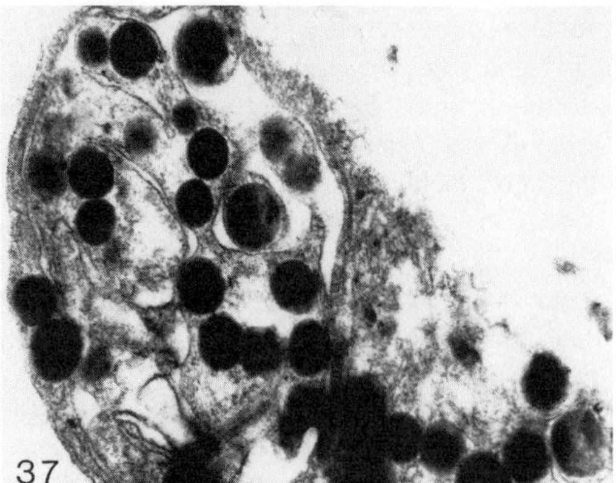
37

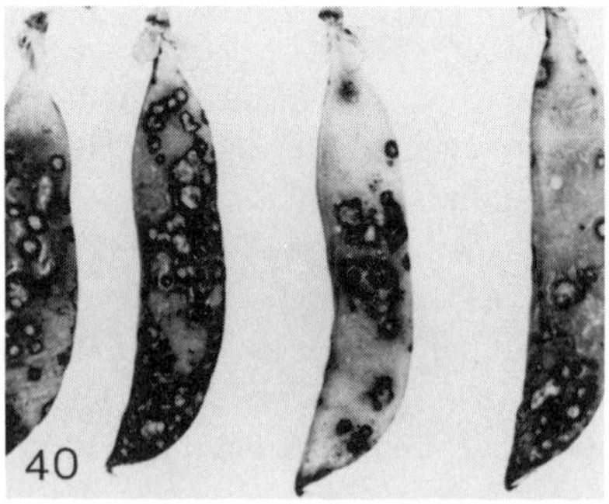
40

PLATE VII. FIGURE 35. Local lesions on a tobacco (cv. Burley) leaf caused by a hypersensitive reaction to inoculation with the tobacco mosaic virus. FIGURE 36. Anthracnose lesions on a melon fruit caused by *Colletotrichum lagenarium*. FIGURE 37. Crossection of a maize stem showing rickettsialike organisms in conductive vessel parenchyma cells. FIGURE 38. Local lesions along and between the veins on a sugar beet leaf caused by inoculation with the beet necrotic yellow vein virus. FIGURE 39. Zoosporangia of *Polymyxa betae* (arrow) in the epidermal root cells of sugar beet. (From Ivanovic, M., Doktaska disertacija, Poljoprivredni fakultat, Zemun, Yugoslavia, 1984.) FIGURE 40. Anthracnose-like lesions on garden pea pods caused by *Ascochyta pisi*.

An indirect influence, which is due to the metabolic by-products of pathogens, may alter the ability of infected cells to produce hormones, or their ability to react to their own hormonal compounds.

Many pathogens are capable of producing hormones. However, most research on this capability in living tissues remains essentially unknown. According to Wood,[15] at least 75 fungal species belonging to 35 genera, as well as numerous bacteria, produce hormones. Among the fungi producing such compounds in culture and also causing changes in plant growth are species of the following fungi: *Gibberella, Exobasidium, Taphrina,* and *Ustilago*. However, according to the same author, several other pathogens produce hormones in culture but do not display adequate activity in diseased plants to result in symptom production.

Research has shown that pathogenic organisms are capable of significantly altering the activity of hormones produced by cells i.e., abscisic acid, auxin, cytokinin, ethlene, gibberellin, and other related compounds.

1. Auxins (Heteroauxin, Beta-Indolacetic Acid)

Auxins are considered the most important group of growth hormones in plant life. The amino acid tryptophan is a precursor of auxin synthesis, giving rise to auxin formation after a series of enzymatic transformations. Auxin is sythesized in young plant tissues, primarily in young leaves of apical shoots, in apical roots, and in ovules; it is also found in large quantities in pollen grains.[21] A constant quantity of auxin is maintained in plants primarily by two mechanisms, one by controlling its rate of synthesis, and the other by controlling its degradation, either by photo- and enzymatic oxidation, or its binding with other compounds in the cell to form inactive complexes.

Primary auxin activities are concerned with their influence on the growth processes involved in ontogenetic plant development. They enhance cell division, influence cell and organ enlargement as well as cell growth and differentiation. Auxins also play an important role in fruit or seed development following pollination. In normal quantities, they prevent senescence by inhibiting ribonuclease activity thus, controlling DNA degradation. Any reduction in auxin content in leaves results in the activation of processes leading to defoliation.

Many investigators have shown that pathogenic and nonpathogenic bacteria and fungi are active auxin producers primarily by growing the microorganisms in culture.[112] The result of this activity is either the increase (hyperauxiny) or decrease (hypoauxiny) of auxin in infected plant tissues resulting in some effect on plant growth and development.

The formation of uneven quantities of auxin has been established in both obligate and facultative pathogenic fungi. The obligate pathogen, *Puccinia graminia tritici*, for example, causes a 24× increase in auxin content in cv. Little Club wheat plants.[113] Otherwise, it is not known to cause hyperauxiny in cereal plants.[114-116] According to the same author, infection by *Erysiphe graminis* in cv. Atlas barley plants resulted in an increase of this hormone ranging from only slight to 5×. The obligate parasitic fungus *Gymnosporangium,* in its aecial stage, causes hyperauxiny in pear, similar to that caused by the fungus *Uromyces* in plants of *Ranunculus*.[117] Hyperauxiny with up to 10× increases has been described on hypocotyls of *Carthamus tinctorius* infected with *Puccinia carthami*.[118] Greater amounts of auxin were recorded in stems and leaves of *Brassica napus* infected with *Albugo candida*[117] as well as in plants of *Capsella bursa-pastoris* infected with *A. candida* and *Peronospora parasitica*.[119] The parenchyma tissue of tumors on potato tubers infected with *Synchytrium endobioticum* contained more indolauxin than uninfected ones.[120] Higher levels of auxin were reported in smut galls on maize infected with *Ustilago maydis* than uninfected tissues.[121-123]

Hyperauxiny also occurred in some plants infected with facultative pathogenic fungi. Increased quantities of this hormone of up to 12× were found in tomato plants infected with *Fusarium* sp.[124] A similar report was made for tomato plants infected with *Verticillium albo-atrum*.[125]

According to Pegg,[114] plant colonization by the phytobacteria *Agrobacterium tumefaciens, Pseudomonas solanacearum, P. syringae* pv. *savastanoi*, and the nitogen-fixing bacteria *Bradyrhizobium leguminorarum*, and *B. melilothi* were characterized by increased levels of beta-indolacetic acid and its derivatives. Changes in auxin quantities also were related to metabolic changes in host plants infected with certain viruses. Jaros[126] reported potato plants infected with potato mosaic virus X had an approximate 2× increase in auxin 9 days after infection followed by a 2× decrease 50 days after infection. The level of auxin was reduced in bean, beet, and tomato plants infected with the beet curly top virus.[127]

Some cell metabolic by-products prevent or stimulate auxin synthesis. Phenolic compounds, for example, influence the activity of auxin oxidase, and it is generally understood that orthodiphenols prevent, while monophenols stimulate activity of this enzyme.[128-130]

Pegg[114] cited the following phenolic compounds as inhibitors of auxin oxidase; caffeic acid, catechol, chlorogenic acid, scopoletin, and others; and the following as activators: amino acid tyrosine, and o-, m-, and p-cumaric acid, and others.

Changes in auxin quantities in infected plants result in abnormal development and growth. These disorders also are accompanied by other symptoms depending upon the pathogen and host involved. It is often difficult to assign particular symptoms to the action of specific auxins, since the changes in infected plants are influenced by other hormones, as well as other factors. Within these limitations, it is possible to relate some pathological changes to altered amounts of these hormones.

Hyperauxiny is related to characteristic changes in plant growth and development. One such change is the appearance of hyperplasmic formations such as cankers, tumors, wart-like, or similar types of neoplastic growth stimulated by any of several pathogen bacteria and fungi. The influence of auxin(s) on the occurrence of hyperplasmic formations is primary, since it is the hormone that stimulates normal cell division. The elongations and/or enlargement of infected plant parts frequently are related to increased auxin levels. For example, abnormal hypocotyl elongation of plants of *Carthamus tinctorius* is symtomatic of infection by the rust fungus *Puccinia carthami*.[118] Plants of *Euphorbia cyparissias* infected with the smut fungus *Uromyces pisi* are elongated and less branchy with shorter and thicker leaves, which fall off prematurely, than uninfected plants.[15] Increased growth of plants of *Capsella bursa-pastoris* infected with either *Albugo candida* or *Peronospora parasitica* and leaf elongation of plants of *Sempervivum tectorum* infected with the fungus *Endophyllum sempervivi*, as well as the enlargement and curling of peach leaves caused by *Taphrina deformans* all are examples of these types of changes. It is assumed that the increase in the size of plum fruits infected with *T. pruni* is induced by the same mechanism(s).

According to some authors, the function of auxins in apical shoots, which controls lateral bud development, may be disturbed in infected plants resulting in premature and abnormal development and branching. Wood[15] pointed out that such a possibility exists in beans infected with either the southern bean mosaic virus or aster yellows mycoplasma.

An increase in auxins can cause changes in various biochemical processes in infected plants. For example, increases in auxins increased the respiration rate in cv. Little Club wheat plants infected with the rust fungus, *Puccinia graminis tritici*. However, some authors stated that these auxins might not be the direct cause of the intensified respiration, but an indirect one influencing other biosynthetic processes requiring energy.[131,132] According to Beckman et al.,[133] auxins formed by *Ceratocystis ulmi*, the oak wilt fungus, influences the change in shape of parenchyma cell walls in the xylem. Following the weakening of the cell walls, tyloses form through the action of pectolytic enzymes.

The fungus, *Omphalia flavida*, degrades auxins through the production of auxin oxidase resulting in hypoauxiny, which causes early defoliation. According to Pegg,[114] the fungus *Diplocarpon rosae*, which causes black spot of rose, is responsible for similar changes in rose leaves. Abnormal flower and seed development in *Brassica napus* were related to decreased auxin content in plants infected with *Albugo candida*. Stunting of plants caused by the tobacco mosaic virus,[134,135] as well as the potato leaf roll virus,[136,137] in their respective hosts, was attributed to reduced auxin content. It should be stressed that the influence of other hormones may be implicated in these and similar changes involving whole plants.

2. Gibberellins

Gibberellins are considered the next most important group of growth hormones. They were discovered during studies on bakanae disease of rice, caused by the fungus *Gibberella fujikuroi* (anamorph *Fusarium moniliforme*). Kurosawa[138] first established the presence of these compounds in fungus filtrates. As a result of these studies, it was established that various strains of *G. fujikuroi* produced at least 18 of the 38 gibberellins. Also, several mono- and dicotyledonous plants are capable of forming seven of the gibberellin compounds.[115] Acetic and mevalonic acids are considered indispensable for gibberellin biosynthesis.[17] The presence of large quantities of these compounds in young plant parts, such as apical meristems and young leaves, suggested that they are formed in these plant parts. Large quantitites of gibberellins also were found in immature and developing seeds. Gibberellins are transported from the point of synthesis, monopolarly, through the xylem and phloem to other plant parts. Gibberellins were found in plants in their free form or bound to sugars as glucosides, and to proteins.[21]

Gibberellins perform a variety of activities which change in infected plants. These hormones stimulate plant cell elongation, which is the most common effect of their action. For example, genetically dwarfed plants elongate under the influence of gibberellins and may reach the height of a nondwarfed plant. Gibberellins influenced the growth and development of apical shoot cells, indicating their role in plant growth.[17] Brian and Hemming[139] suggested the possibility of gibberellins participating in the stimulation of auxin synthesis and thus, increasing their level in plants.

These hormones are capable of changing plant photoperiodism by changing long-day plants to flower under short-day conditions; and changing biennial plants to flower in the first year. Gibberellins are one of the factors that influence the dormancy of vegetative buds and, in some plants, seeds. Large quantities of giberellins have been found in some seeds suggesting their paticipation in seed germination.

Gibberellin synthesis has been studied in many microorganism, but experimentally confirmed in a limited number. Pegg[115] indicated that the presence of gibberellin or gibberellin-like substances has been established, inter alia, in the following Basidiomycetous fungi: *Agaricus Bisporus, Boletus elegans, Grifora frondesa, Phallus impudicus,* and *Phellinus pomaceus*. The same author cited data from other researchers indicating that 26 bacterial species of the genus *Pseudomonas* form gibberellin-like sustances. Regardless of the limited number of studies concerned with the formation of gibberellin and gibberellin-like substances in microorganisms, it may be considered as a biological trait.

Pathogens alter the quantities of gibberellin in infected plants, thus affecting plant development and symptom expression. An excess of gibberellins in infected plants can be considered as a consequence of altered plant metabolism, caused by the influence of the metabolic byproducts of the pathogen involved.

Fusarium moniliforme causes the bakanae symptoms on infected rice plants, which are characterized by rapid plant growth due to the elongation of plant internodes, plant chlorosis, and reduced tillering compared to uninfected plants. According to Wood[15] internode elongation also was characteristic for other plants infected by other fungal pathogens; for example, *Euphorbia cyparissias* infected with *Uromyces pisi*, *Sempervivum hirsutum* infected with *Endophyllum sempervivi*, and *Cirsium arvense* infected with *Puccinia suaveolens*. According to the same author, early flowering of plants of *Primula* spp. infected with *Uromyces primulae integrifoliae* is a symptom possibly influenced by gibberellins.

Abnormal gigantic growth has been described on flowers of tomatoes and other solanaceous plants infected with the stolbur mycoplasma. Flower parts, particularly sepals, elongates covering the other flower parts resulting in the "big bud" symptom. Such growth in infected plants is associated with a disturbed hormone balance in which excess gibberellins play an important role.

In contrast, reduced levels of gibberellin in infected plants causes stunting. Such symptoms appeared on plants of *Silene vulgaris* sub. sp. *maritima* infected with fungus *Ustilago violaceae*.[115] Russell and Kimmins[140] recorded reduced quantities of some gibberellin-like substances in barley plants infected with the barley yellow dwarf virus. Stunted tobacco plants infected with the tobacco etch virus, or sour cherry trees, infected with the prune dwarf virus, tend to have remission of symptoms with the application of gibberellic acid, implying that these disorders were the result of reduced quantities of gibberellin. Reduced levels of gibberellin also have been reported in plants infected with other viruses.

Although the data are not sufficient regarding the role of the lack of gibberellin in plant stunting and dwarfing, reduced quantities of these hormones, nevertheless, are considered to be associated with stunted plant growth. There are no explanations as to the mechanism of this relationship, but it has been suggested that reduced quantities of gibberellin-like substances also may result from decreased plant metabolic activity.

3. Cytokinins

Cytokinins are natural or synthetic compounds which stimulate cell division. Skoog et al.[141] and Srivastava[142] defined them as compounds, which, when acting jointly with auxins, stimulate cell division. Because of this ability to induce cytokinesis, they have been named cytokinins.

Several substances with cytokinin activity have been isolated from microorganisms as well as from animal and plant tissues. Most cytokinins occur in the form of free purine bases, nucleosides, or nucleotides, but may to found bound to transport RNA,[143] or in the form of zeatine glucosides.[144]

Some cytokinins isolated as naturally-occurrng substances from plants, animals, fungi, and bacteria are zeatine, ms-zeatine, dihydrozeatine, 2iP, and ms-2iP. Kinetin is a cytokinin of animal origin which can stimulate plant cell division, although it has not yet been isolated as a natural component of plant tissues.[21]

It is assumed that cytokinins are synthesized in young leaves and meristematic tissues, but according to Dekhuijzen,[143] this synthesis may take place in other types of cells, since they are found in the vascular sap of many plants, then transported by the sap to growing plant parts.

Cytokinins, as well as other plant hormones, have a defined role in plant growth and development. Their primary role is to stimulate cell division. The presence of these hormones ensures cell multiplication and normal formation of tissues and organs. However, auxins also influence cell elongation without causing cell division. Cytokinins also influence bud regeneration in tissue cultures, with organ regeneration from calllus tissue depending on these compounds as well as auxins.[21]

Cytokinins stimulate cell division in root and stem tissues, leaf development, and may intitiate the development of lateral buds, freeing them from the influence of growth-control substances produced by apical bud tissues.[139] The application of cytokinins experimentally induced the proliferation fo lateral buds, whose shoots, together with the central ones, form the witches' broom symptom.

Cytokinins also play a role in delaying senescence and defoliation in healthy plants. For example, leaves removed from plants soon loose their green color due to degradation of proteins, including those to which chlorophyll molecules are bound. However, when such leaves are treated with cytokinins, they maintain their green color for a longer period, because they prevent protein, and thus, chlorophyll degradation.[145] Cytokinins also may induce chlorophyll formation.[146-148] Other compounds tend to accumulate in plant parts, such as young fruits and in some leaf parts, with increased amounts of cytokinin.

By forming these compounds, pathogens change the amount of cytokinins in infected plants. Also, pathogens are capable of altering cytokinin biosynthesis in infected cells and tissues. Increased or decreased quantities of cytokinins alter the normal ontogenetic plant development.

Changes in the quantities of cytokinin in infected plants, at first, either increase or decrease cell division. The consequences of such changes are either hyperplasia or hypoplasia of infected cells and tissues. An increase in cytokinin has been established in hyperplastic tissues, such as tumors, on roots and hypocotyls of some crucifers infected with the fungus *Plasmodiophora brassicae*. Ten to 100× increases in cytokinins above that in uninfected plants have been recorded in turnips infected with this fungus. It is difficult to determine the origin of these compounds with certainty, but it can be assumed that they are either pathogen byproducts or formed by infected plant cells.[143] Norberg[149] believed that cytokinins and auxins caused tumor formation in azalea leaves infected with the fungus *Exobasidium*. The activity of cytokinins in tumors of pine tree stems infected with *Cronartium fusiforme* was 10× greater than in uninfected trees.[150]

Among the phytopathogenic bacteria *Agrobacterium tumefaciens*, cause of crown gall, is an active producer of some cytokinins. In addition, infected cells form their own auxins and cytokinins. Since the presence of both types of these compounds cause cell division and tumor formation, it represents an interesting system for study. Another phytobacterium, *Rhodococcus fascians*, cause of fasciation in dicotyledon seedlings, also is an active producer of cytokinins. However, the mechanism of fasciation is not clear, since increased amounts of auxins and cytokinins associated with the symptoms, may be produced either or both by the fungus and diseased cells. Wilson[151] showed that cytokinin is responsible, in part, for hyperplasia and tumor formation caused by *Pseudomonas syringae* pv. *savastanoi* in oleander plants.

Cytokinins cause lateral bud development and witches‘ broom and may be induced by fungi such as *Taphrina cerasi* in cherry trees and *T. insititiae* in apricot and plum shoots; by phytobacteria such as *Rhodococcus fascians* in sweet peas (*Lathyrus odoratus*); and by mycoplasma in apple shoots and bindweed(*Convolvulus arvensis*).

Cytokinins usually formed in roots are transported to other parts of the plant through the vascular system and govern normal plant growth and development. However, the effect of cytokinins is reduced if conductivity of the vascular elements is restricted due to abiotic or biotic factors. For example, Krikun et al.[152] established a restriction of cytokinin transport by the fungus *Verticillium dahliae* in tomato plants, and Misaghi et al.[153] observed reduced cytokinin activity in cotton plants infected with *Verticillium* sp. These vascular fungal pathogens reduced xylem conductivity, so that the amount of cytokinin transported in the vascular system to other plant parts was reduced. This resulted in reduced plant cell division and growth, and thus, stunted plants. These changes are more detrimental if plants are infected early. Plant stunting associated with infection by *V. albo-atrum* also can be explained in part by this mechanism.

Cytokinins also can enhance chlorophyll synthesis, or prevent or delay its degradation. Such activity may explain the formation or maintenance of "green islands" in plant parts resulting from infection by obligate parasites, such as the causal agents of downy and powdery mildews, and rusts. This group includes *Erysiphe gramins*, cause of cereal powdery mildew; *Uninula aceris*, cause of maple powdery mildew; and *Uromyces phaseoli*, cause of bean rust.

Explanations for the development of green islands in infected plants are inconsistent. Some authors believe that they are composed of undegraded chlorophyll, while others believe that chlorophyll is degraded and then reconstituted in infected tissues. Another possibility is that normal chlorophyll degradation is delayed in the green islands alone and accumulates with synthesis of new molecules. The maintenance of green color in herbaceous plant parts is dependent on cytokinin activity, which may delay senescence and defoliation in infected plants.

The participation of cytokinins and other hormones in virescence of infected plant parts, which are normally not green, was discussed in the section on chromatic changes of chlorophyll, and remains an open question.

4. Ethylene

Ethylene, the first hormone to be described, is different from other plant growth hormones since it occurs in the plant in a gaseous form. The hormonal activity of ethylene was observed at the end of the 19th C when it, due to accidental release from illuninating gas, caused changes in development and growth in certain plants. Initially, ethylene was not considered a hormone because it was not thought to be produced by plant cells. However, using sophisticated research methods, small quantities of ethylene were found in plants, establishing its hormone activity.

Glišić[21] showed that ethylene was formed in several types of plant cells through the degradation of the amino acid, methionine, although other compounds were not excluded. It has been reported to occur in several plant organs. For example, young cotton leaves produce greater quantities of ethylene than older ones; and large quantities of ethylene are produced in flowers particularly following pollination, which may account for petal fall. The formation of this hormone in fruits increases with maturation. Ethylene formation is stimulated by various factors, such as natural and synthetic auxins, ionizing radiation, mechanical injuries, as well as pathogens.

Ethylene formation has been established in a number of pathogenic organisms. It is produced in pure cultures of *Penicillium digitatum*, pathogenic to citrus fruits, and *Pseudomonas solanacearum*.[154] Species of *Erwinia, Pseudomonas,* and *Xanthomonas* form substantial quantities of ethylene in culture.[17] Epinasty of leaves of *Physalis floridana* infected with potato streak virus (potato virus Y) was associated with abnorally high levels of ethylene.[98] Some pathogens do not form ethylene in culture even though it is detected in plants infected by the pathogen(s). This led Williamson[99] to conclude that ethylene formation originated in susceptible host plants as a consequence of cell damage.

Excess ethylene related to plant disease can cause premature symptoms of senescence and defoliation; for example, in black spot of rose, caused by *Diplocarpon rosae,* and cherry leaf spot, caused by *Cocomyces hiemalis*.[15] Both pathogens cause dark, necrotic spots on leaves with limited growth; infected leaves turn yellow beyond the points of infection and drop premature. Toxins also can have a role in early yellowing, but the primary cause has been attributed to excess ethylene. Increased levels of ethylene have been found in tomato leaves infected with *Cercospora personata*, accompanied by leaf yellowing and premature defoliation.[155] Significant increases of ethylene were reported in carnation plants infected with *Botrytis* sp.[156] Freebairn and Buddenhagen[154] recorded large quantities of ethylene in banana fruits infected with *Pseudomnas solanacearum* which was related with early fruit maturation.

Increased quantities of ethylene in plant tissues have been associated with similar symptoms in fruits and leaves damaged by arthropod pests. For example, cherry and rose leaves attacked by spider mites contain higher concentrations of ethylene than leaves that have not been damaged. The apple and pear fruit maggot, *Laspeyresia pomonella*, as well as the peach fruit maggot, *Cydia molesta*, cause permature yellowing, maturation, and fruit drop. Symptoms of maggot damage suggested ethylene production as a possible cause of the symptoms.

The second general symptom caused by excessive ethylene is epinasty of leaves on infected plants. Epinasty is characterized by downward bending of leaf petioles beginning at the leaf base, so that the entire leaf curves downward giving a hooked appearance. This bending is the result of more intense growth of the tissues on the upper portion of the petiole, particularly at the base, than the lower portion. Excess ethylene in plant tissues causing epinasty may occur when pathways of ethylene dissemination are blocked; for example, when vascular bundles are blocked by polysaccaride products of degradation, or stomatal closures in plant wilting.

Epinasty has been described as a symptom for several diseases; for example, tomato leaf epinasty caused by F. *oxysporum* f. sp. *lycopersici*;[77] and epinasty, chlorosis and other malformations of hop plants caused by *Verticillium* sp.[4] Quince sooty ringspot disease causes epinasty of young leaves of cultivars F and C7/1.[35] The same disease, also known as the Spy 227 apple epinasty and decline, causes similar symptoms in that particular apple cultivar.

There are numerous reports on ethylene influencing infected plants in other ways. Ethylene stimulates respiration in plants, which may lead to senescence of plant organs. Ethylene also can stimulate enzyme and phytoalexin formation in infected plants, thus possibly participating in plant disease resistant reactions. If this is true, the study of the effect of ethylene on resistance is of practical and theoretical importance.

5. Abscisic Acid and Abscisic-Like Compounds

In addition to growth stimulating compounds, plants also contain growth inhibitors, which delay or prevent growth processes. Growth inhibitors generally are wide-spread in plant life. Among this group is phenolic compounds which have been shown to inhibit certain metabolic processes in plants. However, abscisic acid is the only true hormone-type of growth inhibitor which is active at low concentrations and induces dormancy in various plant organs.[116] Abscisic acid is an isoprenoid compound which may be formed either by synthesis from mevalonic acid or degradation of some carotenoids.[21] Violaxanthin is such a carotenoid which oxidizes under the influence of light to form xantho toxin, another highly inhibitory compound. Xantho toxin can be a source of abscisic acid, the latter which is found in plant tissues either in a free state or bound to glucose.

Abscisic acid acts in concert with other hormones making it difficult to clearly define its activity or role in diseased or healthy plants. Pegg[116] summarized the results of several investigators who reported a variety of activities associated with this hormone; it caused fruit drop in cotton, defoliation in yellow lupin, bud and leaf dormancy in maple trees, and delayed seed germination and hypocotyl elongation in lettuce and other plants. Abscisic acid influences stomatal opening and closing, thus controlling water loss in several plants.[157]

The role of pathogens in abscisic acid synthesis in infected plants has not been thoroughly investigated. However, it can be expected that certain pathogens alter the levels of this compound thus, affecting ontogenetic plant development.

Limited data indicate that abscisic acid increases in infected plants resulting in either stunting or dwarfing, or stomatal closure, which in the case of vascular disease, directly influences water loss. Steadman and Sequeira[158] established the presence of abscisic acid in tobacco plants infected with *Pseudomonas solanacearum* and attempted to define its influence on stunting in acute stages of the disease. Increased levels of abscisic acid were found in the vascular bundles of bean, tomato and wheat plants that were blocked by vascular pathogens. It was assumed that the pathogens were responsible for the increase. Wiese and DeVay[159] showed that cotton plants infected with a defoliating strain of *Verticillium dahliae*, contained twice the level of abscisic acid than plants infected with a nondefoliating strain. Since this hormone is known to cause defoliation in plants, it was assumed that it was responsible for the defoliation in cotton. However, several pathogens can cause defoliation in many plant species, therefore, the role abscisic acid play in these cases should be considered.

Finally, it should be stressed that all hormones act in concert in normal plant development. By altering the level of any hormone, either by abiotic or biotic agents, can cause pronounced structural and physiological changes in affected plants.

I. CYTOGENETIC CHANGES

Plant growth and development from embryogenesis to sexual reproduction depends on regular nuclear division. In a broad sense cytogenetic changes refer to altered chromosome structure and function in infected plant cells. These changes include changes in the quantity of chromatin, and disturbed nuclear and chromosome division. Genetic changes pertaining to genotype and mutations are not included in this discussion.

1. Changes in Chromatin and Chromosome Division

The reduction or disappearance of chromatin in infected plant cells has been known for some time. For example, Küster[5] observed the reduction of chromatin in maize cells infected with *Ustilago maydis*, and cells emptied of contents in crucifers infected with *Plasmodiophora brassicae*. The loss of chromatic reticulum was recorded in leaf cells of highly susceptible wheat plants infected with *Puccinia graminis tritici;* upon reaching this state, cells decayed and were transformed within 14 days into uniform, evenly colored masses.[3] The fungus vector of beet necrotic yellow vein virus, *Polymyxa betae*, develops inside the epidermal cells of sugar beet roots (Figure 39). In the course of the development of the plasmodia, zoospores, and resting spores; it entirely fills the internal cell spaces. The disappearance of cell constituents in the final stages of its development results in cell death.

Another example of the loss of nuclear constituents is nuclear fragmentation in onion cells infected with *Botrytis cinerea*.[160] Nuclear fragmentation or splitting is irreversible and the nuclear parts are not capable of karyonatic division. Such cells resemble old cells, which may divide amitotically, since they are not capable of any other form of cell division.

Phytopathogenic viruses have a direct effect on nuclei and cell organelles as a result of virus multiplication in living cells of susceptible plants. For example, Gibbs and Harrison[161] showed that RNA of tobacco mosaic virus is synthesized in the nucleus, particularly in the nucleolus, of tobacco plants. Particles of red clover yellow mosaic virus were shown to accumulate first in the nucleolus of that host. Final synthesis of virus particles, the assemblage of nucleic acid and protein coats, usually takes place in the nucleus or the cytoplasm. Thus, virus RNA or whole viruses can be found in the nucleus, significantly changing the composition of that organelle.

Virus particles in a cell and nucleus can either disturb normal nuclear division or prevent it. According to Tarr[40] this has been shown in chrysanthemum, petunia, and tomato infected with the tomato aspermy virus. It was shown that amitotic nuclear division ending with nuclear fragmentation occurred in shoot cells of infected petunia plants.

Little reliable data are available to explain the cause of amitotic division in infected plant cells. It is assumed, *inter alia,* that it is a result of the lack of RNA needed for the formation of the spindle apparaturs, which is indispensable for mitotic division. RNA deficiency may be a consequence of either decreased nucleus biosynthesis activity or caused by virus particle RNA formation. Other anomalies pertain to chromosomes division and movement during nuclear division. Following the destruction or disappearance of the spindle apparatus, chromosomes may scatter in the cell or aggregate to form large or giant nuclei. Tarr[40] suggested that this may explain, in part, sterility in plants infected with certain viruses.

Pejkić and Šutić[162] observed irregular nuclear division in pollen mother cells of the plum cultivar Požegača infected with plum pox virus, resulting in an increased number of polyvalent chromosomes in reduction division and a delay in chromosome movement during the division process. The influence of virus particle multiplication on nuclear division in host cells provides a model for further study.

2. Changes Occurring under Foreign Genetic Control

Metabolic and morphogenetic processes in cells are under the control of DNA, which transmits genetic information from one cell to another and from one organ to another during cell division. Each cell of every organism contains a complete genetic message ensuring continuity for the individual species.

The fundamental role of DNA in the continuation of individuals and species is performed in the controlled snythesis of various proteins indispensible in cell structure and function. Genetic messages are transferred, by transcription, from DNA into messenger RNA (mRNA). Messenger RNA, through the process of translation, bind amino acids in a determined order resulting in polypeptide and protein synthesis (for mechanisms of transcription and translation, see section on protein synthesis in diseased plants). Various parts of the DNA molecule differ functionally and form several types of mRNA, which provides for the formation of corresponding proteins. Genes arranged along DNA molecules form chromosomes, which can be defined as genetic information carriers. In addition to chromosomes, extrachromosomal formations, such as plasmids and episomes whose genetic structure is formed by DNA, also serve as carriers of genetic information in several species of bacteria, particularly the Gram-negative forms. Plasmids are existing chromosomes.[21]

Some plant pathogens are capable of altering the autonomy of host cell genetic information. This information may be impaired by changes in the plant cell genome function or structure. For example, pathological changes in the genome structure were described in plant tumor cells induced by *Agrobacterium tumefaciens* and the clover wound tumor virus. The introduction of tobacco mosaic virus nucleotide sequence into tobacco host cell genome substantially altered the composition and function of this genome.

A. tumefaciens altered the genome composition in tumor cells after penetrating the host cell plasmid DNA in which it reproduced.[13] Tumor cells thus contained a new genome composed of both bacterial and host cell genomes. The new genome formed is the product of two very different organisms, a eukaryote (the host) and a prokaryote (the bacterium).

The mechanism of incorporation of the bacterial DNA into host DNA is not understood completely. It is possible that it is inserted into host chromosomes or other organelles, where it replicates, or plasmid DNA first replicates autonomously and is later incorporated onto different locations in the host cell DNA. The mechanism of DNA plasmid replication also is not completely known, but it probably occurs along with host DNA with the aid of host polymerase DNA.[163]

Plant tissue damage favorably influences the binding of bacterial and host cells, and facilitates the penetration of bacteria plasmids into host protoplasts. Infected cells composed of both genomes further divide without the presence of bacterial cells. The consequence of this division is tumor formation, basically similar to that of animal and human tumors. The amino acids octopin and nopalin were found in such tumors, but not in healthy susceptible plants.[164-166] The synthesis of these amino acids in tumor tissues is encoded in the bacterial DNA plasmid.

The formation of new genomes was described in tumor tissues on clover infected with the clover wound tumor virus.[163] In infected tumor cells, the virus genes combined with host cell genes.[167] Even though the genetic composition and variability of this virus has been studied,[168] the mode of incorporation of the virus gene into host cell genes has not been clarified completely.

Deom et al.[169] successfully introduced the nucleotide sequence gene of the Ls_1 strain of the tobacco mosaic virus (TMV) into the cell genome of *Nicotiana tabacum* cv. Xanth. This nucleotide sequence encoded synthesis of 30-kD protein in the structure of the TMV capside, facilitating the spread of virus particles from one cell to another.

The introduction of the TMV nucleotide sequence into the tobacco cell genome was performed using genetic transformation techniques for RNA viruses. In this process a chimeric gene containing a cloned complimentary DNA of the 30-kD protein gene was introduced into tobacco cells via a Ti-plasmid mediated tranformation system of *A. tumefaciens*. The successful transfer of the virus nucleotide sequence was confirmed by the fact that transgenic plants, regenerated from transformed tobacco plants, expressed the 30-kD protein messenger RNA and accumulated 30-kD protein. In the inoculated transgenic plants, virus particles of the Ls_1 strain moved from cell to cell. In inoculated, genetically normal plants, movement of the virus did not occur. These results confirmed that 30-kD TMV protein was responsible for virus movement in inoculated tobacco plants.

Generally, virus particles are among the most common extrachromosomal entities that can redirect genetic message activity in plant cells. This is borne out by the active participation of host cells in the synthesis of virus particles.[18]

All parts of the virus particle are synthesized in plant cells outside of host cell genome control. Progeny virus nucleic acids representing virus particle genomes are formed under the influence of viral nucleic acid and host polymerase activity. Ribonucleic acids of some viruses such as TMV and brome mosaic virus, directly transform into mRNA. In other plant viruses, mRNA is formed by transcription from the virus DNA or RNA. Messenger RNA also takes part in the translation of virus particle coat protein synthesis. In the relationship between bacteria and bacteriophages, normal bacterial cell genome transcription is hindered facilitating bacteriophage synthesis.[170]

Thus, *A. tumefaciens* and plant viruses alter the genome integrity and autonomous functions of plant cells. These changes in genetic information can be considered as among the most detrimental to a cell because they alter and disturb the normal reproduction mechanisms on which the continuity of life depends.

V. INTRACELLULAR PATHOGENS

Numerous species of pathogens colonize plant cells and live partially or completely within them, and are defined as intracellular pathogens. In contrast to animal cells, intracellular parasitism occurs in plant cells that possess cell walls, which in turn act as a protective skeleton for the inner plant cell constituents. Intracellular pathogens belong to different groups of pathogenic agents which may colonize and live in cells of all plant tissues. These pathogens also can colonize tissues of the vascular system.

Intracellular pathogens develop in the protoplasts and cause specific cytological and structural changes. The type or extent of these changes depend on the pathogen, the host cell and its sensitivity to the pathogen, and environmental conditions during disease development.

A. FUNGI — INTRACELLULAR PATHOGENS

Several pathogenic fungi both facultative and obligate parasites live in plant cells. Intracellular obligate pathogenic fungi are characterized by thalli in the form of plasmodia, and since they lack an outer wall, survival is possible only in host cell protoplasts. Therefore, these obligate intracellular pathogens are an ecological group of organisms for which the living cell represents the indispensable life environment. All stages of these pathogens develop in the living cells. This group stimulates cell development resulting in hypertrophy and hyperplasia in affected tissues.

Such changes in host tissues are distinctive and represent the effect by a group of microorganisms that affect the hormone balance and composition of infected cells. These pathogens also represent the principle of self-preservation in phyllogenetic development, since spread within an infected host is limited to the immediate area of penetration.

Obligate intracellular fungi develop in the living cells of epidermal, parenchyma, and meristematic tissues, but not in phloem tissues, which is composed of both living and dead cells. Common representative of this group are *Plasmodiophora brassicae* and *Synchytrium endobioticum,* causal agents of cytoplasmic and cytotrophic changes resulting in the production of clubs, warts, and other neoplastic formations (Figures 9 and 10) (see section on hyperplasia and hypertrophy).

Among some of the facultative parasitic fungi that colonize plant cells there is a group that live exclusively in xylem cells and are designated as vascular pathogens. Other fungi belonging to this group also develop in cells of other tissues, primarily parenchyma tissue, and are considered as less specialized intracellular pathogens.

Intracellular facultative parasites infect host cells and complete their development following the death of these cells. In contrast to obligate parasites, these fungi affect infected cells primarily by stimulating enzyme and toxin production (see section on vascular necrosis). The consequences of these specialized fungi are chlorosis, stunting and wilting of plants, and cell necrosis in infected plant tissues.

The less specialized vascular pathogens include the facultative fungus *Pythium debaryanum*. *P. debaryanum* is one of the most destructive pathogens in herbaceous and woody plant seedling production. This fungus develops in and between parenchyma, epidermal, and conductive tissue cells of the lower stems of seedlings of many plant species. Hyphae of this fungus lack haustoria, thus feed directly on infected cell constituents, causing rapid cell decay and death resulting in lodging of young plants.

The vascular fungi are polyphagous and develop in the xylem of a large number of herbaceous and woody hosts. This group includes *Fusarium oxysporum, F. vasinfectum, Verticillium albo-atrum, V. dahliae* and others. *V. albo-atrum* cause of green pepper wilt[25] and *F. vasinfectum* cause of cotton wilt[171] are of economic importance in Yugoslavia. The fungus *Phomopsis helianthi,* following hydathode penetration, colonizes lead vessels, spreads along midribs and into the vascular bundles of sunflower stems. Infected plants desiccate rapidly resulting in scald-like symptoms.

Vascular pathogens of herbaceous plants often are soilborne since they survive in infected crop debris for several years. Such pathogens are particulary destructive and difficult to control under normal plant production conditions.

Specialized vascular pathogens of woody plants include the facultative fungi such as *Ceratocystis ulmi* cause of Dutch elm disease, and *C. fagacearum* cause of oak wilt. These pathogens cause extensive stem decay and are a serious problem in the production and maintenance of these hosts. As specialized vascular pathogens of woody plants, these fungi are transmitted by insect vectors (see section on plant vascular necrosis). The identification of the vectors and subsequent control of them has resulted in better control of the diseases.

B. BACTERIA — INTRACELLULAR PATHOGENS

All phytobacteria are facultative parasites and usually are associated with parenchyma and vascular tissues. Most develop between cells of parenchyma, the least differentiated of plant tissues. Such pathogenesis is considered as primitive compared to the more advanced obligate parasitism. Other phytobacteria survive intracellularly in xylem cells and may complete their entire life cycle in xylem elements. Similar to pathogenic vascular fungi, they stimulate the production of various enzymes and metabolites which are responsible for cellular degradation. The symptoms generally associated with vascular bacteria are stunting, wilting, and significant yield reductions. A diagnostic symptom is vascular discoloration accompanied by a slimy, semifluid exudate in the vessels.

Vascular bacteria are found in all genera of phytopathogenic bacteria. However, the most important belong to the genera *Corynebacterium* and *Curtobacterium,* which are specialized for living in vascular tissues. Being nonflagellate, they move passively, transported in the xylem fluids. Examples of some phytopathogenic vascular bacteria are *Corynebacterium michiganense* pv. *michiganense* cause of tomato canker and wilt, *C. sepedonicum* cause of potato ring rot and wilt, and *Curtobacterium flaccumfaciens* pv. *flaccumfaciens* cause of bean wilt. Examples of other bacterial vascular pathogens are *Erwinia stewartii* cause of maize wilt, *Pseudomonas solanacearum* cause potato brown rot and wilt, and *Xanthomonas campestris* pv. *campestris* cause of black rot of crucifers.

The vascular bacteria represent a specialized group of plant pathogen adapted to living in the dead cells of xylem vascular elements.

C. MYCOPLASMA-AND RICKETTSIALIKE ORGANISMS — INTRACELLULAR PATHOGENS

Mycoplasmas were first described in 1967 as causal agents of plant diseases that has been considered caused by viruses.[172-174] Over 90 plant mycoplasmas have been described. In some disease situations, the symptoms and the presence of what appears to be mycoplasmas in phloem tissues is suggestive, but the association has not been shown conclusively, therefore, the causal agent it referred to as "mycoplasmalike" (MLO).[175,176]

Mycoplasmas belong to the class Mollicutes.[177] The term "mycoplasma" includes all species belonging to this class as well as plant parasitic spiroplasmas.[178] Mycoplasmas are among the smallest of prokarotes. They have no cell wall, but are protected by an envelope. They contain ribosomes with RNA and double-stranded DNA.

Mycoplasma and MLO's have only been described in phloem tissues, and therefore, are considered as intracellular pathogens. They move passively through the phloem being transported by the flow of the sap, pass through the sieve pores of the sieve plates and spread to other sieve tubes and phloem tissues. Since they develop exclusively in phloem cells, they are considered as obligate parasites.

Since plant mycoplasmas are difficult to grow in pure culture on artificial media, their indentification can be uncertain. However, one of them, *Spiroplasma citri,* cause of citrus stubborn disease, has been identified in pure culture and is recognized as an established plant pathogen.

Mycoplasmas multiply and become distributed in sieve tubes rapidly inhibiting the flow of sap and the transport of photosynthates, which results in destruction of chlorophyll and tissue chlorosis.[179] In addition to chlorosis, infection results in phloem necrosis, red discoloration, and leaf rolling (see section on plant vascular necrosis). A variety of physiologic and pathologic changes take place in sieve tubes and companion cells, as well as phloem parenchyma.

Symptoms associated with mycoplasmas are related to different disease mechanisms; yellowing and stunting with disorder in phloem function and to metabolites; witches' broom and bud proliferation with hormone inbalances; and greening or virescence of floral parts with hormone imbalance. The manifestation of symptoms in infected plants depends on the host, insect vector, and environmental conditions during pathogenesis. For example, stolbur mycoplasma causes leaf yellowing, wilting and stunting on pepper plants.[25] The same pathogen causes big bud in tomato and witches' broom on bindweed (*Convolvulus arvensis*). Pears infected with the pear decline mycoplasma have yellow, malformed, and stunted leaves. The apple proliferation mycoplasma causes witches' broom on shoots (see section on virescence).

Rickettsias belong to a group of organisms with a cell wall. Two types of rickettsias related plant diseases have been described, one associated with phloem and one with xylem.[176] Rickettsias are prokaryotes and are classified in the class Schizomycetes.[180] However, Maramorosch[175] suggested that the organisms found in the phloem may be very small bacteria or representatives of *Chlamydia,* and not rickettsias.

Rickettsias associated with xylem cause a number of diseases, for example Pierce's disease of grape, phony peach, and plum leaf scard. Those associated with phloem cause diseases such as carrot proliferation, citrus greening and wheat chlorosis, which resemble diseases caused by mycoplasmas.

Mycoplasma- and rickettsialike organisms differ taxonomically, but many have common plant and insect vector hosts. Maramorosch[181] suggested that the insect vector may represent the original hosts of these organisms and that plants became hosts later.

D. VIRUSES — INTRACELLULAR PATHOGENS

Viruses are obligate parasites which develop exclusively in host cells and persist indefinitely in infected cells. Some viruses stimulate the formation of other constituents, known as inclusions or inclusion bodies, in infected cells. In contrast to other plant pathogens, viruses generally cause systemic infections in susceptible plants, which is achieved by a rapid spread of the virus throughout the plant, both intercellulary and through the vascular system. The virus spreads intracellularly through the plasmodesmata system. A more rapid spread to greater distances within a plant occurs in the phloem and xylem, with phloem being the more favorable.

All organs of systemically infected plants may not become infected. For example, seeds resist infection by many viruses; approximately 45 of 300 economically important plant viruses are known to be seed transmitted. Systemic infection does not always involve meristematic cells, because of some protective mechanism of the immature cell, or the intercellular spread of the virus is slower than the rate of cell division. The isolation of virus-free meristematic cells from apical meristems of infected plants allows for the production of virus-free plants in tissue culture.

A general trait of most plant viruses is their ability to infect all types of cells and plant tissues. The primary characteristic of this large nonspecific group is that they cause symptoms of various mosaics, mottling, and leaf deformaties followed by abnormal growth and development of other plant organs. This group comprises the largest number of virus and includes, *inter alia,* alfalfa mosaic, apple mosaic, cucumber mosaic, plum pox, potato mosaic (potato X), and tobacco mosaic viruses (Figure 43).

Viruses that exclusively or predominantly develop in phloem tissue belong to a smaller, specific group, which cause characteristic leaf rolling, curling and stunting, cell and tissue necrosis, internode shortening, and plant dwarfing as well as rapid plant death. This group includes the beet curly top and potato leaf roll viruses, and probably the beet necrotic yellow vein and citrus tristeza viruses.

Another specific group of viruses are adapted to living in the phloem as well as cells of other tissues. These viruses cause leaf brittleness, yellowing and thickening, phloem necrosis, and plant stunting. Typical representatives of this group are the beet yellows and potato streak (potato Y) viruses.

Cell inclusions are unique foreign substances occurring in cells of plant infected with certain plant viruses (Figures 41,52). They may be found in the cytoplasm, nucleus, or in some cases, nucleolus. Belli[182] classified such inclusion into three types based on their shape and structure: (i) crystals, composed only of virus particles arranged in the form of hexagonal, rhomboid, or elongated prism-like crystals. This type is frequently associated with cells infected with tobacco mosaic virus (TMV); (ii) amorphous or X-bodies, composed of virus particles and cell constituents, such as endoplasmic reticula, mitochondria or ribosomes. This type also is associated with plants infected with TMV (Figure 43); and (iii) pinwheels, composed mostly of protein with a central tubule from which curved "arms" radiate giving a pinwheel appearance (Figure 41). This type develops first in association with the plasma membrane at sites lying over plasmodesmata. Later they are dissociated from the plasmodesmata and move into the cytoplasm. Pinwheel inclusions are associated with potyviruses, such as the plum pox viruses and potato virus Y.

The type of inclusion body formed usually is characteristic of a given virus, although some viruses may form more than one type of an inclusion body, even in the same cell. Because of this specificity, inclusion bodies may be used as an internal symptom for the rapid identification of a virus present in diseased cells (Figure 41).

VI. STRUCTURAL CYTOPATHOLOGICAL CHANGES

The changes that occur in infected plant cells are primarily related to changes in cell structure

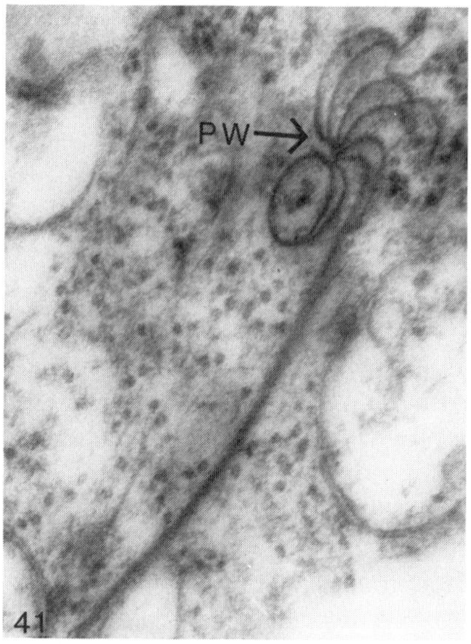

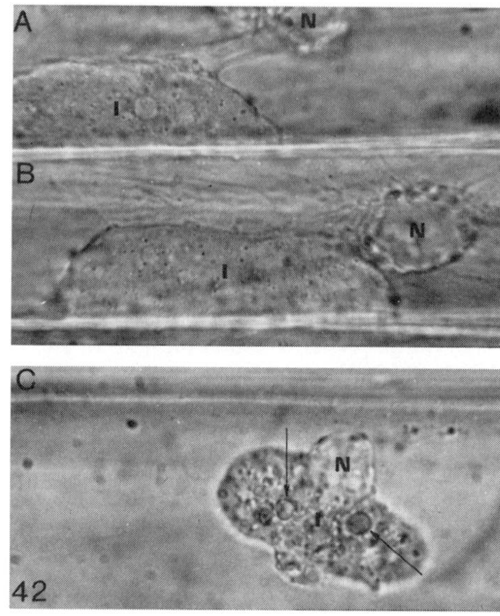

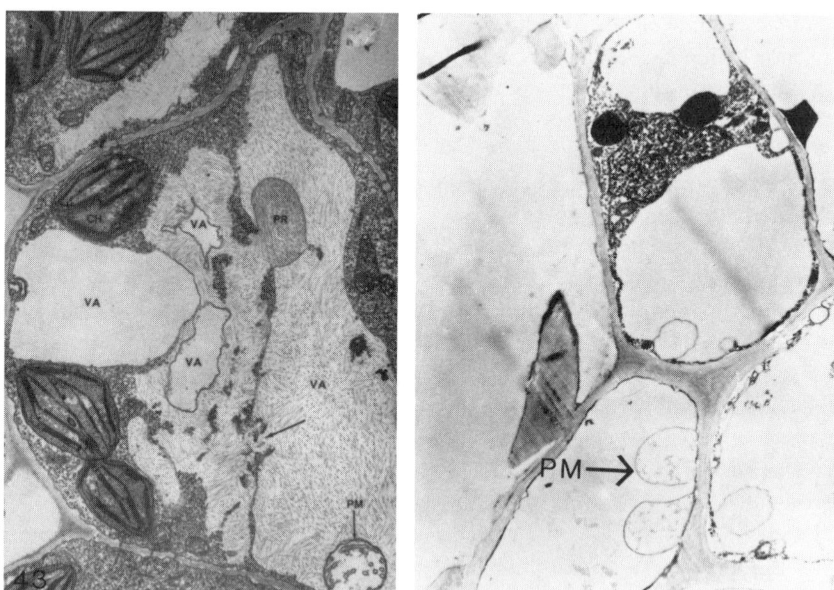

PLATE VIII. FIGURE 41. Pinwheel-shaped inclusion (PW) in a parenchyma leaf cell of *Stellaria media* infected by the plum pox virus. FIGURE 42. Turnip leaf epidermal cells infected with the European strain radish mosaic virus showing inclusions (I) close to the nucleus (N) of two cells (A,B), and a granulated inclusion and host cell nucleus an aggregate (C). Oil droplets are shown (arrows). (From Štefanac, Z. Habilitation's thesis, Belgrade1973.) FIGURE 43. Tobacco mosaic virus particles in a parenchyma cell of tobacco (*Nicotiana tabacum*) showing plasmalemma invagination (PM) into the cell vacuole (VA) and tonoplast breakage (arrow) tonoplast invagination (PR) (left) (From Esau, University of Wisconsin Press, Madison, 1968) and xylem cells of sugar beet infected with the beet necrotic yellow vein virus showing invagination of the plasmalemma (PM) (right).

and physiological activities. These changes are interlinked and are a product of other biochemical and physiological activities affected by these changes. Those changes in the structure of cells, cell walls, or protoplasts that can be observed with the unaided eye or with bright-field microscopy are termed as structural changes. Those that are observed with an electronmicroscope are termed as ultrastructural changes.

Infection of plant cells takes place as soon as a pathogen begins using the protoplast or protoplast constituents released from affected cells as a nutrient base. In order to reach the protoplast and accomplish infection, the pathogen must pass through the cell wall, which is a complex and highly organized protective barrier. Therefore, structural changes may be classified according to the stages of pathogensis, i.e. inoculative or infectious.

Inoculative changes take place in the course of inoculation and/or pathogen penetration into the cell and include all changes occurring in the degradation of the cell wall, and may be referred to as preinfection alterations. Inoculative changes can be caused by pathogenic bacteria and fungi, which possess the ability of locomotion. These pathogens initially establish contact with the epidermal cells, cells in the substomatal cavity, or in parenchyma tissue.

Upon establishing contact with the cell wall, pathogenic fungi use mechanical pressure and/or enzymatic activity for its degradation. In contrast, most plant pathogenic bacteria degrade host cells chemically and are incapable of mechanical penetration. Lacking the ability of active movement, viruses as well as the obligate parasitic mycoplasm- and rickettsialike organisms cannot cause inoculative changes. They must be introduced directly into the protoplast by some other means, such as an insect vector and other natural or artificial means.

Immediately upon protoplast infection, a series of changes occur, which are caused primarily by the action of enzymes, hormones, and/or toxins. Only viruses differ in this respect causing changes in the metabolism of the host cell protoplasts by their genetic influence.

Following colonization of the host cell protoplast and suppression of the defensive mechanisms of the cell, the infected cell must adjust to the burden of being colonized. In most cases, a breakdown of the cell system takes place. The dynamics of such changes, particularly immediately after infection, occasionally appear stimulatory, but it is of such short duration that it does not alter the depressive nature on the life of infected cells. A limited number of pathogens cause changes of a stimulatory nature such as hypertrophy and hyperplasia following infection. Disorders of a stimulatory nature in infected cells are considered in other sections and will not be discussed here.

A. CELL WALL — THE BARRIER FOR PATHOGEN PENETRATION

The cell wall is a barrier limiting or preventing penetration of pathogens into the protoplast and therefore, plays an important role in the protection of the cell from infection. This protective nature of the cell wall is due in part to its chemical composition and physical characteristics.

B. CHEMICAL CELL WALL COMPOSITION

The chemical composition of cell walls differ widely depending on the cell type and function in a tissue. Different chemical compounds which provide partial or full protection against pathogens may be included in the composition of a cell wall. These substances are products of metabolic protoplast activity and can be classified as lignins, lipids, polysaccharides, or proteins.

The most important lipids of a cell are cutin, suberin, and wax, highly resistant to various influences, and are found in cells forming protective tissues on the plant surface. Polysaccharide cell wall constituents are pectic substances, such as celluloses and hemicelluloses. Protein constituents of cell walls are referred to as structural proteins.

Wax. Wax occurs in the outermost parts of epidermal cell walls. It permeates the cuticle layer and also may occur in cutinized parts of cells walls beneath the cuticle. Special waxy deposits consisting of wax crystals of different shapes are formed on the surface of the cuticle. Chemically, waxes consist of a long series of paraffins, higher alcohols, ketones, esters, and various acids.[183]

Cutin. Cell wall cutin is found in the outer portions of cell walls and cuticle. In its pure state, cutin covers the external cell surface; a process called cell wall cuticularization. As an impermeable substance, cutin is wide-spread in epidermal cell walls, but is present in the structure of other types of cells.[17] A cutin envelope has been described along the margin of guard cells as well as in the walls of cells in the substomatal cavity, which represents the invagination of the epidermal tissue with its outer cuticle. Cutin also covers all cells lining intercellular spaces as well as root hairs.[184-186] By being part of the epidermal cell wall structure, cutin becomes wide-spread in all herbaceous plant parts covered with this tissue.

Chemically, cutin is a monocarboxylic acid polyester with a series of 16 and 18 carbon atoms and two to three hydroxyl groups. Binding of hydroxyl groups give rise to three-dimensional forms.[183]

Suberin. Suberin is a lipid precipitated in the cell wall after reaching the cell has reached maturity. Cell wall suberinization involves the placement of this corky material in the form of fine plates on the inner side of the primary wall, giving rise to secondary wall thickening. Differing from cutinization, suberinization involves the formation of a special corky layer associated with cell walls, but not permeating the cell wall structure. Suberinization commonly occurs in peridermal cells involved in wound healing. Corking also is observed in some specialized cell layers, such as the endoderm and exoderm, where suberinization plays an important role in root development.

Suberin chemistry is not fully understood. Džamić[187] stated that it may represent a mixture of compounds consisting primarily of condensed and/or oxidized products of certain unsaturated fatty acids, their esters or soaps, such as corticinic, phellonic, phloinic, phloinolic and suberic acids.

Pectic substances. Cell wall pectic compounds occur in three forms; protopectin, pectin, and pectic acid. Basically, all are composed of araban, galactan, and polymethylgalacturonic acid.[188] Pectic compounds are the primary constituents of the middle lamella and when bound to cellulose, are part of the primary and secondary cell wall composition.

Hemicelluloses. Hemicelluloses are a heterogenous group of polysaccharides partially composed of glucans, mannas, and xylans. They are the main constituents of primary and secondary walls binding cellulose and pectin molecules.[189,190]

Cellulose. Cellulose (alpha-cellulose) is a characteristic polysaccharide found in most plant cells and is the most abundant compound in plant life. It consists of long chains of molecules linked with 1,4-beta-glucoside bonds.[17] It occurs in its elemental filamentous form as microfibrills and is the primary constituent of primary and secondary cell walls. In the primary cell wall, it is arranged more or less at random, whereas in secondary cell walls it occurs in parallel plates with the layer oriented in different directions.

Lignins. Lignins are aromatic macromolecular compounds formed by the polymerization of coniferyl, sinapiryl, and p-hydroxycinnamic alcohol.[191] Lignin occurs in cells walls permeating its parts through a process called lignification. The process usually begins in the middle lamella and later encompasses primary and secondary cell walls, accumulating in the primary wall. It is the main constituent of the middle lamella in lignified tissues. Cowling and Brown[2] stated that it represents 20 to 30% of the primary and secondry cell wall constituents. Lignin generally occurs in the older cells of woody plants, but can be found in epidermal and hypodermal cells of some xerophytic plants.[17] It is the second most important substance in plant life, after cellulose. As a structural constituent of cell walls, lignin acts as a barrier to cell wall polysaccharide-degrading enzymes.

Injuries or death of parasitized cells often result in polyphenol oxidation or formation of lignin both in infected and noninfected cells surrounding points of infection.[192] The formation of lignin to resist pathogen enzyme acivity represents of the most important protective mechanisms in plant cells and tissues.

Structural proteins. Structural proteins are present in cell walls, but data concerned with their occurrence and function differ among authors. According to some researchers, the proteins are glycoproteins rich in hydroxyprolines not found in the composition of protoplasma proteins. Sugars of this protein are primarily arabinose and galactose. Hydroxproline proteins may crosslink with the carbohydrates, forming specific cell wall structures.

According to some authors, protein constituents are located in the cuticle and middle lamella,[193,194] while other authors believe that the proteins may be deposited in the primary cell wall during maturation.[195,196] The presence of proteins in cell walls has raised the issue of whether or not the cell walls can be considered living or nonliving.

In summary, aromatics, lipids, polysaccharides, and proteins take part in cell wall structure. These compounds, differing in chemical and physical characteristics, are bound into a common structural and functional organization. Cell wall structure may be represented in the form of a two-phase system, in which the cellulose "woof" is woven into the continuous "warp" (base or matrix) which is composed of other constituents. The plant cell wall has achieved in the course of evolution, a structural and functional unity assuring a high level of protection of cells against pathogen penetration.

C. STRUCTURAL CHANGES OF THE CELL WALL

In their struggle for survival, pathogens have adapted by developing mechanisms for cell wall penetration and protoplast colonization and infection. The surmounting of cell walls is characteristic of actively motile bacterial and fungal pathogens. Extracellular pathogenic fungi, which form the largest group of plant pathogens, have developed specialized structures for the establishment of infection, such as appressoria, haustoria, infection or inoculation hyphae or pegs and other structures. Collectively, these structures often are referred to as the "infection apparatus".

Pathogenic fungi use mechanical and enzymatic action for cell wall degradation and penetration, and depending on the tissues affected, the fungi are classified as either epiphytic or endophytic. Epiphytic fungi affect primarily epidermal tissue cells, and endophytic ones, depending upon their life cycle, affect epidermal, substomatal cavity (after penetrating through stomata aperatures), parenchyma, or other living tissue cells.

The first barriers to pathogen penetration are cutin, suberin (periderm), and wax. These hydrophobic compounds are resistant both to mechanical and enzymatic action of pathogens.

1. Waxes Degraded by Pathogens

Until recently, it was believed that pathogens lack the ability of enzymatic degradation of these compounds. Cooper[197] said that some pathogens alter or eliminate wax crystals, which is clearly visible underneath germ tubes, hyphae, and appressoria. Disappearance of wax around colonies of an epiphytic yeast also has been observed. Changes in waxes underneath the germ tubes of *Erysiphe graminis* penetrating wheat cuticle were observed and it was assumed that wax degradation represented a condition preliminary to plant infection.[198]

According to some data, germinating hyphae may become attached to the cuticle in order to grow and locate suitable points for penetration. For example, the arrangement of wax crystals in wheat provides direction for hyphae of uredospores of *Puccinia graminis* f. sp. *tritici* towards leaf axes, where the possibility of reaching stomata was greater.[199]

Such investigations have shown that waxes as inert materials in the pathway of pathogen penetration may be subject to chemical degradation even though their ether and ester bonds generally are considered as unsuitable for pathogen action.

2. Cuticle Degradation by Pathogens

The cuticle may be affected mechanically by plant pathogens, although there is increasing evidence that enzymatic action by pathogens also affects this external protected layer. Mechanical breakdown or penetration of the cuticle results from mechanical pressure exerted by the apical growth of hyphae of haustoria. Gäumann[3] observed the penetration of the epidermal cuticle of broad bean leaves by *Botrytis cinerea,* which created a pressure of up to 7 bars; and that of young tomato fruits by *Alternaria tomato* with a pressure up to 3 bars. However, the fungus was incapable of penetrating the thicker cuticle of older tomato fruits even when exerting a pressure of 5 bars. Powdery mildew fungi were shown to penetrate cuticles via haustoria. Hyphae produced from conidia of *Cladosporium* sp. penetrated the cuticle and middle lamella between epidermal cells.[200]

Other pathogens capable of penetrating cuticle by mechanical means include *Ascochyta pisi, Colletotrichum lindemuthianum, Septoria helianthis,* and others.[11] Wood[15] cited data showing that rose leaves infected with *Sphaerotheca pannosa* contained one-fourth less cutin than that present in noninfected leaves, and that the cutin content of apple leaves infected with *Venturia inaequalis* was significantly reduced, proving that the activity of these fungi affected cutin production. The same author referred to other results establishing cutinase activity by the following fungi: *Botrytis cinerea, Cladosporium cucumerinum,* and *Pyrenophora graminea.* Mount[201] stressed that pathogens may use cutinesterase and other oxidative enzymes to penetrate cuticle and to reach the cell wall polysaccharides. Enzyme activity in cutin degradation is more frequent that once believed.

Pathogenic fungi may penetrate cuticle using either mechanical or enzymatic action or both, with enzymatic action being characteristic of a small number of pathogens. The development of perforations or pores in cuticle may be considered as evidence of structural alteration caused by pathogens in the course of epidermal penetration, and the number of perforations formed depending upon the number of hyphae involved.

3. Suberin Degradation by Pathogens

Suberin is the substance considered most resistant to pathogen activity in plants. It is part of the cell wall structure of the nonliving peridermal tissues. However, some pathogens must penetrate such tissue on its way towards living cells. Such penetration is performed, for example, by rhizomorphs of soilborne fungi, such as *Armillaria mellea,* which is capable of infecting roots and root crowns of many coniferous and deciduous plants. In the case of uninjured tissues, pathogen rhizomorphs become attached to the tissues with the aid of a slimy matrix and hyphae. Then, together with the infection peg in a mycelial sheath, it mechanically penetrates the peridermal tissues. Such a mode of penetration is accompanied by the splitting of peridermal cells and the development of openings in the peridermal tissue, suggesting structural damage. The fungus also has the ability to degrade cork.[202]

4. Cell Wall Degradation by Pathogens

Following penetration of wax, cutin, and suberin layers, pathogen hyphae continue their mechanical and/or enzymatic action on the middle lamellae, as well as the primary and secondary cell wall constituents. As summarized in the section on cytolytic changes, pathogens are capable of degrading lignin, polysaccharides, and proteins to their basic units through the action of several hydrolytic enzymes.

Cell wall penetration proceeds with a combination of chemical and/or mechanical activities of the pathogen. The consequence of such an action is, at first, the formation of pores both in the cuticle and cell walls through which the pathogen hyphae reach the cell protoplast. The appearance of pores corresponds to the shape of the penetrating haustoria or hyphae, with sharp and irregular margins, depending upon the degree of chemical or mechanical activity. The formation of such pores is the primary visible symptom of structural cell wall change or damage.

Cell wall penetration and degradation may be accomplished simply by growth of hyphae, or in a more complex manner by means of a special inoculation or penetration apparatus. One of the most simple means of inoculation or penetration is characteristic of the fungal pathogen, *Cladosporium cucumerinum,* whose conidia germinate to form hyphae which penetrate directly through the cuticle reaching the middle lamella between epidermal cells.[200] The conidia of *Botrytis cinerea* germinate on the epidermal cuticle to form simple appressoria. Infection hyphae grow from the appressoria and, either chemically or mechanically, further degrades the cell wall through which it forms a duct enabling it to reach the interior of the cell and then through the same mechanisms penetrate deeper into the tissues.[3]

Powdery mildew fungi, as epiphytic pathogens, form appressoria for penetration of the host cell wall and then haustoria, which colonize the plasmalemma (ectoplast), thus establishing a close linkage with the host cell constituents. Bracker[203] reported that the cell wall of barley plants were penetrated by appressoria and haustoria of *Erysiphe graminis,* which formed a ring-like collar around the neck of the penetrating haustoria. The haustorium was enveloped with a 5u-thick coat linked continuously with the surface of the plasmalemma. Several haustoria may develop independently and remain separate in a single cell. An infected cell may contain one or more than 200 haustoria.[204]

Wynn and Staples[205] described a characteristic sequence for the penetration of host tissue by the rust fungi. Uredospores germinate on the surface of herbaceous plants. Initially, the hypha grows and upon reaching a stomata, forms an appressorium. Penetrating hyphae develop from these appressoria and upon reaching the substomatal cavity, enlarge and form vesiculated hyphae, which give rise to infection hyphae that grow and reach the surface of the mesophyll cells. These hyphae then form haustoria which penetrate the host cell protoplast. The formation of haustoria is the final stage of pathogen colonization. Frequently, 50 to 90% of the hyphae from conidia form appressoria in leaves, of which 99% of these will be found associated with stomata.[206,207]

Another example is *Plasmopara viticola* whose zoospores are released from zoosporangia and transported in water drops. Where the zoospores reach the stomata of grape leaves, they absorb their flagella and encyst. These spores then germinate within 12 minutes forming hyphae which penetrate the substomatal cavities.[3] The hyphae then penetrate further between the parenchyma cells and form haustoria in the cells. Similarly, the zoospores of *Synchytrium endobioticum* move through the soil and when they encounter potato tubers, encyst, germinate, and form pores in the epidermal cell walls, which lack a cuticle. They infect the tissue by allowing their protoplast to enter into the host cell.[3]

These examples show the different and specific modes by which fungal pathogens penetrate host cell wall structure have developed in the course of phyllogenic development and pathogen adaptation to host cells.

Phytopathogenic bacteria represent another group of pathogens which must surmount the outer protective tissues of various hosts to reach susceptible host cells. These bacteria are incapable of overcoming the protective substances associated with outer tissues, such as the waxy coats, cuticles, and corky materials by mechanical means. Therefore, most bacteria, in association with water, penetrate plant tissues through natural openings; stomata, hydathodes, nectary glands, and lenticels. Having penetrated the host, these bacteria develop in the intercellular spaces or, in the case of vascular bacteria, in the xylem or phloem.

Many bacterial species are found in large numbers in the environment and on plant surfaces, and are capable of penetrating plants through their natural openings. However, infection takes place only if a virulent bacterium invades sensitive or compatible plant tissues. Avirulent bacteria, although they may penetrate intercellular spaces of compatible hosts are incapable of initiating infection.

Virulent bacteria upon penetrating susceptible plants and after colonizing the intercellular spaces or vascular tissues, degrade pectic substances through the action of pectolytic and in som cases, cellulolytic enzymes. With the degradation of cell wall constituents, plasmolysis of the cells occurs in response to toxic substances released during the degradation. Increased permeability of the cell membrane occurs, and additional substances are released which become subject to the enzymatic activity of the bacteria. Phytopathologenic bacteria thus differ substantially from phytopathogenic fungi.

Some avirulent bacterial strains, although they do not establish infection in nonhost cells, stimulate host cell necrosis, a type of hypersensitive reaction, in inoculated plant parts. Klement and Goodman[208] first observed this type of reaction in tobacco plants inoculated with bacteria. Sequeira[209] induced this reaction on tobacco leaves with *Pseudomonas syringae* pv *pisi,* avirulent strains of *P. solanacearum,* heat-killed bacterial cells, and saprophytic bacteria.

These hypersensitive reactions are characterized by rapid host cell death, decreased bacterial populations, and persistence of the bacteria on inoculated plants.[209] More detailed studies have shown that such bacteria cells become compressed to plant cell walls and either individually or in groups, become surrounded with granular or filamentous materials, assume to be products of the host cells. Characteristic structural changes in plant cells occur at these points where the bacteria are located, such as invagination of the plasmalemma and formation of vesicles on the inner cell walls. It is assumed that the material surrounding the bacteria is a byproduct of these vesiculated excrescences. The mechanism by which avirulent bacterial cells are surrounded by the adjacent plant cells, thus preventing further spread in the intercellular spaces, has not been investigated thoroughly. Sequeira[209] supported the opinion that the lectins in plant cells walls bind as receptors with glycolipids in the external walls of Gram-negative bacteria. This mechanism should represent a general form of resistance enabling plants to survive in the presence of a variety of bacterial species.

D. STRUCTURAL PROTOPLAST CHANGES

All plant pathogens, upon penetration of the cell wall influence the protoplasts. Infection is established when the pathogen involved begins to use host cell protoplast constituents for its own metabolism. Since viruses have no metabolism of their own, the beginning of infection for this group of pathogens is associated with the replication of virus particles.

In order to initiate infection, pathogens must establish links with the protoplast in a specific manner, depending upon the extent of parasitic development. For example, intercellular pathogenic fungi, i. e. the biotrophes and necrotrophes, penetrate protoplasts and establish their connection with the aid of haustoria. Intracellular biotrophic fungi completely colonize and develop within cell protoplasts. Intracellular necrotrophic fungi develop only their vegetative organs primarily in the protoplast.

Phytopathogenic bacteria, except for the vascular invaders, develop in intercellular spaces, chemically degrade cell wall constituents, and indirectly affect the protoplast. Plant viruses and mycoplasmas develop as obligate parasites intracellulary with their entire life cycle taking place in the host cell.

Changes corresponding to the needs of the pathogens and the preservation of these species take place in the protoplast of infected, susceptible host cells.

Obligate parasites establish a relationship in the protoplast that is best suited to their nature and survival. For example, the biotrophic fungi *Peronospora tabacina, Plasmopara viticola,* and *Puccinia graminis,* as well as the epiphytic biotropic fungi, *Erysiphe graminis* and *Leveillula taurica,* stimulate protoplast activity at the initiation of infection and later cause decay of the parasitized cells in order to preserve their source of nutrients as long as possible.

Intracellular biotrophic fungi, such as *Plasmodiophora brassicae* and *Synchytrium endobioticum,* cause hypertrophy of host cells to ensure the intracellular development of their vegetative and reproductive organs. Replicated virus particles spread in susceptible host cells establishing a network of systemic infection, which represents an advanced form of parasitism.

Necrotrophic pathogens as facultative parasites can represent different levels of parasitism. One of the lowest forms is represented by *Botrytis cinerea, Pythium debaryanum,* and *Rhizoctonia solani,* which cause a rapid death of susceptible cell protoplasts and then develop intensively in the dead cells and tissues. Those fungal pathogens that belong to a group with an advanced degree of parasitism are represented by species of *Mycosphaerella, Septoria, Venturia,* and others. These fungi cause a rapid decay of the cell protoplast followed by a short-term stimulatory effect. The group of pathogenic fungi that represent one of the highest forms of parasitism include *Nectria galligena, Taphrina* spp., and *Ustilago maydis.* These fungi are distinguished by their stimulation of protoplast division and hyperplastic formations. The influence that phytopathogenic bacteria have on protoplasts represent an advanced form of facultative parasitism.

In summary, the course of protoplast infection, plant pathogens use a variety of mechanisms to establish infection, such as mechanical, chemical, toxin, and hormonal activities. The extent to which each mechanism is used depends on the mode of parasitism adapted to each host cell. Structural and physiological changes occur in infected protoplast. These changes differ, being a consequence of a variety of factors in infected cells. Due to the diversity of these changes, it is difficult to systematize them with certainty. However, they can be classified as changes in the cell membranes, the cell organelles, and cytoplasm.

E. STRUCTURAL CHANGES OF CELL MEMBRANES

Cell membranes are part of the protoplast structure and form the basis for intercellular space organization and include the plasmalemma (ectoplast or cytoplasmic membrane) and tonoplasts (vacuole membranes); and according to some authors,[210] the endoplasma reticula, dictyosomes (Golgi bodies), and other structures. These membranes primarily consist of lipoproteins and are closely linked structurally and functionally within the cell. In addition, the endoplasmic reticula spread between cells with the aid of the plasmodesmata, thus linking them into a unique entity. The fundamental biochemical enterity indispensable for all plant cell life is based on these membranes and their functions, such as regulation of permeability in and outside of the cell, various synthetic and nutrient transport processes in the cell, and others.

These membranes are highly sensitive to pathogen action. Changes in their structure and function are among the first reactions to protoplast infection. They are sensitive to all types of pathogen influence, including enzymatic, mechanical, and/or toxic. The consequence of such influences is membrane degradation and destruction, which is easily determined in late stages of infection.

Ultrastructure changes in cell membranes has been documented in the exposure of wheat cells to victorin, a toxin produced by *Helminthoporium victoriae,* and other toxins.[211-213] Various changes in the ultrastructure of cell walls, plasmalemmas, dictyosomes, and endoplasmic reticula in oat cells were described following a 24-hour exposure to victorin. Others described barely detectable changes in the ultrastructure of plasmalemmas and vacuole membrane tonoplasts. Goodman[214] stated that victorin caused ultrastructure changes by the laceration of the internal membrane systems (exclusive of the mitochondria) in the cells of the oat rootcap and green leaf tissues following a 24-hour exposure to victorin. Irregularities in dictyosomes characterized by the formation of vesicles in the external rootcap were recorded. Structural changes in the plasmalemma also were described as a reaction to victorin.[215]

Structural changes in the tonoplast were caused both by victorin and by the presence of other pathogens. Destroyed tonoplasts have been described in maize leaves infected with *Bipolaris maydis.*[216] The degradation of the tonoplast accelerated cytoplasm degeneration and decay.

Wheeler[217] stated that tonoplast degradation in infected oat cells may be related to decreased mechanical strength of cell walls. Enzymatic degradation of the cell wall results in vacuole swelling, ending with tonoplast disintegration.. This mechanism attributed to pectic enzymes could have an important influence on membrane system disorders.

Vaculoes, in addition to other constituents, contain organic acids which are used by pathogens only after tonoplast disintegration, which occurs under direct influence of pathogens or during the degradation process of normal senescence.[215]

Thus, cell membrane damage in susceptible plant cells may be caused by pathogenic fungi,[218] bacteria,[214] or viruses.[210] Structural and morphological changes of the plasmalemma are evident by the invagination at the point of haustorium penetration and spreading into the protoplast. Such invaginations occur under the pressure exerted by haustorial tips and therefore, their shape corresponds to that of the haustorium. Such changes are caused by all biotrophic and necrotrophic fungi.

The structure of haustoria and therefore, the changes that occur in the plasmalemma and cytoplasm, differ depending on the pathogenic fungus involved and particularly on the extent of parasitic development. The fundamental structure of haustoria consists of two main parts, the neck, located at the point of cell wall penetration; and the body, formed in the host cell cytoplasm. Haustoria of *Erysiphe graminis* in epidermal barley cells are separated from the host protoplast with a sheath up to 5µm thick.[203] The sheat is surrounded with a membrane which frquently protrudes toward the haustorium and form invaginations that are continuous with host protoplast. The invaginations of simple or complex shape are usually found around the haustorial neck on top of the collar. The membrane which surrounds the haustorium and its parts maintains a connection between the haustoria and host cytoplasm indispensable for the host cell parasitization.

The haustoria of *Puccinia graminis tritici,* which cause plasmalemma invagination, are enveloped by a capsule 80 to 340 µm thick, the latter being of a composition similar to that of the haustorium.[15] The envelope also is protected with a membrane 12 to 24 µm thick, possibly originating from the plasmalemma or may represent a new formation. The haustorial envelope sometimes is interrupted by small vesiculated formations, which spread into the host cell cytoplasm.

Formation of haustoria and processes of their encapsulationin the host cell have been described for other fungi including those that cause downy mildew and white rust of crucifers.

Mechanisms of haustoria nutrition in host cells differ depending upon their structure and colonization of protoplasts. According to Bracker[203] nutrition for *Puccinia graminis* comes from the transport of nutrients from the host protoplasts through ducts in the haustorial walls, and that for *Erysiphe graminis* through the capsule on the surface of the haustorium. *Peronospora manchurica* obtains its nutrition through the activity of secretory bodies formed on the haustorium.

Thus, morphological and structural changes in plasmalemma take place in the course of haustoria formation and nutrient uptake. Structural changes may occur in the form of subtle injuries and interruptions, or more important as degenerative changes in the cytoplasm and cell organelles. An important alteration discovered in some diseases is the separation of the plasmalemma from the internal part of the cell wall. Other significant changes in the form of vesiculated formations may occur on the plasmalemma beneath the inner surface of the cell wall, which has been defined at lysosomes or lysosome-like structures.[219]

Plasmalemma invagination also may be caused when nonpathogenic bacteria are introduced into intercellular spaces and become attached to the outer cell wall.[209,220] Plasmalemma invagination also occurs in and around these points on the inside of the cell wall. Special vesiculated formations also develop on the inside of the cell wall, and it is assumed that their products take part in the structure of the external bacterial envelopes.

There are limited data on the role of plant pathogen enzymes in cell membrane degradation.

Pathogens produce enzymes which are capable of degrading cell membrane constituents. The role of such enzymes in cell membrane degradation needs more study.

Phytopathogenic viruses also influence membrane systems during the processes involving their replication in the cell protoplasts. It is assumed that the replication of RNAs of the barley yellow dwarf virus is performed in vesicles surrounded by the endoplasmic reticulum.[221] It is also assumed that the RNA of the turnip yellow mosaic virus is replicated in the vesicles developed in chloroplasts from parts of the membrane and that the occurrence of the vesicles are related to virus infection.[222]

Finally, it should be stressed that all structural changes in cell membranes caused by one or several means are characterized by a common and general consequence, i.e. the increase in cell membrane permeability, and if irreversible, leading to cell death.

F. STRUCTURAL CHANGES IN CELL ORGANELLES

Structural changes in various organelles, such as chloroplasts, mitochondria, and nuclei occur in infected host cell protoplast leading to partial or complete dysfunction.

1. Structural Changes in Chloroplasts

Chloroplasts contain chlorophyll a, b, c, and d, and other pigments, such as carotene and xanthophyll. Several metabolic processes take place in the chloroplasts; photosynthesis as well as the snythesis of low and high molecular weight compounds, such as carbohydrates, amino acids, peptides, nucleotides, and lipids. They contain a large number of enzymes which take part in these processes. The chloroplasts are complex structures. The chloroplast body is in the form of a disk or grain, separated from the surrounding cytoplasm by a double membrane. The primary structure of the inner membrane is a stroma with numerous lamellae. Short lamellae are arranged as one about the other and form grana, which are interlinked with longer lamellae, defined as stromatal lamellae.

The development of chloroplasts may be delayed or impared by various abiotic factors. For example, chloroplast development may be disturbed by the presence of smaller chlorophyll granules, which is a consequence of irregular cell nutrition; vacuoles may form in stromata of degenerated chloroplasts,[5] or water penetrating into chloroplast can cause swelling.

Plant pathogens as biotic factors are capable of causing disorders in chloroplast structure, number, and biochemical activity. As has already been noted, plant pathogens cause changes in pigment synthesis in chloroplasts. As a general rule, infection by pathogens reduce chlorophyll quantities either by preventing chlorophyll synthesis or by chemical degradation of the lipoproteins in the lamellar chloroplast membranes. A decrease of chlorophyll generally occurs in all infected plants (see section on pathogens causing green color reduction).

Chlorophyll degradation in mature chloroplasts results in the loss of structure followed by deformation. Loss of structure is followed by chloroplast dysfunction, since only whole or uninjured chloroplasts are biologically active.

Rubin et al.[223] have shown that fungi associated with Verticillium wilt caused 26% reduction in quantities of chlorophyll a and b in chlorotic cotton leaves by disturbing protoplast structure. These changes are characterized by chloroplast swelling, injuries to the enveloping membrane, and degradation of the internal lamellar constituents.[224] Degradation of the lipoprotein constituents of the membrane results in the release of free lipids which accumulate in the chloroplast in the form of osmophile granules. At the time of the appearance of necrotic spots on diseased leaves, cells contain disintegrated organelle parts, among which are found remnants of chloroplast lamellae.

Rahmankulova[225] established that in the course of disease development, the fungi associated with Verticillium wilt caused a reduction in the quantity of grana and lamellae, as well as a disintegration of the stromata leading to the complete destruction of the chloroplasts in cotton plants. The result of such structural changes in chloroplast function, primarily in energy transport.

Rubin and Artsikhovskaya[226] referred to data obtained by several authors pertaining to chloroplast destruction in infected plant cells. Also, chloroplast disintegration was observed in chestnut and lupine leaves infected with *Phytophthora cambivora*.[227] In wheat plant cells infected with *Puccinia graminis*, a gradual reduction and finally the complete disappearance of chloroplasts was reported.[228] Chloroplast destruction in plants infected with *Vaccinium macrocarpon* was caused by pathogen lipase degradation of chloroplast lipoproteins.

In studies on oat plants infected with *Helminthosporium victoriae*, it was shown that the toxin victorin caused swelling of plastides and disorganization of the grana and intergrana network of these organelles in oat leaves.[214] Also, the occurrence of chlorosis and yellow striping in plants infected with the fungus was explained on the basis of the destruction of the chloroplasts. The same author showed that the halo toxin produced by *Pseudomonas syringae* pv. *tabaci* caused chloroplast swelling and a decrease in stromatal content as well as a reduction in the number of ribosomes in chloroplasts and cytoplasm of susceptible plants.

Plant viruses, similar to other pathogens, usually reduce chlorophyll in infected plant cells. Viruses can bring about a reduction in chlorophyll in two ways: by disturbing the processes of pigment formation in infected cells by reducing the amount of porphyrin needed for synthesis; or by damaging the chloroplast structure, resulting in the release of chlorophyllase, which degrades chlorophyll and separates it from the protein constituents.[229,230] Such released chlorophyll is subject to hydrolysis by chlorophyllase. Such reductions in chlorophyll quantities results in the incomplete formation or disintegration of chloroplasts. Tu et al.[231] recorded a 19.2% reduction in the number of chloroplasts in maize and sorghum plants infected with maize dwarf mosaic virus; and the average size of the chloroplasts was 3.6 x 2.4 μm in infected compared to 4.4 x 3 μm in uninfected plants.

Chloroplasts from Chinese cabbage plants infected with turnip yellow mosaic virus were rounded and grouped in aggregated forming cuplike depressions with their openings usually pointing towards the cell walls.[232] Numerous vesicles bound to their membranes occurred in these chloroplasts. Some strains of this virus also caused fragmentation of the aggregated chloroplasts, and the so-called "white strains" of the virus caused degeneration of the grana in the chloroplasts.

Hršel and Brčak[233] observed vesicles formed as depressions in membranes of chloroplasts and the cytoplasm of tobacco plant cells infected with the tobacco mosaic virus. Abnormal chloroplast vesiculation also was observed in tobacco plants infected with the broad bean mottle virus.[161] Chloroplasts in chlorotic barley leaves infected with the barley strip mosaic virus contained a reduced number of smaller grana than noninfected ones.[234]

Changes in chloroplast structure due to pathological causes depend on the stage of viral infection. For example, chloroplasts of necrotic tissue cells caused by the tobacco mosaic virus when compared to noninfected cells, usually lack membranes, are surrounded by osmophile granules, contain virus particles, and contain a large number of starch granules between the disorganized lamellae.[40]

An increased quantity of starch also has been found in leaves of susceptible plants infected with *Puccinia striiformis*.[235] Increases in the quantity of starch has been described in mesophyll cell chloroplasts of susceptible oat plants exposed to victorin.[236] There is evidence that cell plastides of hypersensitive wheat plants of the cultivar Little Club infected with *Puccinia triticina*, frequently contain more starch than noninfected ones.[3] Thus, the accumulation of starch in infected cell protoplasts frequently occurs as a consequence of the interruption of the enzymatic translation of insoluable starch into soluble forms suitable for transport through the plant system.

In conclusion, several pathogens causing disorders in chloroplasts of susceptible and hypersensitive plants have been described. The general characteristic of this disorder is the laceration of the membranes, resulting in the loss of the chloroplast structure, which results in the free movement of enzymes and various chloroplast metabolites into the cytoplasm disturbing the metabolism of the infected cell.

2. Structural Changes of Mitochondria

Mitochondria are cytoplasmic organelles which may be granular, cylindrical, or rod-like in shape and are composed of three parts: an enveloping membrane separating it from the cytoplasm, an inner membrane, and the matrix of the interior. Cristae in the form of incomplete septa-like structures extend from the inner membrane into the matrix toward the mitochondrial axis, thus increasing the surface of the inner membrane.

Mitochondria primarily are cell respiratory organelles composed of lipoproteins. The membrane lipids are made up of more than 90% phospholipids. These membranes also contain respiratory, tricarboxylic acid cycle, and phosphorylation enzymes. Biological oxidation of cell substrates, primarily metabolites of the tricarboxylic acid cycle, as well as synthesis of energy to accumulate the phosphatic compound, adenosin triphosphate, take place within them. Thus, mitochondria take part in the exchange of energy, which is the primary activity for the maintenance of cell activity and life in general.

According to Glišić,[21] mitochrondria are highly changeable cell organelles susceptible to morphological, structural, and functional alterations caused by various external and internal factors. The less pronounced changes often are reversible, while the more profound changes are irreversible, resulting in destruction of these organelles.

The influence of abiotic factors, such as temperature and water, may cause mitochondria to swell to twice or more of their original size. Conversely, degenerative changes result in the reduced size or growth of mitochondria matrix and cristae. Mechanical injuries induced by the artificial cutting of tissues, resulted in an increased number and the activity of mitochondria, as well as the microsomes of the ribosomes and endoplasma reticulum.[192]

The penetration of plant cells and tissues by pathogens can cause substantial disruption in the structure and the biological activities of mitochondria. The extent of these changes depends upon host plant susceptibility and the extent of parasitic development.

Several authors have established that mitochondrial membrane injuries occurred in infected cells. Detrimental consequences of membrane injuries are the complete disorganization of mitochondria. Several authors have recorded increased number of mitochondria and volume at the beginning of infection as a consequence of intensified energy consumption for the synthetic processes activated in the cell at that time.

Ultrastructure investigations of damaged mitochondrics have shown swelling and deformation of internal structures as well as changes in the electroconductivity of the matrix and the internal structure, and the formation of abnormal vacuoles within diseased mitochondria.[237-239]

According to Akseenova and Brinza[240] infection of cabbage by *Botrytis cinerea* increased the volume of mitochondria in susceptible plants by 8.4%, but not in resistant ones; and that infection of cucumber by *Erysiphe cichoracearum* caused a swelling of mitochondria in susceptible plant cells. The formation of new mitochondria was reported in sweet potato plants infected with *Ceratocystis fimbriata* as well as up to 60% more in tomato plants infected with the Fusarium wilt fungus compared to uninfected plants.[226]

The degradtion of mitochondria results in the separation of respiration processes from phosphorylation. Such changes have been recorded in wheat plants affected by the rust and powdery mildew fungi, in cabbage plants infected with *Botrytis cinerea*, and in virus-infected tobacco plants.[241]

Changes in mitochondrial structure have been observed in cells of tissues exposed to the T-toxin of *Bipolaris maydis*, race T. The T-toxin caused swelling of mitochondria and reduced the matrix density in root cells of susceptible maize plants after 15 minutes exposure.[242] Also, the halo-toxin produced by *Pseudomonas syringae* pv. *tabaci* caused reduced matrix volume in mitochondria of susceptible plant cells.[214] Thus, this and other experimental evidence showed that toxins can play a significant role in pathological changes in mitochondria.

Changes in the number and structure of mitochondria also occur in plant cells infected with plant viruses. The extent of these changes depends upon the susceptibility of the host and the virulence of the virus strain involved. In cases of severe infection, an increase in the respiration rate was accompanied by an increase in the number of mitochondria. For example, following infection by tobacco mosaic virus, tissues of *Nicotiana glutinosa* surrounding necrotic areas showed an increase number of mitochondria;[243,244] and a greater number of cristae within the mitochondria.[238] The increased number of mitochondria and the intensified metabolic activity are examples of the reaction of a hypersenative host. Weintraub et al.[86] considered the intensified respiration as a result of both increased number of mitochondria and intensified enzyme activity. A similar increase in the number of mitochondria was recorded in virus-infected apple tissues.[237] Hatta et al.[239] described abnormal mitochondria in cells of cucumber infected with the cucumber green mottle virus.

Harrison and Roberts[245] discovered that the long particles of the tobacco rattle virus (TRV) were closely linked with mitochondria membranes. However, it was not known whether this relationship was established to induce virus synthesis or was a consequence of it. According to other data reported by Harrison,[161] some strains of the TRV not only caused changes in the mitochondria, but they were found aggregated into cell inclusion bodies. MacLeod et al.[246] showed that cell inclusion bodies associated with infection by the potato yellow dwarf virus included mitochondria.

In conclusion, the effect of virus infection on mitochondria differs according to the host and pathogen involved. An increase in the number and changes in the size of the mitochondria in infected cells usually results in and may be related to the intensified respiration during the first stages of infection. The deformation and incorporation of mitochondria into cell inclusions results in a reduction in their number and, thus, their respiratory activities. These changes come during the later stages of infection.

3. Structural Changes in Nuclei

The nucleus is an organelle found in plant cells in either a discrete form or in a diffused state depending upon the nature of the cell. It is located in the cytoplasm and incapable of surviving elsewhere. The most frequent form of nuclei is globular or ellipsoid. Nuclei of other shapes are found in cells performing special functions. The nucleus is separated from the cytoplasm by the plasma reticulum and is filled with nuclear plasma in which one or several nucleoli may be located. The nuclear envelope is composed of several membranes. Pores in these membranes enable direct contact between the nuclear plasma and the cytoplasm of the cell, which ensures a constant exchange of nutrients on which the physiology unity of the cell is based. The size of the nucleus and the volume of the cytopasm are found to be in a certain relationship called the nucleoplasmatic ratio.

Proteins and nucleic acids, the primary constituents of nuclei, are present in greatest quantities compared to other components. Other compounds are found in their structure such as enzymes, lipids, and mineral salts.Deoxyribonucleic (DNA) and ribonucleic (RNA) acids are located within the nucleus. The quantity of DNA found in the chromosomes is constant for any given chromosome set. It is the carrier of genetic messages and directs protein synthesis. In contrast, the quantity of RNA, located in the nucleolus and chromosomes during mitosis, is changeable.

Nuclear activity is manyfold and indispensable for cellular organization and function. It ensures protoplasm synthesis and the maintenance of balance among cell functions. It influences growth as well as cell wall and plant morphogenesis. The nucleus plays an important role in the synthesis of cell protein and in the control of cell respiration. As the primary carrier of genetic messages, it ensures the transmission of hereditable characters. The nucleolus takes part in protein and nucleoprotein synthesis and represents the center of RNA synthesis in the cell.

Various nuclear changes that occur in the course of development may be a consequence of either normal cytogenesis or detrimental influences of external abiotic or biotic factors. Depending upon the intensity of the change, they may be reversible or irreversible, with irreversible changes leading to cell death, since the nucleus is an unregenerative organelle.

Several changes in the nucleus may occur under the influence of nonpathogenic factors.[5] For example, insufficient cell nutrition results in a reduction of the cytoplasm and in nuclear development with a tendency for rapid degeneration. Chromatic and nuclear substances disappear from such "starved" cells. The loss of water may result in the formation vacuoles in the nucleus with meristemattic cell nuclei being more susceptible to such changes than are those of older cells. Vacuoles also may appear in the nucleolus and may be absorbed upon reaching the nuclear vacuoles. The return of water into a cell may result in nuclear swelling and bursting. Nuclear swelling and deformation have been observed following cell plasmolysis. Several chemical compounds, water deficiency, and other nonpathogenic factors may cause granular degeneration.

Many investigators have shown that pathogens cause various structural and physiological disorders in host cell nuclei. Some pathogens may alter the size and the shape of nuclei and thus, the functions.

Nucleus enlargement has been recorded in some plants infected with intracellular obligate parasitic fungi. For example, *Synchytrium mercurialis* caused a 250-fold increase in the size of nuclei in cells of infected plants of *Mercurialis perennis*.[5] In addition, the enlarged nuclei were permeated by slightly ramified systems of ducts spreading in the direction of the pathogen. *Plasmodiophora brassicae* caused an 8- to 10-fold enlargement of susceptible cabbage cells which was accompanied by a 6-fold enlargement of nuclei and a 30-fold enlargement of nucleoli.[17]

According to some authors, the intracellular biotrophic fungi caused nuclear enlargement in the initial stages of infection, while in later stages, there was a reduction in nuclear size. Nuclei of rust-infected wheat leaf cells were found to be larger than those of normal cells at 6 days after infection, whereas at 15 days they became smaller and crinkled.[15] In cells of the susceptible wheat cultivar Little Club infected with *Puccinia triticina,* the nuclei appeared to be drawn to the pathogen haustoria and to swell to form rings.[3] Similarly, leaf cells of wheat plants infected with *Puccinia graminis tritici,* the nuclei were drawn to the infectious hyphae, became flattened, and simultaneously with nucleolus enlargement, doubled in size.

Müller et al.[247] recorded the enlargement of nuclei in potato plant tissues infected with *Phytophthora infestans.* They divided the course of infection into five stages. Enlargement of the nucleus started in the second and continued into the third stage. Shrinking and crinkling of some nuclei accompanied by a loss in cell turgor occurred in the fouth stage.

Nucleus enlargement at the beginning of the infection progress with this group of pathogens is connected with an increase in RNA. Accompanying an increase in RNA is an increase in protein synthesis, which usually occurs in infected plant cells.

Increased dimensions of nuclei were observed in neoplasmatic tissue cells induced by *Agrobacterium tumefaciens*.[226] In contrast, the necrotrophe *Botrytis cinerea* reduced the size of nuclei in infected onion tissues.[226] Crinkling of the nuclei in infected bulb cells occurred during the first stage of disease. With continued disease development, the remnants of the nucleus divided into separate parts, with the nucleus eventually disappearing in affected cells near the hyphae. It was shown also that the toxin produced by *B. cinerea* as well as its fractions, caused a more pronounced reduction in nuclei in susceptible than in resistant cabbage cultivars. Two hours after the infusion of the toxin the dimensions of nuclei was reduced 11% in susceptible plants and by 13%, 24 hours later; and in the same periods, an organic acid reduced dimensions of the nuclei by 14 and 17%, respectively.

Thus, as a general rule, necrotrophic fungi, depending upon the extent of parastism, may be expected to cause a more rapid disintegration of the nucleus than other fungi (see section on cytogenetic changes).

4. Structural Changes in Cytoplasm

The cytoplasm includes all the constituents of a cell outside of the nucleus. It represents the internal cell environment in which all organelles as well as numerous particles, differing in shape and size, are located. It is composed of protein molecules of a granular appearance, among which are several enzymes. The physical properties of cytoplasm are viscosity, elasticity, and mobility within the cell. A physiochemical trait of the cytoplasm is its behavior as a complex colloidal system in which hydrophylic colloids, which play an important role, are composed primarily of protein macromolecules. The hydrophylic nature of these colloids is based on their eletrical charge, which attracts the bipolar water molecules, forming a water envelope about them. This water disappears with the loss of the electrical charge, resulting in cytoplasm coagulation and cell death.

The cytoplasm directs many processes within the cell, such as growth, morphogenesis, extranuclear heredity, and others. Together with its organelles, it takes part in most biochemical reactions. For example, the first step of respiration (glycolysis) takes place in the cytoplasm, while the remaining stages, such as the tricarboxylic acid cycle, electron transport, and oxidative phosphorylation, take place in the mitochondria. The cytoplasm coordinates the activities of all its constituents. The size and composition of cytoplasm changes during cytogenesis; some cell constituents are more abundant in younger than older cells, and in meristematic than in somatic cells. In addition, numerous disorders caused by abiotic and biotic factors also occur in the cell.

Nutritional deficiences, mechanical injuries, and other types of cell damage, for example, result in the formation of numerous vacuoles in the cytoplasm; and some chemicals, water deficiences, and other agents cause cytoplasm granulation. Lipoprotein degradation and the release of lipid constituents cause fatty cytoplasm degeneration.

The cytoplasm also represents the living environment for may plant pathogens. Most of the processes of pathogenesis from the beginning of infection to cell death occur here. These processes are accompanied by a variety of changes manifested in disease symptoms.

All cytoplasmic changes can be either general or specific ones. Specific changes result from disorders of the cytoplasm (Figures 44 to 47) and are discussed in more detail in sections pertaining to different types of cytoplasmic changes.

Ribosomes are granular macromolecules of ribonucleoproteins synthesized as a product of cell metabolism. They are found free in the cytoplasm or in chloroplast endoplasmic reticulum performing identical biochemical activities. Several authors have shown that various plant pathogens are capable of influencing the number and activity of ribosomes in cells. For example, a significant number increase was reported in resistant sweet potato plants infected by *Ceratocystis fimbriata*.[226] Also, a 30% increase in ribosome numbers was calculated occurring in cells of resistant and a 24% increase in susceptible cucumber plants infected by the biotrophe *Erysiphe cichoracearum*.[248] The same authors showed a 40% and 15% increase in ribosome numbers in resistant and susceptible plants infected by the necrotrophe *Botrytis cinerea*. The authors found such an increase in ribosomes was accompanied by an increase in the number of polysomes.

Biotrophic pathogens were found to stimulate an increase in ribosome number of the endoplasmic reticulum and other organelles in infected cells.[197] In contrast, necrotrophes reduced ribosome populations in cytoplasm and its organelles. The introduction of the halo-toxin from *Pseudomonas syringae* pv. *tabaci* into susceptible tobacco plants caused a reduction in ribosome number both in the cytoplasm and chloroplasts.[214] An increase in the number and activity of ribosomes was observed in tissues damaged by cutting, suggesting that such a change represented an atypical phenomenon.[192]

Šutić[18] referred to data showing that phytopathogenic viruses influenced the number and activity of ribosomes, with the influence being more pronounced in the chloroplasts than in the cytoplasm. Hirai and Wildman[249] observed a reduction in ribosome number other than in the cytoplasm of tobacco plants infected with the tobacco mosaic virus. Plants infected with turnip yellow mosaic virus had a reduced number of ribosomes in chloroplasts of infected leaves with little influence on those in the cytoplasm.[250]

Changes in the content and activity of ribosomes rank among the general changes in infected plants. The dynamics of these changes depends upon the host plant susceptibility and characteristics of the pathogen. In the initial stage of infection, and in the case of mechanical injuries, the increased number of ribosomes center around the synthesis of new proteins as a defensive reaction. The degree of plant resistance and damage depends upon the ability of the infected or damaged plant to form new ribosomes and intensify protein synthesis.

The dynamics of pathological changes in cytoplasm depend on the host cell, pathogen, and extent of parasitism. Those associated with biotrophic pathogens usually influence both cytoplasm and its organelles.[40,197] Some intracellular biotrophes cause hypertrophy and hyperplasia, some of which are irreversible and persist until cell death. Intercellular biotrophes influence cytoplasm constituents in the initial stages of infection and later cause degenerative changes leading to cell death. The initial stimulation usually is evident by the increase in the size of cytoplasmic mitochondria and ribosome number accompanied by an accumulation of starch in plastides, temporary emlargement of nuclei and other changes.

Some necrotrophes, similar to biotrophes, cause hyperplasia. However, most cause stimulative changes in cytoplasm constituents for a short time with a rapid degeneration leading to cell death. This is characterized in later stages of development by intensive cytoplasma degradation, nucleus degeneration, and accumulation of phenolic compounds and tannins.

Cooper[197] stated that several pathogens cause changes in the cytoplasm resembling those that occur during normal cell senescence, such as disintegration of stromata and chloroplast lamellae, degeneration of starch granules, decrease in ribosome number, appearance of vesicles on the endoplasmic reticulum and dictyosomes, degradation of membranes, and crinkling and deformation of nuclei.

Host cell protoplasts play an important role in the final result of the host-parasite relationship. Akseenova et al.[248] observed that protoplasts of susceptible plants are more labile to the influence of pathogen metabolites than are resistant ones. For example, protoplasts of susceptible cabbage plants infected with *B. cinerea* were more prone to injury, rapid damage of cytoplamic membranes, disturbances of the cytoplasm colloidal state, and other changes than resistant ones.

Partial or complete cell membrane and organelle disintegration of the cytoplasm results in the formation of a variety of foreign compounds as well as in uncontrolled cytoplasmic chemical reactions. Reduction of the colloidal particles and changes in the isoelectric points, cell sap acidity, and other reactions take place. Due to these disorders, the colloidal particles precipitate resulting in a granular-appearing cytoplasm.

The structure of granulated cytoplasm has been described in cells of plants infect with *B. cinerea* and other pathogens; and such altered protoplasts lose their ability for plasmolysis.[3] The granular protoplasm later becomes fibrillated as was observed in cells of susceptible potato plants infected with *P. infestans*.[247]

Some authors described granulation and rapid collapse of protoplasts in plant tissues exposed to pectic enzymes and pathogen toxins as well as in cells penetrated by pathogens.[219] Further degradation occurred in the form of coagulation and aggregation of the cytoplasm constituents, followed by their death.

In conclusion that structural cytopathological changes partially or completely disturb the

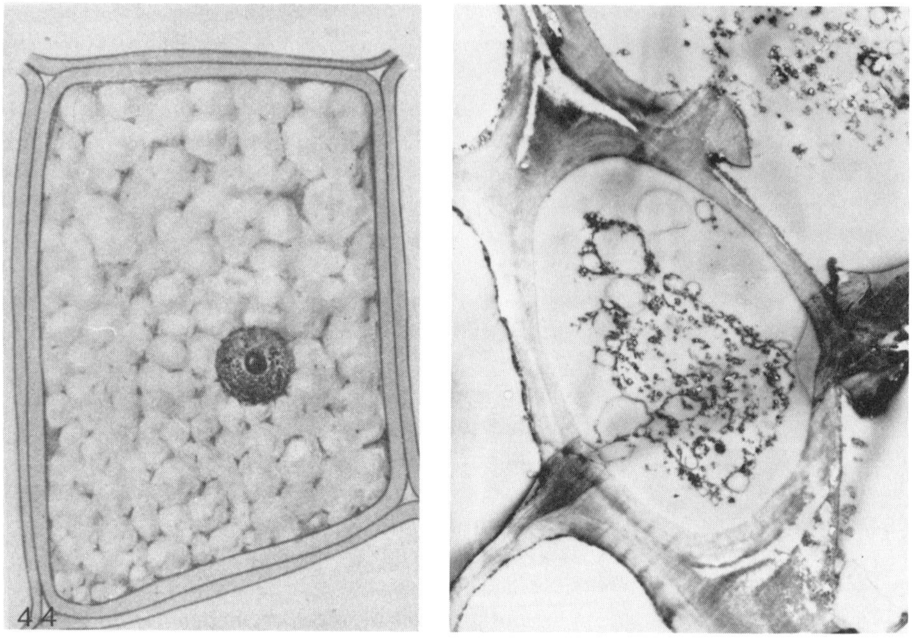

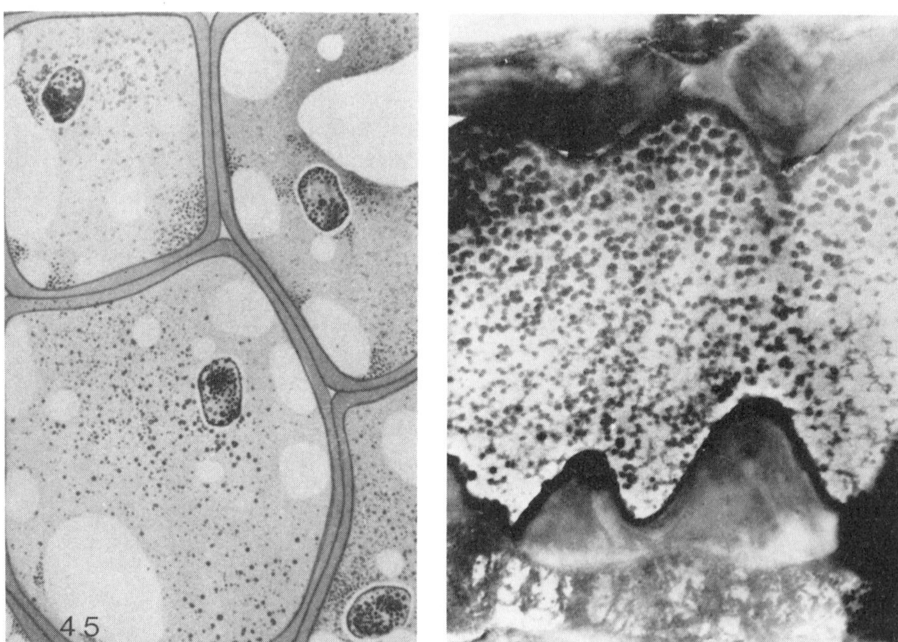

PLATE IX. FIGURE 44. Vesiculated cytoplasm with vacuoles in a parenchyma cell (left) and a xylem cell (right) of sugar beet infected with the beet necrotic yellow vein virus. (From Miloranović, M., Ph. D. thesis, Belgrade, 1986.) FIGURE 45. Granular degeneration of the cytoplasm in a parenchyma cell (left) and in a xylem cell (right) of sugar beet infected by the beet necrotic yellow vein virus. (From Miloranović, M., Ph. D. thesis, Belgrade, 1986.)

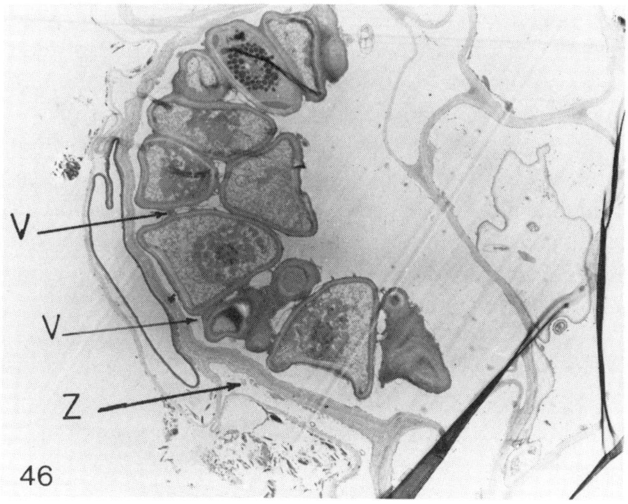

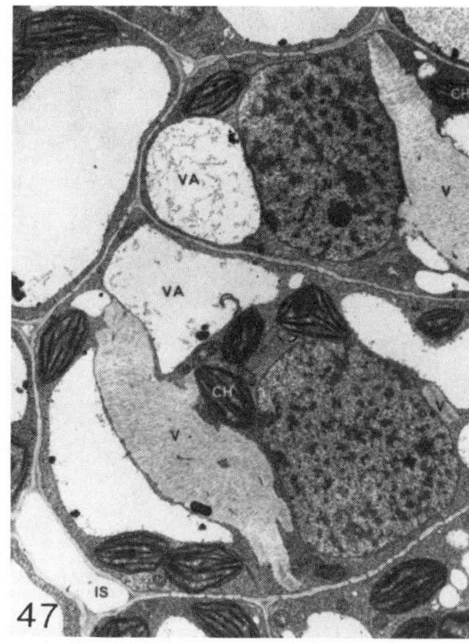

FIGURE 46. Crossection of a sugar beet root epidermal cell infected with *Polymyxa betae* showing cystospores interlinked (V) with the cystosorus and collapse of cytoplasm from the cell wall (Z). (From Ivanovic, M., Doktorska disertacija, Poljoprivredni fakultet, Beograd (Yugoslavia), 1984.) FIGURE 47. Cells of tobacco (*Nicotiana tabacum*) in the process of cell division infected with the tobacco mosaic virus showing aggregates of the virus (V), vacuoles (VA), intercellular space (IS) and chloroplasts (CH). (From Esau, K., University of Wisconsin Press, Madison, 1968.)

basic organization of plant cells. For example, cell membrane and organelle degradation results in disturbances of the principle of the internal cell space division. Changes in cell walls, in cytoplasmatic membranes and organelles alter the principle of controlled permeability in and outside the cell. This also disturbs the principle of maintenance of internal pressure (turgor) on which the colloidal state of protoplasts and the structural unity of cell constituents are based. Changes in cell structure lead to disorders in the division of labor among cell constituents (membranes, plastides, mitochondria, ribosomes, nuclei). Structural changes prevent the most efficient energy exchange, which in healthy cells, according to the current level of development, is performed in the most perfect manner.

GENETIC ENGINEERING AND PLANT GENOTYPE CHANGES

I. INTRODUCTION

Genetic engineering pertains to the transfer of genes from one cell genome to another cell genome resulting in the creation of new cell genotype capable of autonomous reproduction. Genomes belonging to taxonomically very distant organisms may be combined. The developement of such genomes with new features open extraordinary and unimagined prospectives in human medicine, veterinary medicine, and agriculture.

Human genes coding the synthesis of proteins is extremely important in medicine. Various enzymes, such as insulin and interferon for example, have been introduced into the *Escherichia coli* bacterium genome.[252] Plant genes also have been introduced into the *E. coli* genome.[252] However, the synthesis of plant compounds has been limited in genetically altered bacterial cells. If this is achieved, and it is considered highly probable, bacteria replicating in fermentors could become plant compound production factories.

Gene recombination and new genotype development are based on the transfer of deoxyribonucleic acid (DNA) as the main gene component. An important discovery in manipulatory gene recombination is that, in addition to DNA, messanger ribonucleic acid (mRNA) may also be included into genetic recombination. mRNA is included into genome recombination by the way of its complementary synthetized DNA (cDNA, DNA transcript). cDNA is obtained as a mRNA copy through the mediation of the reverse transcriptase enzyme, followed by the reconstruction of the double-stranded cDNA according to a given method.[253] Double-stranded cDNA is introduced into the vector DNA (plasmid) and then into bacterial cells. Genetic information for the synthesis of insulin, for example, has been transferred and cloned with the aid of cDNA in *E. coli.* cells.[254] The cDNA method is of prime importance in genetic engineering with plant viruses, since they are approximately 75% single-stranded RNA.

II. FOREIGN GENE VECTORS IN PLANTS

Vectors are organisms or agents capable of transferring genetic material from a donor to an acceptor organism, enabling it to survive and express its influence in the acceptor cell (Table 1).

1. Bacteria as Plant Gene Vectors

The Ti-plasmid (tumor-inducing) of *Agrobacterium tumefaciens* is the most suitable, if not the only plant gene vector. Upon contact with an injured plant cell, the Ti-plasmid is introduced into a susceptible plant cell. A portion of the Ti-plasmid is further separated, introduced into the plant cell nucleus and becomes part of the nucleus genome, giving rise to a new cell type. Plasmid T-DNA also introduces new genes the modified cell genotype being expressed by the synthesis of new products and proliferation division of genetically altered cells. The gene conferring resistance to *A. tumefaciens* has also been tranferred through the mediation of Ti-plasmids into tobacco plants.[255]

A portion of the inner T-DNA plasmid may be substituted with a small *E. coli* plasmid, for example, pBR322. Thus, foreign DNA genes may be cloned in the small plasmid and through the mediation of Ti-plasmids introduced into the plant cell genome. Based on the gene transfer mechanism, the Ti-plasmid may be termed as an integration vector, since it directly introduces foreign genes into plant cell genomes, thus bringing about the greatest possible changes in plant cells and plants.

2. Plant Viruses as Plant Gene Vectors

Bacterial viruses (bacteriophages) are the most active vectors in the transfer of genetic material into bacteria and animals. Plant viruses may also be a valuable vector for active gene transfer among plants. However, the potential mechanism of their direct vector role has not been established. However, in the view of rapid research development this aim is expected to be achieved within the next decade.

Plant viruses may be used as possible vectors. However, virus particles develop independently in infected plant cells and do not become part of the cell genome. Viruses could introduce foreign genes into plant cells which could replicate independently and spread throughout the plant simultaneously with the plant virus genome. Thus, plant viruses could deliver foreign genes in plant cells but in a manner sustantially different from that of the Ti-plasmid.

In order for plant viruses to act as vectors, engineered viruses acting as vectors of plant genes should be capable of infecting the plant cell and replicating in it, accompanied by a simultaneous multiplication fo the introduced foreign genetic material. The genes being transferred must not cause disease or a reduction in plant yield.

The cauliflower mosaic virus (CaMV) has been successfully used in plant gene transfer. It is a small isometrical virus with double-stranded DNA. The DNA has been cloned in plasmids of *E. coli* and preserved its infectivity. However, CaMV could not be used as a vector because

TABLE 1.
Vectors Capable of Transferring Genetic Material From a Donor to Acceptor Organisms.

Vectors	Acceptors
Plasmids, viruses (bacteriophages for bacterial and SV_{40} for animal cells) Cosmids (specially engineered plasmids from lambda bacteriophages)	Bacteria, yeasts, animals
A. tumefaciens plasmids, double-stranded cauliflower mosaic virus DNA	Plants
Single-stranded tobacco mosaic virus RNA	Plants
Single-stranded geminivirus DNA, etc.	Plants

with the degradation of its small dimension DNA, the virus loses its infectivity and therefore its role as a vector. This problem is solved with the application of an auxiliary virus system in which the viral genome is expanded with foreign DNAs, while infectivity is expressed with normal viral DNA.

The potential of gemini viruses plant virus vectors has been investigated. Gemini viruses are isodiametrical with single-stranded DNA. They penetrate and replicate in plant cell nuclei, forming an infectious double-stranded DNA in the nucleus which may be used by bacterial plasmids *in vivo*. The virus infects both mono- and dictotyledonous plants, which is of particular importance for gene transfer between a wide variety of hosts.

RNA viruses, particularly multicomponent and satellite viruses, might play an important role in recombination genetics in the future. In some multicomponent viruses, such as the brome mosaic virus, the smallest component contains RNA responsible only for coat protein synthesis which may be completely substituted with foreign genetic material (RNA of cDNA) without disturbing virus replication and infectivity.

The tricomponent brome mosaic virus RNA has been used to transfer DNA into barley protoplasts in which the foreign DNA proved to be more active. This virus could be used as a gene vector since it is virulent to mono- and dicotyledonous plants. Satellite virus RNA is not indispensible for basic virus development and maybe completely or partially substitiuted with foeign genetic material (RNA or cDNA) which may be introduced into the plant following infection.

III. VIRUS GENE TRANSFER INTO PLANTS AND PLANT RESISTANCE TO VIRAL INFECTIONS

Plant virus genes determining (coding) viral (capsid) protein synthesis may be introduced into plant cells using gene transfer techique. Virus protein formation is a product of the mRNA activity formed by trransformation from viral RNA (RNA viruses) or by transcription from viral DNA (DNA viruses). cDNA may be synthetized by reverse (irreversible) transcription from individual mRNA. By introducing cDNA with a vector system these nucleic acids replicate autonomously and spread through the plant. Plants containing such foreign genetic material are termed "transgenic plants". By replication transgenic plants confer these features to subsequent generations.

The transfer of virus genes into transgenic plants is performed by the transfer system with the aid of *A. tumefaciens* Ti-plasmids. In this transfer system virus cDNA is introduced into intermediate of *E. coli* plasmids where it is bound to certain plasmid promoters, which results in a new chimeric gene. With the aid of the genetic transfer techniques, using an intermediate plasmid, the chimeric gene is built into the Ti-plasmid. Bacteria with the modified Ti-plasmid transfer the chimeric genes into the plant cells. The plants in which the introduced chimeric gene replicates and spreads, obtain the transgenic characters.

Viral genes in transgenic plants effect on both, plant resistance to viral infection and viral particle spreading. The viral protein, synthetized by the activity of the gene increases plant resistance virus from which the gene was originally isolated.

1. Examples of Transgenic Plant Resistance to Viral Infections

Transgenic tobacco plants resistant to tobacco mosaic virus (TMV) have been obtained using the viral gene transfer technique.[256] Inoculated transgenic tobacco plants inocluated with TMV have not manifested systemic symptoms, nor accumulation fo virus particles.

Similarly, tansgenic tobacco and tomato plants, containing alfalfa mosaic virus capsid protein remained resistant to this virus.[257-259] These authors reported delayed symptoms in transgenic plants; the degree depending on accumulation of the capsid protein.

Transgenic tomato plants with capsid protein of TMV strainU_1 were resistant to infection and development of symptoms caused by TMV and tomato mosaic virus strains L, 2 or L^2.[260] In field trials 5% of inoculated transgenic plants were infected, whereas infection occurred on 99% of control plants. The yield of transgenic tomato plant inoculation in the field with TMV was not affected, whereas the yield on control plants was reduced by 26-35%. These trials have shown that the use of transgenic plants may be used as an effective means of protecting tomatoes against TMV.

Transgenic tobacco plants with the potato mosaic virus capsid protein were protected against infection by this virus as evidenced by a decrease in the number of lesions on inoculated leaves, delayed or absence of systemic symptoms, and reduced accumulation of viral particles.[261]

Transgenic tobacco plants with the cucumber mosaic virus capsid protein was resistant to this virus. Viral particle development and accumulation symptoms were reduced or absent in transgenic plants.[262]

Viral proteins are primarlily responsible for the spread form cell to cell in transgenic plants.[263] For example, using the chimeric gene transfer technique, cDNA for 30-KD TMV protein synthesis was introduced into transgenic plants. Transgenic plants generated from transformed tobacco cells, as well as their seedlings, expressed virus movement of the L_s1 strain in the plants, confirming that the 30-KD protein was responsible for virus movement from one cell into another.

2. Transgenic Plant Resistance Features

Capsid protein may protect transgenic plants from infections by the virus from which the gene was isolated. In all described cases, this protection was due to the accumulation of the capid protein in transgenic plants, showing that this protein plays an important role in plant protection against its mRNA.

In view of the viral particle replication mechanism, it is possible to assume that transgenic plant protection is achieved in the early stages of infection when virus uncoating takes place. Infection by TMV on trangenic alfalfa leaves was prevented at a stage prior to viral RNA release from its particles.[264] Thus, protection is a result of competition between the previously formed capsid protein and the virus introduced to the site of virus uncoating.

In conclusion, transgenic plant resistance is achieved by mechanisms in processes of plant cross protection. In classical cross protection, prior inoculation is performed with a less virulent strain offering protection against a later infection with a more virulent strain of the same virus. In transgenic plants, protection is obtained by the introduction of certain genes for viral protein production, so that this type of protection may be termed genetic cross protection.

REFERENCES

1. **McClendon, J. H.,** Evidence for the pectic nature of the middle lamella of potato tuber cell walls based on chromatography of macerating enzymes, *Am. J. Bot.*, 51, 628, 1964.
2. **Cowling, E. B. and Brown, W.,** Structural features of cellulosic materials in relation to enzymatic hydrolysis, in *Cellulases and Their Applications,* Advances in Chemistry, Series 95, Gould, R. F., Ed., Amer. Chem. Soc., Washington, DC, 1969, 152.
3. **Gäumann, E.,** *Pflanzliche Infeckitionslehre,* Verlag Birkhauser, Basel, Germany 1946, 611.
4. **Talboys, P. W.,** Dysfunction of the water system, in *Plant Disease, An Advanced Treatise,* III, Horsfall, G.J., and Cowling, B.E., Eds., Academic Press, Inc., New York, 1978, 141.
5. **Küster, E.,** *Pathologische Pflanzenanatomie,* Dritte Auflage, Verlag von Gustav Fisher, Jena, Germany, 1925.
6. **Tošić, M. and Mišović, M.,** Uticaj viroznog mozaika na porast i prinos nekih sorti i hibrida kikiriza, *Zastita bilja* 93, 173, 1967.
7. **Pop, I. and Tusa, C.,** Influence of maize mosaic on the growth and yield of maize hybrids, *Proc. Int. Conf. Plant Viruses,* Wageningen, 1966, 170.
8. **Tošić, M.,** Virus diseases of wheat in Serbia II, Some changes in wheat plants infected with wheat streak mosaic virus (WSMV) and brome mosaic virus (BMV), *Phytopathologesche Zeitschrift,* 71, 327, 1971.
9. **Šutić, D. and Milovanović, M.,** Pojava i znasčaj kržljavosti korena šecérne repe, Agrohemija, 9, 363, 1978.
10. **Kegler, H.,** Untersuchungen über Virosen des Kernobstes IV, die Flachätigkeit des Apfels. *Phytopath. Z,* 50, 297, 1964.
11. **Josifović, M.,** *Poljoprivredna Fitopatologija* III, Naučna Knjiga, Belgrade, 1964.
12. **Butler, J. E. and Jones, G. S.,** *Plant Pathology,* Macmillan and Co., Ltd., London, 1949, 979.
13. **Merlo, J. D.,** Crown gall — a unique disease, in *Plant Disease,* An Advanced Treatise III, Horsfall, J.G. and Cowling, B.E., Eds., Academic Press, Inc., New York, 1978, 201.
14. **Lippincott, A. J. and Lippincott, B. B.,** Morphogenic determinants as exemplified by the crown-gall disease, in *Physiological Plant Pathology,* Heitefuss, R. and Williams, P.H., Eds., Springer-Verlag, Berlin, 1976, 356.
15. **Wood, R. K. S.,** *Physiological Plant Pathology,* Blackwell Scientific Publ., Ltd., Oxford, 1967.
16. **Cincović, T., Čanak, M., and Kojić, M.,** *Botanika.* Naučna *Knjiga,* Belgrade, 1980.
17. **Goodman, R. N., Kiraly, Z., and Zaitlin, M.,** *The Biochemistry and Physiology of Infectious Plant Disease,* D. van Nostrand Co., Inc., London, 1967.
18. **Šutić, D.**, Biljni virusi, *Nolit,* Belgrade, 1980.
19. **Scaramuzzi, G.**, Una virosi con deformazione, maculatura verde e suberosi interna dei frutti de cotagne (*Cydonia oblonga* Mill.), *Phytopath.* Z. 30, 259, 1957.
20. **Popović, Z.**, *Fiziologija Bilja,* Naučna Knjiga, Belgrade, 1976.
21. **Glisić, Lj.**, Opśta Citologija, *Unija Bioloskih Naučnih Društava Jugoslavije,* Belgrade, 1980.
22. **Kastori, R.**, Uloga elementa u ishrani biljaka, *Matica Srpska,* Novi Sad, 1983.
23. **Wallace, T.**, The diagnosis of mineral deficiencies in plants by visual symptoms, 3rd Ed., Chemical Publishing Co., New York, 1961, 125.
24. **Neuffer, G. M., Loring, G., and Zuber, S. M.**, *The Mutants of Maize,* Crop Sci. Soc. Amer., Madison, 1968, 74.
25. **Aleksić, Z., Aleksić, D., and Šutić, D.**, Bolesti povrca i njihovo suzbijanje. *Nolit,* Belgrade, 1980.
26. **Gassner, G. and Goeze, G.**, Assimilationsverhalten, Chlorophyllgehalt und Transpirationsgrosse von Getreideblattern mit besonderer Berucksichtung der Kalium- und Stickstoffenährung, *Zeitsch f. Bot.* 27, 257, 1934.
27. **Kokin, A. J. and Tumarinson, S. H.**, Fiziologicheskoe obsnovanie vrednosti rzhavchini ovsa *Puccinia coronifera* Kleb., *Tr. Po. Zasc. Rast.*. II ser. 6, 5, 1934.

28. **Tumarinson, H. S.,** K fizioologicheskomu obosnovaniyu shkal uchetu vrednosti rzhavchini,*Tr. Po. Zaskchite Rast* II ser. 6, 35, 1934.
29. **Oknina, E. Z.,** Ispitanie virulentnogo deystviya verticilov na hlopchatnika, *Tr. Inst. Fiziol. Rast. im. Timiryazeva* 3, I, 109, 1973.
30. **Dvoretskaya, I. E. and Skurtul, M. A.**, O vliyanii porazheniya tomatov gribom *Cladosporium fulvum* na soderzhanie pigmentov i aktivnosty protolicheskih fermentov, *Biohemija plodov i ovoshey,* 20 F, Moscow, 1964.
31. **Kuprevich, F. V.**, Fiziologiya bolynovo rasteniya v svyazi s obshimi vaprosami parazitizma,*Izdat. Akad. nauk SSSR,* Moscow, 299, 1947.
32. **Bawden, C. F.**, *Plant Viruses and Virus Diseases,* Chronica Botanica Book Co., Waltham, Massachusetts, 1956.
33. **Pešić, Z., Jovanović, Z., Mitrović, G., and Pantić, S.**, Promene u količini hlorofila i skroba u biljakama kukrurza zarazenih virusom mozaika, in *Symposium IV — Savremeni Problemi Virusnih Infeckija,* Vrnjačka Banja, 1977.
34. **Wang, D.**, The nature of starch accumulation at the rust infection site in leaves, of pinto bean leaves, *Can. J. Bot.*, 39, 1595, 1961.
35. **Šutić, D.,** Viroze Biljaka, *Nolit,* Belgrade, 1983.
36. **Kojić, M.**, Botanika, *Naučna knjiga,* Belgrade, 1984.
37. **Sullian, J. T. and Chilton, S. J. P.**, The effect of leaf rust on the carotene content of white clover, *Phytopathology,* 31, 554, 1941.
38. **Elmer, O. H.**, Transmissibility and pathological effects of the mosaic disease, Iowa Agricultural Experiment Station Research Bulletin 82, 39, 1925.
39. **Guthrie, J. D.**, Effect of environmental conditions on the chloroplasts pigments. *American Journal of Botany* 16, 716, 1929.
40. **Tarr, J. A. S.**, *The Principles of Plant Pathology,* MacMillan Press, Inc., London, 1972.
41. **Matthews, R. E. F.**, *Plant Virology,* Academic Press, Inc., New York, 1981.
42. **Ladigina, M. E. and Edreeva, M. A.,** Biohimicheskie mehanizmi patogeneza pri peresporonoze i tochechnom nekroze u tabaka, in *Patologicheskaya fiziologiya i imunitet rasteniy,* Izdatelstvo Moskovskogo Univerziteta, 1976, 60.
43. **Bossi, R.**, Über die Wirkung der Fusarinsäure auf die Polyphenoloxydase, *Phytopathologesche Zeitschrift,* 37, 273, 1959.
44. **Džamić, M.**, *Osnovi biohemije II,* ICS, Belgrade, 1978.
45. **Stević, B.**, Tehnološka mikrobiologija stočnih proizvoda ishrane stoke, *Naučna knjiga,* Belgrade, 1962.
46. **Bateman, F. D. and Basham, G. H.**, Degradation of plant cell walls and membranes by microbial enzymes, in *Physiological Plant Pathology,* Heitefuss, R. and Williams, P. H., Eds., Springer-Verlag, Berlin, 1976, 315.
47. **Heitefuss, R. and Williams, H. P.**, *Physiological Plant Pathology,* Springer-Verlag, Berlin, 1976.
48. **McMurray, W. C. and Magee, W. L.**, Phospholipid metabolism. *Ann. Rev. Biochem.*, 41, 129, 1972.
49. **Gatti, S. and Barenholz, Y.**, Enzymes of complex lipid metabolism, *Annual Review of Biochemistry,* 42, 61, 1976.
50. **Hancock, J. G., Millar, R. L., and Lorbeer, J. W.**, Pectolitic and cellulolytic enzymes produced by *Botrytis allii, B. cinerea,* and *B. squamosa in vitro* and *in vivo, Phytopathology*, 54, 928, 1964.
51. **Uritani, I. and Stahmann, M. A.,** Changes in nitrogen metabolism in sweet potato with black rot, *Plant Physiology,* 36, 770, 1961.
52. **Strider, D. L. and Winstead, N. N.**, Production of cell-wall dissolving enzymes by *Cladosporium cucumerinum* in cucumber tissue and in artificial media, *Phytopathology,* 51, 765, 1961.
53. **Cole, M. and Wood, R. K. S.**, Pectic enzymes and phenolic substances in apples rotted by fungi., *Annuals of Botany,* 25, 435, 1961.
54. **Winstead, N. N. and McCombs, C. L.**, Pectinolytic and cellulolytic enzyme production by Pythium aphanidermatum, *Phytopathology,* 51, 270, 1961.
55. **Bateman, D. F.**, Pectolytic activities of culture filtrates of *Rhizoctonia solani* and extracts of *Rhizoctonia*-infected tissues of bean, *Phytopathology*, 53, 197, 1963.
56. **Hancock, J. G. and Millar, R. L.**, Relative importance of polygalacturonate trans-eliminase and other pectolytic enzymes in southern anthracnose, spring black stem, and Stemphylium leaf spot of alfalfa, *Phytopathology,* 55, 346, 1965.
57. **Foley, D. C.**, The presence of cellulase in corn stalks infected with *Fusarium moniliforme, Phytopathology,* 49, 538, 1959.
58. **Husain, A.**, Production of cellulolytic enzymes by *Sclerotium rolfsii, Phytopathology,* 48, 338, 1958.
59. **Sumere, C. F., Van Sumere-De Preter, C., and Ledingham, G. A.**, Cell-wall-splitting enzymes of *Puccinia graminis* var. *tritici., Can. J. Microbiol.*, 3, 761, 1957.
60. **Linskens, H. G., Heninen, W., and Stoffers, L. A.**, Cuticula of leaves and the residue problem, in *Residue Reviews,* Guther, F. A., Ed., 8, 1965, 136.

61. **Linskens, H. F. and Haage, R.**, Cutinase Nachweis in Phytopathogen pilzen. *Phytopathologesche Zeitschrift*, 48, 306, 1963.
62. **Kuć, J.**, Production of extracellular enzymes by *Cladosporium cucumerinum*, *Phytopathology*, 52, 961, 1962.
63. **Goto, M. and Okabe, N.**, Two kinds of pectin-depolymerase secreted by *Erwinia carotovora* (Jones) Holland, II, Comparative studies of endo-polymethygalacturonase and endopolygalacturonase, *Annals. of Phytopathological Society of Japan*, 28, 1, 1963.
64. **Smith, W. K.**, A survey of the production of pectic enzymes by plant pathogenic and other bacteria. *Journal of General Microbiology*, 18, 33, 1958.
65. **Wood, R. K. S.**, Studies in the physiology of parasitism. XVIII. Pectic enzymes secreted by *Bacterium aroideae.*, *Annals of Botany*, 19, 1, 1955.
66. **Gehring, E.**, Untersuchungen über die Infektionsverlauf einer durch *Pectobacterium parthenii* (Starr) Hellmers var. *dianthicola* Hellmers verursachten Nelkenbakteriose sowie über enzymatische Eigenschaften dieses Bakteriums im Vergleich mit *Pseudomonas caryphylli* (Burkholder) Starr et Burkholder und einigen typischen Nassfäuleerregern, *Phytopathologesche Zeitschrift*, 43, 383, 1962.
67. **Dye, D. W.**, Pectolytic activity in *Xanthomonas.*, *New Zealand Journal of Science*, 3, 61, 1960.
68. **Abo-El-Dahab, M. K.**, Production of petic and cellulolytic enzymes by *Xanthomonas malvacearum*, *Phytopathology*, 54, 597, 1964.
69. **Amman, A.**, Über die Bildung von Zellulase bei pathogenen Mikroorganismen, *Phytopathologesche Zeitschrift*, 18, 416, 19652.
70. **Goto, M. and Okabe, N.**, Studies on the celluloytic enzymes of phytopathogenic bacteria. Part 2. Relations between the activity of cellulase "cx" and pH, reaction temperature or salts., *Annals of Phytopathological Society of Japan*, 24, 182, 1959.
71. **Friedman, B. A.**, Physiological differences between a virulent and a weakly virulent radiation-induced strain of *Erwinia caratovora*, *Phytopathology*, 52, 328, 1962.
72. **Krstić, M. and Usčuplić, M.**, Najvažnije parazitske glijive šumskog biljja, str. 224, *Sarajevo*, 224, 1965.
73. **Erikson, D. and Montonery, H. B. S.**, Certain aspects of resistance of plum trees to bacterial canker. Part III. The action of cell-free filtrates of *Pseudomonas mors-prunorum* Wordmald and related phytopathogenic bacteria on plum trees., *Annals of Applied Biology*, 32, 117, 1945.
74. **Arsenijević, M.**, *Bakterioze Biljaka*, Poljoprivredni Fakultet, Novi Sad, Yugoslavia, 1975.
75. **Goidanich, G.**, *Manuale di Patologia Vegetale*, Edizione Agricola, Bologna, Italy, 1959
76. **Gottlieb, D.**, The presence of a toxin in tomato wilt, *Phytopathology*, 33, 126, 1943.
77. **Dimond, A. E. and Waggoner, E. P.**, The cause of epinastic symptoms in Fusarium wilt of tomatoes, *Phytopathology*, 43, 663, 1953.
78. **Braun, A. C.**, A study of the mode of action of the wildfire toxin, *Phytopathology*, 45, 659, 1955.
79. **Pringle, R. B. and Scheffer, R. P.**, Host-specific toxins, *Annals Review Phytopathology*, 2, 133, 1964.
80. **Scheffer, R. P. and Samaddar, K. R.**, Host-specific toxins as determinants of pathogenicity, *Recent Advances Phytochemistry*, 3, 123, 1970.
81. **Scheffer, R. P. and Yoder, O. C.**, Host-specific toxins and selective toxicity, in *Phytotoxins in Plant Diseases*, Woods, R. K. S., Ballio, A. and Graniti, Eds., Academic Press, Inc., London, 1972, 251.
82. **Pringle, R. B.**, Chemistry of host-specific phytotoxins, in *Phytotoxins in Plant Diseases*, R.K.S. Wood, Ballio, A. and Graniti, Eds., Academic Press Inc., London, 1972, 39.
83. **Turner, N. C.**, Stomatal behavior of *Avena sativa* treated with two phytotoxins, victorin and fussiococcin, *American Journal of Botany*, 59, 133, 1972.
84. **Scheffer, R. P.**, Host-specific toxins in relation to pathogenesis and disease resistance, in *Physiological Plant Pathology*, Heitefuss, R. and Williams, H. P., Eds., Springer-Verlag, Berlin, 1976, 245.
85. **Farkas, G. L., Kiraly, A., and Solymosy, F.**, Role of oxidative metabolism in the localization of plant viruses, *Virology*, 12, 408, 1960.
86. **Weintrub, M., Ragetli, J. W. H., and Dwurazna, M. M.**, Studies on the metabolism of leaves with localized virus infections: Mitochondrial activity in TNV-infected *Nicotiana glutinosa* L., *Canadian Journal of Botany*, 42, 541, 1964.
87. **Parish, C. L., Zaitlin, M., and Siegel, A.**, A study of a necrotic lesion formation by tobacco mosaic virus, *Virology*, 26, 413, 1965.
88. **Weintrub, M., Kemp, G. W., and Ragetli, J. W.**, Studies on the metabolism of leaves with localized virus infection: I Oxygen uptake, *Canadian Journal of Microbiology*, 6, 407, 1960.
89. **Yamaguchi, A. and Hirai, T.**, The effect of local infection with tobaccco mosaic virus on respiration in leaves of *Nicotiana glutinosa*, *Phytopathology*, 49, 447, 1959.
90. **Müller, K. O. and Behr, L.**, Mechanism of *Phytophthora* resistance of potatoes, *Nature*, 163, 498, 1949.
91. **Tomiyama, K., Sakai, R., Takase, N., and Takakuwa, M.**, Physiological studies on the defense reaction of potato plant to the infection by *Phytophtohora infestans:* IV. The influence of preinfectional ethanol necrosis upon the physiological reaction of potato tuber in the infection by *P. infestans*, *Annals of the Phytopathological Society of Japan*, 21, 153, 1957.

92. **Maine, E. C.**, *Physiological Responses in the Tobacco Plant to Pathogenesis by Pseudomonas solanacearum, Causal Agent of Bacterial Wilt,* Ph.D. Thesis, Dept. of Plant Pathology, North Carolina State University, Raleigh, 1960.
93. **Sequeira, L. and Kelman, A.**, The accumulation of growth substances in plants infected by *Pseudomonas solanacearum, Phytopathology* 52, 439, 1962.
94. **Perdrizedt, E. and Macquaire, G.**, Étude des variations due métabolisme des feuilles de pommes de terre provoquées par le virus l'enroulement durant les premiers stades dé l'infection, *Compt. Rend. Acad. Sci.*, Paris, 257, 3208, 1963.
95. **Diener, T. O.**, Free amino acids and amides in healthy and virus-infected cherry and peach leaves, *Phytopathology,* 50, 141, 1960.
96. **Bozarth, R. F. and Diener, O. T.**, Free amino acid concentrations in plants singly and doubly infected with potato virus X and potato virus Y, (Abst.), *Phytopathology* 53, 871, 1063.
97. **Selman, I. W., Brierley, R. M., Pegg, G. F., and Hill, A. T.**, Changes in the free amino acids and amides in tomato plants inoculated with tomato spotted wilt virus, *Annals of Applied Biology,* 49, 601, 1961.
98. **Ross, A. F. and Williamson, E. C.**, Physiologically active emanations from virus-infected plants, *Phytopathology,* 41, 431, 1951.
99. **Williamson, C. E.**, Ethylene, a metabolic product of diseased or injured plants, *Phytopathology*, 40, 205, 1950.
100. **Meyer, G.**, Zellphysiologische und anatomische Untersuchungen über die Reaktion der Kartoffelknolle auf den Angriff von *Phytophthora infestans* bei Sorten verschiedener Resistenz, *Arb, Biol. Reichsanst,* 23, 97, 1940.
101. **Tomiyama, K.**, Cytological studies on the resistance of potato plants to *Phytophthora infestans* II: The death of the intercellular hyphae in the hypersensitive cells, *Annals of Phytopathological Society of Japan,* 19, 149, 1955.
102. **Ward, H. M.**, On the relations between host and parasite in the bromes and their brown rust, *Puccinia dispersa (Erikss.), Annals of Botany,* 16, 233, 1902.
103. **Allen, R. F.**, A cytological study of infection of Baart and Kanred wheats by *Puccinia graminis tritici., Journal of Agricultural Research,* 23, 131, 1923.
104. **Allen, R. F.**, A cytological study of orange leaf rust, *Puccinia triticina* physiologic form 11, on Malaoff wheat., *Journal of Agricultural Research,* 34, 697, 1927.
105. **Thiers, H. D. and Lister, M. B.**, A histological study of bacterial blight of cotton, *Phytopathology,* 41, 499, 1951.
106. **Talboys, P. W.**, The possible significance of toxic metabolites of *Verticillium albo-atrum* in the development of hop wilt symptoms. *Transactions of the British Mycological Society,* 40, 415, 1957.
107. **Kiessig, R. and Haller-Kiessig, R.**, Beitrag zur Kenntnis einer infektiösen Welkekrankheit der Luzerne (*Verticillium albo-atrum* R. et B.), *Phytopathologesche Zeitschrift,* 31, 185, 1957.
108. **French, W. J. and Stassi, L. D.**, Further observations on plum leaf scard and peach phony: Rickettsia-associated disease of stone fruits, (Abstr.) *Abstr. of Papers, 3rd International Congress of Plant Pathology,* 79, 1978.
109. **Goheen, C. A., Nyland, G., and Lowe, K. S.**, Association of a rickettsia-like organism with Pierce's diseases of grapevines and alfalfa dwarf and heat therapy of the disease in grapevines, *Phytopathology,* 63, 341, 1973.
110. **Nyland, G., Gooheen, C. A., Lowe, K. S., and Kirkpatrick, C. H.**, The ultrastructure of a rickettsia-like organisms from a peach tree affected with phony peach, *Phytopathology*, 53, 1275, 1973.
111. **Schneider, H.**, Cytological and histological aberrations in woody plants following infection with viruses, mycoplasmas, rickettsias, and flagelates, *Annual Review of Phytopathology,* 11, 119, 1973.
112. **Gruen, H.**, Auxin and fungi, *Annual Review Plant Physiology,* 10, 405, 1959.
113. **Shaw, M. and Hawkins, A. R.**, The physiology of host-parasite relations V. A preliminary examination of the level of free endogenous Iaa in rusted and mildewed cereal leaves and their ability to decarboxylate exogenously supplied readioactive IAA. *Canadian Journal of Botany,* 36, 1, 1958.
114. **Pegg, F. G.**, Endongenous auxins in healthy and disease plants, in *Physiological Plant Pathology*, Heitefuss, R. and Williams, H. P., Eds. Spring-Verlag, Berlin, 1976, 560.
115. **Pegg, F. G.**, Endogenous gibberellins in healthy and diseased plants, in *Physiological Plant Pathology,* Heitefuss, R. and Williams, H.P. Eds., Spring-Verlag, Berlin, 1976, 592.
116. **Pegg, G. P.**, Endogenous inhibitors in healthy and diseased plants, in *Physiological Plant Pathology*, Heitefuss, R. and Williams, H.P., Eds., Springer-Verlag, Berlin, 1976, 607.
117. **Hirata, S.**, Studies on the phytohormone in the malformed portion of the diseased plants I: The relatin between the growth rate and the amount of free auxin in the fungus galls and virus infected plant. *Annals of the Phytopathological Society of Japan,* 19, 33, 1954.
118. **Daly, J. M. and Inman, R. E.**, Changes in auxin levels in safflower hypocotyles infected with *Puccinia carthami, Phytopathology,* 48, 91, 1958.
119. **Kiermeyer, O.,** Papierchromatographische Untersuchungen über der Wuchsstoffgehalt von *Capsella bursa-pastoris* nach Infektion mit *Albugo candida* und *Peronospora parasitica, Öst. Botan. L.*, 105, 515, 1958.
120. **Reingard, T. A. and Pashkar, S. I.**, __________, *Fizol. Rasteniy Akad. Nauk*, USSR, 5, 501, 1958.
121. **Turian, G.**, Recherches anatomo-physiologiques sur les cecides, *Review Géneral Botany* 65, 279, 1958.

122. **Turian, G.**, Accumulation of 3-indole acetic acid and activation of enzymes in the *Ustilago* induced tumor of maize, *Conference on Scientific Problems in Plant Protection*, Budapest, 1, 35, 1961.
123. **Turian, G. and Hamilton, H. R.**, Chemical detection of 3-indolylacetic acid in *Ustilago zeae* tumors, *Biochem. Biophys. Acta* 41, 148, 1960.
124. **Matta, A. and Gentile, I.**, Variation in auxin content induced in tomato by *Fusarium oxysporum* f. sp. *lycopersici*, *Review of Applied Mycology* 44, 233, 1964.
125. **Pegg, G. F. and Selman, I. W.**, An analysis of the growth response of young tomato plants to infection by *Verticillium albo-atrium* II: The production of growth substances, *Annals of Applied Biology* 47, 222, 1959.
126. **Jaros, H.**, Studies on the phases of development in healthy and virus Z, Y and Z- and Y-infected potatoes. *Acta Biol. Cracov Ser. Bot.*, 6, 75, 1963.
127. **Smith, S. H., McCall, S. R., and Harris, J. H.**, Auxin transport in curly top virus-infected tomato, *Phytopathology*, 58, 1169, 1968.
128. **Vendrig, J. C. and Buffel, K.**, Growth-promoting activity of caffeic acid, *Nature*, 193, 1204, 1962.
129. **Henderson, J. H. and Nitsch, J. P.**, Effect of certain phenolic acids on the elongation of *Avena* first internodes in the presence of auxins and tryptophan, *Nature*, 195, 780, 1962.
130. **Zenk, M. H. and Muller, G.**, *In vivo* destruction of exogenously applied indoyl-3-acetic acid as influenced by naturally occurring phenolic acids, *Nature*, 200, 761, 1963.
131. **Daly, J. M., Krupka, L., and Bell, A. A.**, Influence of hormones on respiration metabolism of healthy and rust-affected tissues, *Plant Physiology*, 37,130, 1962.
132. **Galston, A. W. and Purves, L. W.**, The mechanism of action of auxin, *Annual Review of Plant Physiology*, 11, 239, 1960.
133. **Beckman, C. H., Kuntz, J. E., Riker, A. J., and Berbee, J. G.**, Host respones associated with development of oak wilt, *Phytopathology*, 43, 448, 1953.
134. **Beaucamp, C.**, La constitution auxinique des tabacs, sains ou virosés, *Review of General Botany*, 65, 477, 1958.
135. **Pavillard, J.**, Recherches sur la croissance des plantes virosés: Virus et auxins, *Compt. Rend.* 235, 87, 1952.
136. **Lucas, H.**, Weitere Untersuchungen über den Wuchsstoffhaushalt abbaukranker Kartoffeln, *Phytopathologesche Zeitschrift*, 12, 234, 1939.
137. **Söding, H. and Funke, H.**, Über den Wuchsstoffhaushalt abbaukranker Kartoffeln, *Phytopathologesche Zeitschrift* 13, 351, 1941.
138. **Kurosawa, E.**, Experimental studies on the secretion of *Fusarium heterosporum* on rice plants, *Transactions of the Natural History Society of Formosa*, 16, 213, 1926.
139. **Brian, P. W. and Hemming, B. H.**, Complimentary action of gibberellic acid and auxin on pea internode extension, *Annals of Botany*, 22, 1, 1958.
140. **Russel, S. L. and Kimmins, W. C.**, Growth regulatory and the effect of barley yellow dwarf virus on barley (*Hordeum vulgare* L.), *Annals of Botany* 35, 1037,1971.
141. **Skoog, F., Strong, F. M., and Miller, C. O.**, Cytokinins, *Science*, 148, 532, 1965.
142. **Srivastava, R. I. S.**, Cytokinins in plants, *International Review of Cytology*, 22, 350, 1967.
143. **Dekhuijzen, M. H.**, Endogenous cytokinins in healthy and diseased plants, in *Physiological Plant Pathology*, Heitefuss, R. and Williams, H. P., Eds., Springer-Verlag, Berlin, 1976, 526.
144. **Letham, D. S.**, Cytokinins from *Zea Mays*, *Phytochemistry*, 12, 2445, 1973.
145. **Richmond, A. and Lang, A.**, Effect of kinetin on protein content and survival of detached *Xanthium* leaves, *Science*, 125, 1957.
146. **Fletcher, R. A. and McCullagh, D.**, Cytokinin-induced chlorophyll formation in cucumber cotyledons, *Planta* 101, 88, 1971.
147. **Fletcher, R. A. and McCullagh, D.**, Benzyladenine as a regulator of chlorophyll synthesis in cucumber cotyledons, *Canadian Journal of Botany*, 49, 2149, 1971.
148. **Fletcher, R. A., Teo, C., and Ali, A.**, Stimulation of chlorophyll synthesis in cumcumber cotyledons by benzyadenine, *Canadian Journal of Botany*, 51, 937, 1973.
149. **Norberg, S. O.**, Studies in the production of auxins and other growth stimulting substances by *Exobasidium*, *Symbolae Botan. Upsalienses*, 19, 1, 1968.
150. **Rowan, S. J.**, Fusiform rustgall formation and cytokinin of loblolly pine, *Phytopathology*, 60, 1225, 1970.
151. **Wison, E. E.**, Pathological histogenesis in oleander tumors induced by *Pseudeomonas savastanoi*, *Phytopathology*, 55, 1244, 1965.
152. **Krikun, J., Chorin, N., and Vazaadia, Y.**, Kinetics of the hormonal imbalance of tomato plants infected with *Verticillium dahliae*, *Israed Journal of Agricultural Research* 21, 149, 1971.
153. **Misaghi, I., Devay, J. E., and Kosuge, T.**, Changes in cytokinin activity associated with the development of Verticillium wilt and water stress in cotton plants, *Physiological Plant Pathology*, 2, 187, 1972.
154. **Freebairn, H. T. and Buddenhagen, W. I.**, Ethylene production by *Pseudomonas solanacearum*, *Nature*, 202, 313, 1964.
155. **Swamy, R. N.**, Gaseous emanation from groundnut infected by *Cercospora personata* (Berk. et Curt.) Ell. et Everh., *Phytopathologesche Zeitschrift*, 40, 245, 1961.

156. **Smith, W. H., Meigh, D. F., and Parker, F. D.**, Effect of damage and fungal infection on the production of ethylene by carnations, *Nature,* 204, 92, 1964.
157. **Little, C. H. A. and Eidt, D. C.**, Effect of abcisic acid on budbreak and transpiration in woody species, *Nature*, 220, 448, 1968.
158. **Steadman, J. R. and Sequeira, L.**, Abscisic acid in tobacco plants. Tentative identification and it relation to stunting induced by *Pseudomonas solanacearum, Plant Physiology*, 41, 691, 1970.
159. **Wiese, M. V. and deVay, J. E.**, Growth regulator changes in cotton associated with defoliation caused by *Verticillium albo-atrum. Plant Physiology,* 45, 304, 1970.
160. **Rubin, B. A. and Artsihovskaya, E. B.**, *Biochimiya i Fizologiya Imuniteta Rasteniy,* Izdatelstvo "Vishays skola", Moscow, 1960.
161. **Gibbs, A. and Harrison, B.**, *Plant Virology - The Principles,* Edward Arnold, Ltd., London, 1976.
162. **Pejkić, B. and Šutić, D.**, Neke promene pri reduckecionoj deobi materinskith célija polena u mikrosprogenezi kod šjive požegace zaražene virusom šarke, *Zastita bilja,* 106, 347, 1969.
163. **Wison, C. L.**, Plant teratomas - Who's control of them? in *Plant Disease An Advanced Treatise* III, Horsfall, J.G. and Couling, B.E., Eds., Academic Press, Inc., New York, 1978, 215.
164. **Goldmann, A., Tempé, J., and Morel, G.**, Quelques particulartés de diverses souches d'*Agrobacterium tumefaciens,* C.R. *Seances Soc. Biol. Ses Fil* 162, 630, 1968.
165. **Goldmann, A., Thomas, D. W. , and Morel, G.,** Sure la structure de la noplaine, métabolite anormal de certaines tumeurs de crown-gall, *C. R. Hebd. Seances Acad. Sci. Ser.* B 268, 852, 1969.
166. **Petit, A., Delhaye, L., Tempé, J., and Morel, G.**, Recherches sur les guanidines des tissues de crown gall. Mise en évidence d´une relation biochemique spécifique entre les souches d´*Agrobacterium tumefaciens* et les tumeurs qu'elles induisent, *Physiol, Veg.* 8, 205, 1970.
167. **Reanny, D.**, Extrachromosomal elements as possible agents of adaptation and development, *Bacteriology Review*, 40, 552, 1976.
168. **Reddy, D. V. R. and Black, L. M.**, Isolation replication of mutant populations of wound tumor virus lacking certain genome segments, *Virology*, 80, 336, 1977.
169. **Deom, M. C., Oliver, J. M., and Beachy, R. N.**, The 30-kilodalton gene product of tobacco mosaic virus potentiates virus movement, *Science*, 237, 389, 1987.
170. **Samborski, J. D., Rohringer, R., and Kim, W. K.**, Transcription and translation in diseased plants, in *Plant Diseases An Advanced Treatise,* III, Horsfall, J.G. and Cowling, B.E., Eds., Academic Press, Inc., New York, 1978, 375.
171. **Šutić, D.**, Propadanje pamuka u nekim mestima NR Makedonije, *Zaštita bilija*, 26, 53, 1954.
172. **Doi, Y., Terenaka, M., Yora, K., and Asuyama, H.**, Mycoplasma or PLT group-like microorganisms found in the phloem elements infected with mulberry dwarf, potato witches´ broom, aster yellows, or paulownia withches´ broom , *Nippon Shokubutsu Byori Gakkaiho* 33, 259, 1967.
173. **Ishii, T., Doi, Y., Yora, K., and Asuyama, H.**, Suppressive effects of antibiotics of tetracycline group on symptom development of mulberry dwarf disease, *Nippon Shokubutsu Byori Gakkaiho,* 33, 267, 1967.
174. **Nasu, S., Sugiura, M., Wakimoto, T., and Iida, T. T.**, On the etiologic agent of rice yellow dwarf disease, *Nippon Shokubutsu Byori Gakkaiho*,33, 343, 1967.
175. **Maramorosch, K.**, Plant mycoplasma diseases, in *Plant Physiology* IV, Williams P.H., and Heitefuss, R., Eds., Springer-Verlag, Berlin, 1976, 150.
176. **Maramorosch, K.**, How mycoplasmas and rickettsias induce plant disease, in *Plant Disease An Advanced Treatise* IV, Horsfall, J. G. and Cowling, B. E., Eds., Academic Press, Inc., New York, 1979, 203.
177. **Tully, J. G. and Razin, S.**, The mollicutes, in *CRC Handbook of Microbiology*, 2nd ed. Laskin, A. I. and Lechevalier, H. A., Eds., CRC Press, Inc., Boca Raton, Florida, 1977, 417.
178. **Saglio, P., L´Hospital, M., LaFlèche, D., Dupont, G., Bové, J. M., Tuffy, F. H., and Freundt, E. A.**, *Sprioplasma citri* gen & sp. nova, A mycoplasma-like organism associated with stubborn disease of citrus, *Int. J. Syst. Bacteriol.* 23, 191, 1973.
179. **Geiger, D. R.**, Effects of translocation and assimilate demand on photosynthesis, *Canadian Journal of Botany,* 54, 2337, 1976.
180. **Auger, J. G., Shalla, T. A., and Kado, C. I.**, Pierce´s disease of grapevine: Evidence for a bacterial etiology, *Science* 184, 1375, 1974.
181. **Maramorosch, K.**, Origin and classification of viruses and mycoplasmas, *Symposium on Ecosystematics*, Univ. Arkansas Press, Fayetteville, 1972, 195.
182. **Belli, G.**, *Virologia vegetale, CLESAV*, Milan, Italy, 1977.
183. **Hoffmann, M. G., Nienhaus, G. M., Schönbeck, F., Weltzien, H. C., and Wilbert, H.**, *Lehrbuch de Phytomedizin,* Paul Parey, 1976.
184. **Scott, F. M.**, Internal suberization of tissues, *Botanical Gazette,* 111, 378, 1950.
185. **Scott, F. M., Barbara, G., Bystrom, E., and Bowler, E.**, Root hairs, cuticle and pits, *Science* 140, 63, 1963.
186. **Scott, F. M., Hamner, C. K., Baker, E., and Bowler, E.**, Electron microscope studies of the epidermis of *Allium cepa. American Journal of Botany*, 45, 449, 1958.

187. **Džamić, M.**, *Podaci iz rukopisa,* (Manuscript) 1984.
188. **Aspinall, G. O.,** *Polysaccharides,* Pergamon Press, New York, 1970.
189. **Bauer, W. D., Talmadge, K. W., Keegstra, K., and Albersheim, P.**, The structure of plant cell walls, II. The hemicellulose of the walls of suspension, cultured sycamore cells. *Plant Physiol.*, 51, 174, 1973.
190. **Northcote, D. H.**, Chemistry of the plant cell wall, *Annual Review of Plant Physiology,* 23, 113, 1972.
191. **Kirk, T. K.**, Effects of microorganisms on lignin, *Annual Review of Phytopathology*, 9, 185, 1971.
192. **Uritani, I.**, Protein metabolism, in *Physiological Plant Pathology*, Heitefuss, R. and Williams, H. P., Eds., Springer-Verlag, Berlin, 1976, 509.
193. **Priestly, J. H.**, The cuticle in angiosperms, *Botanical Review* 9, 593, 1943.
194. **Ginzburg, B. Z.**, Evidence for a protein gell structure crosslinked by metal cations in the intercellular cement of plant tissue, *Journal of Experimental Botany,* 12, 85, 1961.
195. **Sadawa, D. and Chrispeels, M. J.**, Hydroxyproline-rich cell wall protein (extension): Role in the cessation of elongation in excised pea epicotyles, *Developmental Biology*, 30, 49, 1973.
196. **Sadawa, D., Waller, F., and Chrispeels, M. J.,** Hydroxproline-rich cell wall protein (extension): Biosynthesis and accumulation in growing pea epicotyls, *Developmental Biology*, 30, 42, 1973.
197. **Cooper, M. R.**, Pathogen-induced changes in host ultrastructure, in *Plant Disease Control*, Staples, C. R. and Toenniessen, H. D., Eds., John Wiley & Sons, Inc., New York, 1981, 105.
198. **Kunoh, H., Tsuzuki, T., and Ishizaki, H.**, Cytological studies of early stages of powdery mildew in barley and wheat IV. Direct ingress from superficial primary germ tubes and appressoria of *Erysiphe graminis hordei* on barley leaves, *Physiology Plant Pathology*, 13, 327, 1978.
199. **Staples, C. R. and Toenniessen, H. G.**, Eds. *Plant Disease Control,* John Wiley and Sons, Inc., New York, 1961, 339.
200. **Paus, F. and Raa, J.**, An electron microscope study of infection and disease development in cucumber hypocotyls inoculated with *Cladosporium cucumerinum, Physiology Plant Pathology*, 3, 461, 1973.
201. **Mount, S. M.**, Tissue is disintegrated, in *Plant Disease An Advanced Treatise,* III, Horsfall, J. G. and Cowling, B. E., Eds., Academic Press, Inc., New York, 1978, 279.
202. **Swift, M. J.**, Loss of suberin from bark tissue rotted by *Armillaria mellea*, Nature, 207, 436, 1965.
203. **Bracker, E. C.**, Ultrastructure of the haustorial apparatus in *Erysiphe graminis* and its relationship to the epidermal cell of barley, *Phytopathology*, 58, 12, 1968.
204. **Hirata, K. and Kojima, M.**, On the structure and the sack of the haustorium of some powdery mildews, with some considerations on the sack, *Transactions of the Mycological Society of Japan*, 3, 43, 1962.
205. **Wynn, W. K., and Staples, C. H.**, Tropisms of fungi in host recognition, in *Plant Disease Control*, Staples, C. R. and Toenniessen, H. G., Eds., John Wiley & Sons, Inc., New York, 45, 1981.
206. **Heath, M. C.**, A comparative study of non-host interaction with rust fungi, *Physiology Plant Pathology*, 10, 73, 1977.
207. **Wynn, W. K.**, Appressorium formation over stomates by the bean rust fungus; response to a surface contact stimulus, *Phytopathology*, 66, 136, 1976.
208. **Klement, Z. and Goodman, R. N.**, The hypersensitive reaction to infection by bacterial plant pathogens, *Annual Review of Phytopathology*, 5, 17, 1967.
209. **Sequeira, L.**, Induction of host physical responses to bacterial infection: A recognition phenomenon, in *Plant Disease Control*, Staples, C. R. and Toenniessen, H. G., Eds., John Wiley and Sons, Inc., New York, 1981, 143.
210. **Esau, K.,** *Plant Anatomy,* 2nd ed., John Wiley and Sons, Inc., New York, 1965.
211. **Hanchey, P. and Wheeler, H.**, Pathological changes in ultrastructure: Effects of victorin on oat roots, *American Journal of Botany*, 55, 53, 1968.
212. **Hanchey, P. and Wheeler, H.**, Pathological changes in ultrastructure: False plasmolysis, *Canadian Journal of Botany,* 47, 675, 1969.
213. **Luke, H. H., Warmke, E. H., and Hanchey, P.**, Effects of the pathotoxin victorin on ultrastructure of root tissue and leaf tissue of *Avena* spp., *Phytopathology* 56, 1178, 1966.
214. **Goodman, R. N.**, Phytotoxin - induced ultrastructural modifications of plant cells, in *Phytotoxins in Plant Diseases*, Wood, R. K. S., Ballio, A. and Graniti, A., Eds., Academic Press, Inc., New York, 1972.
215. **Aist, R. J.**, Cytology of penetration and infection fungi, in *Physiological Plant Pathology,* Heitefuss, R. and Williams, H. P., Eds., Springer-Verlag, Gerlin, 1976, 197.
216. **White, J. A., Calvert, D. H., and Brown, M. F.**, Ultrastructure changes in corn leaves after inoculation with *Helminthosporium maydis* race T, *Phytopathology*, 63, 296, 1973.
217. **Wheeler, H.**, Disease alterations in permeability and membranes, in *Plant Disease an Advanced Treatise,* III, Horsfall, J. G. and Cowling, B. E., Eds., Academic Press, Inc., New York, 1978, 327.
218. **Ehrlich, M. A. and Ehrlich, H. G.**, Fine structure of host-parasite interfaces in mycoparasitism, *Annual Review of Phytopathology*, 9, 155, 1971.
219. **Wheeler, H. E.**, Permeability alterations in diseased plants, in *Physiological Plant Pathology*, Heitefuss, R. and Williams, P. H., Eds., Springer-Verlag, Berlin, 1976, 413.

220. **Huang, J. S. and van Dyke, G. C.**, Interaction of tobacco callus tissue with *Pseudomonas tabaci, P. pisi,* and *P. fluorescens, Physiology Plant Pathology*, 13, 65, 1978.
221. **Gill, C. C. and Chong, J.**, Differences in cellular ultrastructural alterations between variants of barley yellow dwarf virus, *Virology*, 75, 33, 1976.
222. **Hamilton, R. I.**, Replication of plant viruses, *Annual Review of Phytopathology*, 12, 223, 1974.
223. **Rubin, B. A., Voronkov, L. A., and Perova, I. A.**, Transport elektronov v hlopastov hlopchatnika pri zarazheniy verticilieznim viltom, *DAN SSSR*, 3, 1207, 1972.
224. **Voronkov, A. L., Perova, A. I., and Sevireeva, V. V.**, Vliyanie zarazheniya verticilieznim viltom na strukturu i funkcii hloroplastov hlopchatniku, *Patologicheskaya Fiziologiya i Imunitet Rasteniy*, 172, 1976.
225. **Rahmankulova, M. E.**, Vliyanie griba *Verticillium dahliae* na strukturu i funkcii hloroplastov hlopchatnika, *Uzb, Biol. Zurn*, 3, 1973.
226. **Rubin, A. B. and Artsikhovskaya, E. B.**, Biohimiya i fiziologiya imuniteta rasteniy, *Izdat. Akademiy Nauk SSSR*, Moscow, 1960.
227. **Allain, A.**, Contribution à l étude du *Phytophthora cambivora, Morphologie, Cytologie et Action Pathogene du Parasite*, Paris, 1935.
228. **Allen, R. F.**, Cytological studies of forms 9, 21 and 27 of *Puccinia graminis tritici* on Khapli Emmer, *Journal of Agricultural Research*, 32, 701, 1926.
229. **Ladigina, E. M. and Tukeeva, I. M.**, Biosintez hlorofila pri virusnom patogeneze u rasteniy. *Patologicheskaya Fiziologiya i Imunited* Rasteniy, Izdatelstvo Moskovshogo Universiteta, 1976, 43.
230. **Bailiss, K. W.**, Infection of cucumber cotyledons by cucumber mosaic virus and the participation of chlorophylase in the development of chlorotic lesions, *Annals of Botany*, 5, 34, 1970.
231. **Tu, J. C., Ford, R. E., and Kras, J. C.**, Comparison of chloroplasts in maize dwarf mosaic virus infected and healthy corn, sorghum and Johnson grass, *Phytopathology* 57, 834, 1967.
232. **Matthews, R. E. F.**, *Plant Virology*, Academic Press, Inc., New York, 1970.
233. **Hršel, I. and Brčak, J.**, Ultrastructure changes in chloroplasts and cytoplasm caused by local infection of tobacco with tobacco mosaic virus and cucumber virus 4, *Virology*, 23, 252, 1964.
234. **Gardner, W. S.**, Electron microscopy of barley stripe mosaic virus; comparative cytology of tissues infected during different stages of maturity, *Phytopathology*, 57, 1315, 1967.
235. **MacDonald, P. W. and Strobel, G. A.**, Adenine diphosphate glucose pyrophosphorylase control of starch accumulation in rust-infected wheat leaves, *Plant Physiology*, 46, 126, 1970.
236. **Wheeler, H.**, __________, in *Plant Pathogenesis*, Thomas, G.W., Sabey, R. B., Vaadea, Y. and van Vleck, D. L., Eds., Springer-Verlag, Berlin and New York, 1975, 54.
237. **Weintraub, M. and Ragetli, H. W. J.**, A mitochondrial disease of leaf cells infected with an apple virus, *Journal of Ultrastructure Research*, 36, 669, 1971.
238. **Ladigina, M. E. and Balnokin, J. V.**, Elektro-mikroskosuspicheskaya harakteristika energetichishkih centrov tabaku v svyazi s imunitetom k virusu tabachnoy mozaiku, *Fiziologya i Biohemiya Zdorovoyo i Bolynogo Rasteniya*, Izdat., MGU, 1970.
239. **Hatta, T., Makamoto, T., Takagi, Y., and Ushiyama, R.**, Cytological abnormalities of mitochondria induced by infection with cucumber green mottle mosaic virus, *Virology*, 45, 292, 1971.
240. **Alseenova, A. V. and Brinza, I. A.**, Vliyanie infekciy na metabolicheskie processi mitohondriy rasteniy, *Patologicheskaya Fiziologiya i Imunitet Rasteniy*, Izdat. Moskovshovo Univerziteta, 98, 1976.
241. **Voronkov, A. L. and Guzova, V. N.**, Energoobmen hiopchantnika pri zabolevaniy viltom, *Patologicheskaya Fiziologia i Imunitet Rasteniy*, Izdatelstvo Moskovshogo Univerziteta 164, 1976.
242. **Aldrich, H. C., Gracen, E. V., York, D., Earle, D. E., and Yoder, O. C.**, Ultrastructural effects of *Helminthosporium maydis* race T toxin mitochondria of corn roots and protoplasts, *Tissue Cell*, 9, 167, 1977.
243. **Weintrub, M. and Ragetli, H. W.**, An electron microscope study of tobacco mosaic virus lesions in *Nicotiana glutinosa* L., *Journal of Cell Biology*, 23, 1964.
244. **Rubin, B. A., Akseenova, V. A., and Ladigina, M. E.**, Novoobrazovanie mitohodriy v tkanych rasteniy pod vliyaniem patogennih agentov, *DAN USSR*, 166, 1966.
245. **Harrison, B. B. and Roberts, I. M.**, Association of tobacco rattle virus with mitochondria, *Journal of General Virology*, 3, 121, 1968.
246. **MacLeod, R., Black, L. M., and Meyer, F. H.**, The fine structure and intercellular localization of potato yellow dwarf virus, *Virology*, 29, 540, 1966.
247. **Müller, K. O., Meyer, G., and Klinkowski, M.**, Die *Naturwiss.*, 46, 765, 1939.
248. **Akseenova, A. V., Nguen, D. G., Kozhanova, O. N., and Sultanov, J.**, Nekatoriye osobennosti processa sinteza belka v inficirovannih tkanyah rasteniy, *Patologiceskaja Fiziologija i Imunitet Rastenij*, Izdat. Moskovskovo Univerzaiteta, 1976, 79.
249. **Hirai, A. and Wildman, S. G.**, Effect of tobacco mosaic virus multiplicaion on ribounucleic acid and protein synthesis in tobacco chloroplasts, *Virology*, 38, 73, 1969.
250. **Reid, M. S. and Matthews, R. E. F.**, On the origin of the mosaic induced by turnip yellow mosaic virus, *Virology*, 28, 563, 1966.

251. **Asahi, T., Kojima, M., and Kosuge, T.**, The energetics, pathogenism and resistance, in *Plant Disease, An Advanced Treatise* IV, Horsfall, J. G., and Cowling, B. E., Eds., Academic Press, Inc., New York, 1979, 17.
252. **Brill, W. J.**, Selskohazjajstvenaja mikrobiologija, in *Industrial Microbiology and the Advent of Genetic Engineering*, Scientific American, Inc., New York 1984, 148.
253. **Hull, R. and Davies, W. J.**, Genetic engineering with plant viruses and their potential as vectors, *Advances in Virus Research* 28, 1, 1983.
254. **Kaczorek, M.**, Clonage de gènes: Des méthodes classiques aux sondes DNA synthétique, Biofutur, Janiver, France, 1985, 13.
255. **Agrios, N. G.**, *Plant Pathology*, 3rd Ed., Academic Press, Inc., New York, 1988.
256. **Powell Abel, P., Nelson, R. S., De, B., Hoffman, N., Rang, Rogers, S. G., Fraley, R. T., and Beachy, R. N.**, Delay of disease development in transgenic plants that express the tobacco virus coat protein, *Science* 232, 738, 1986.
257. **Turner, N. W., O'Connel, K. M., Nelson, R. S., Sanders, P. R., Beachy, R. N., Fraley, R. T., and Shah, D. M.**, Expression of alfalfa mosaic virus coat protein gene confers cross-protection in transgenic tobacco and tomato plants, *EMBO Journal* 6, 1181, 1987.
258. **Loesch-Fries, L. S., Merlo, D., Zinnen, T., Burhop, L., Hill, K., Drahn, K., Jarvis, N., Nelson, S., and Halk, E.**, Expression of alfalfa mosaic virus RNA 4 in transgenic plants confers resistance. *EMBO Journal* 6, 1845, 1987.
259. **Van Dun, C. M. P., Bol, J. F., Van Vloten-Doting, L.**, Expression of alfalfa mosaic virus and tobacco plants, *Virology*, 159, 299, 1987.
260. **Nelson, R. S., McCormick, S. M., Delannay, X., Dube, P., Layton, J., Anderseon, E. J., Kaniewska, M., Proksch, R. K., Horsch, R. B., Rogers, S., Fraley, R. T., and Beachy, R. N.**, Virus tolerance, plant growth and field performance of transgenic tomato plants expressing coat protein from tobacco mosaic virus, *Biotechnology*, 6, 403, 1988.
261. **Hemenway, C., Feng, R-X., Kaniewski, W. K., Chua, N-H., and Tumer, N. E.**, Analysis of the mechanism of protection in transgenic plants expressing the potato virus X coat protein or its antisens RNA, *EMBO Journal* 7, 1273, 1988.
262. **Cuozzo, M., O'Connel, K. M., Kaniewski, W., Fang, R-X., Chua, N-H., and Tumer, N. E.**, Viral protection in transgenic tobacco plants expressing the cucumber mosaic virus coat protein or its antisens RNA, *Biotechnology*, 6, 549, 1988.
263. **Deom, M. C., Oliver, J. M., and Beachy, N. R.**, The 30-kilodalton gene product of tobacco mosaic virus potentiates virus movement, *Science* 245, 389, 1987.
264. **Beachy, N. R.**, Virus cross-protection in transgenic plants, *Department of Biology*, Washington University, St. Louis, Missouri, 1988.

Chapter 2

HISTOPATHOLOGY

I. INTRODUCTION

Plant histopathology is the scientific discipline devoted to the study of pathological changes in diseased plant tissue structure and, indirectly, in function. Histopathological changes arise from changes occuring in the cells. Any changes that occur depend upon the sensitivity of the infected cells or tissues to the pathogen, the type of pathogen involved, and the conditions under which infection and disease development take place.

II. DISEASED PLANT TISSUE

Regardless of the type of causal agent, damaged or diseased tissues are the consequence of disorders within the cells. However, the structure and function of such a tissue may change indirectly due to these disorders, or disorders in other tisues, since all are functionally and organizationally interdependent and interlinked.

Pathological changes in cells, and thus the tissues with which they are associated, are caused by external abiotic and biotic factors. Abiotic factors include all those causing mechanical injury, extremes in environmental factors, chemical and noxious gas injury, etc. Mechanical injuries can occur during normal phytotechnical practices, such as pruning, grafting, cutting of shoots for vegatative propagation, and others. Some factors, such as insects, rodents and other harmful mammals, and other plant pests are similar to mechanical injuries.

Among the biotic factors, various living and nonliving entities of special importance as agents of various disorders in plant tissues and include bacterial, fungal, mycoplasmalike, nematode, rickettsia, and virus pathogens. Most plant pathogens in addition to mechanical damage, cause injuries to cells often employing a variety of mechanisms during the course of their development.

Protective reactions by plants to injuries, whether of a mechanical or pathological nature, generally are manifested in similar manners. Cells and tissues do not recognize different types of causal agents; since all causal agents are foreign, they react with similar defense mechanisms. For example, hypertrophy and hyperplasia may be the result of a variety of traumas, such as mechanical injury or infection with various phytopathogenic agents.

III. INFECTION IN HISTOGENESIS

Infection may take place at any time during the course of its formation and development in susceptible plant tissues. The stage of development at which a tissue becomes infected can influence the type of changes that take place in their structure and function.

Plant tissues represent the first specialized multicellular associations taking part and performing determined functions in the structure of higher plants. As cells represent the fundamental units of life, tissues are the first organized cell associations, giving rise to the organization and function of higher plants. Disorders in individual tissues, depending upon their location, result in more or less pronounced deformations of the internal and morphological structure.

All plant tissues are formed by meristematic cells and tissues. Meristematic cell activity provides for plant growth and development, and gives rise to the entire cell mass forming tissue structure, organs, and whole plants. Thus, any meristematic disorders, regardless of the causal agent, represent a great danger to plant development. To counter the effects of the disorganization or destruction of primary meristematic tissues, plants have developed the ability to form secondary meristematic tissues, which can take over the function of the primary ones. All living somatic cells of all tissues, particularly parenchyma tissue, can be transformed into secondary meristematic cells.

Cells formed from meristematic activity become interlinked and differentiated depending upon the tissues to which they belong. They form all constitutent cell parts and acquire a determined size and shape depending on the function to be performed. Like ontogenetic plant cell development, normal histogenesis may be disrupted by pathogens and abiotic factors at any stage of the developmental process.

All cells belonging to a given tissue are approximately identical. Thus, the structure and function in histological differetiation and specialization are defined as types of cells. Their ability to react similarly to the influence of various factors also is based on the similarity of tissue structures.

For example, conductive tissues differ from all other types by being complex in structure, composed of cells belonging to different tissues, such as the conductive elements, parenchyma and mechanical tissues. Conductive tissues are the only ones that permeate throughout the entire plant without interruption. Thus, pathological changes in the conductive tissues will differ depending upon the type of cells damaged or infected.

From the histopathological point of view, the fact that all tissues are interlinked to form the organized system in plants needs to be stressed. On the basis of the division of labor and association within plants, this organization ensures the structural and functional unity of the plant. Therefore, pathological changes in individual tissues endanger the function of the entire system and, in the case of irreversible damage or changes, will eventually cause plant death.

IV. TYPES OF HISTOPATHOLOGICAL CHANGES

All histopathological changes result from alterations in cells, which in turn, alter the tissues to which they belong. Each cytopathological change is in a sense a histopathological one. Thus, histopathological disorders may be classified into the same or similar categories as those used for describing cytopathological changes.

In discussing types of histopathological changes, particular emphasis will be placed on characteristic forms of disorders in individual tissues and on the sensitivity of these tissues to various disorders. Examples of such disorders, as well as the mechanism of their occurrence, will be described in detail in corresponding cytopathological changes and will not be analyzed separately.

A. HISTOPLASMIC CHANGES

Histoplasmic changes are a consequence of cytoplasmic disorders and can be manifested in the tissue structure in two opposite ways, hypo- or hyperplasmic changes. Histoplasmic changes occur as a result of the disruption of the primary or secondary meristematic tissues with a detrimental impact on the formation and differentiation of parenchyma and conductive tissues. Hyperplasmic changes usually take place in cells previously affected by hypertrophic processes.

B. HISTOTROPHIC CHANGES

Histotrophic changes result from cytotrophic changes characterized by changes in cell size and volume. Processes of atrophy, hypertrophy, and hypotrophy are part of such changes. Although all living plant tissues are susceptible to trophic changes, parenchyma tissue is the most sensitive. Hypertrophic changes, generally, are related to hyperplasmic changes, since hypertrophic processes usually precede hyperplastic ones.

C. HISTOMORPHOLOGICAL CHANGES

Histomorphological changes are a consequence of cytomorphological changes. Although such changes can occur in all cell forms, these disorders are most evident in isodiametric and prosenchymatous cells. These changes take place within the frame work of either existing (i.e. quantitative changes) or altered (qualitative changes) cell forms. Quantitative and qualitative morphological changes are characteristic of all living tissues. Parenchyma tissue is most susceptible to such changes during cell differentiation and transformation. Histomorphological changes are linked with histotrophic disorders.

D. HISTOTRANSFORMATIVE CHANGES

Histotransformative changes in cells occur during the processes of inversion and metaplasia. Meristemization processes in which somatic cells, frequently parenchyma cells, form secondary meristematic tissue, such as phellogen, are characteristic of the reversion process. Meristemization processes also may occur in collenchyma tissue, which is composed of living somatic cells. For example, meristemization of somatic cells occurs in leaves of certain plants at points of injury resulting in the formation of somatic tissue, which can be used for vegetative propagation.

The process of metaplasia cell formation generally would follow repeated differentiation transforming the cells to form another type of tissue. Some examples of this process are vascularization, sclerenchymization and viruescence of tissue normally without vascular elements, without sclerides, and not green in color, respectively.

All transformed tissues must be considered as secondary and also may be defined as induced tissues, since they occur under the influence of external agents. Parenchyma tissue generally is the most susceptible to the transform processes. These transformation responses represent a protective reaction by plants to increase their ability to survive under unfavorable conditions.

E. HISTOCHROMATIC CHANGES

Histochromatic changes pertain to tissue coloration or decoloration as a consequence of cytochromatic changes. Color changes take place in all plant tissues and are manifested in a multitude of different ways.

Histochromatic changes frequently occur in tissues bearing chlorophyll, such as chlorenchyma or photosynthetic parenchyma, often differentiated into palisade and spongy tissues such as found in leaves. Decoloration as well as abnormal coloration of these tissues is a result of processes which alter the quantity of permanent pigments in plastides, such as chloroplasts and chromoplasts, and in vacuoles containing anthocyanins and flavonoids.

Pathogens and abiotic factors generally decrease chlorophyll in photosynthetic tissues. Following partial or complete destruction of chlorophyll, previously masked colors of other pigments become visible. Similar changes may occur in epidermal, in stem and aboveground fruit parenchyma, and in clorenchyma tissues. Mottling of these tissues, for example, is a result of the partial or irregular destruction of green and other plant pigments.

A characteristic histochromatic change is tissue darkening as a result of the accumulation of dark plant pigments. Darkening occurs in all living plant tissues in and around points of damage regardless of its cause. In mesophyll tissue it occurs in the form or spots of various sizes, following either pathogenic or mechanical damage. A similar color change takes place in damaged parenchyma tissues of potato tubers and various fruits. Darkening often is evident in xylem elements infected with a vascular pathogen.

White coloration in various green or colored plant tissues, may occur under the influence of a variety of causal agents. In green tissues it is a result of either the disruption of chlorophyll synthesis or its destruction. It is manifested locally or, more generally, spread in leaf mesophylls, green fruits, green stems, and other such plant tissues. Similar pathological changes can occur in pigmented tissues of flower petals, resulting in color-breaking.

Parenchyma color changes in wody perennials are of particular interest. Agents of these changes include a variety of wood-rotting pathogens. Color changes in this tissue are more or less typical of the fungus involved, and the disease is named after the color given to the infected wood, such as blue stain, white wood rot, brown root rot, etc.

F. HISTOLYTIC CHANGES

Histolytic changes include those that partially or completely degrade tissues resulting from cytolytic disorders. In most cases, they are manifested in different types of decays and rots, gummoses, and resinoses.

Parenchyma tissue of herbaceous and woody plants is most susceptible to cytoloytic processes. Wet rot of pathogenic origin, for example, takes place in parenchyma tissue of the organs and tissues of many plants, including that of leafy vetgetables, fleshy friuts and vegetables, potato tubers and various root crops.

Dry rot of pathogenic origin occurs in parenchyma tissue of fleshy fuits and vegetables, including potato tubers and pome fruits resulting in mumification of affected tissues. Parenchyma tissue in perennial woody plants is susceptible to a variety of wood decays and rots caused by pathogenic fungi.

Gummosis and resinosis, caused by biotic and abiotic agents, occur predominantly in parenchyma tissues, but may occur in medullary ray cells, conductive elements, and mechanical tissues. The occurrence of gummosis and resinosis is wide-spread in plants becoming detrimented when occurrring in the aboveground woody organs of stone fruit trees, for example. However, resinosis is induced in confer trees for commercial resin production.

G. HISTONECROTIC CHANGES

Histonecrotic changes are characterized by the death of portions of individual tissues as a consequence of cytonecrotic changes. Cytonecrotic changes take place in all living tissues, but primarily in parenchyma tissue. Histonecrotic changes are one of the most frequent of symptoms caused by plant pathogens. Necrosis of all epidermal, mesophyll and conductive tissues initially takes place in localized areas, and can be caused by a variety of pathogens.

For example, necrosis of parenchyma tissue appears as a leaf spot in most anthracnose diseases of foliage; and stem anthracnose of herbaceous plants affects epidermal, parenchyma, and mechanical tissues in such crops of beans and peas. In the case of anthracnose of fleshy fruits, all parts of the fruit pericarp are susceptible. Necrotic processes are most pronounced in multilayered mesocarp parenchyma cells in which various nutrients accummulate, such as bean and pea pods, and watermelon and other succulent, fleshy fruits. Necrosis accompanied with gummosis occurs in Požegača cultivar plum fruits infected with plum pox virus.

Necrosis also results in the formation of cankers in fruit trees and other woody plants involving all bark tissues; and in the case of cankers formed by *Nectria galligena* on apple trees, necrosis is accompanied by secondary cambium activity resulting in the development of neoplastic formation around the points of infection.

Necrosis caused by vascular pathogens may affect either xylem or phloem or both. Early necrosis of vascular tissue eventually spreads throughout the plant significantly reducing plant growth and development.

Necrosis of terminal shoots or other apical meristems occurs in many plants as a hypersensitive reaction to plant viruses. For example, hypersensitive potato cultivars react in this manner to potato viruses A, S, X and Y.

H. STRUCTURAL HISTOPATHOLOGICAL CHANGES

Structural histopathological changes result from corresponding cytopathological changes and occur in all infected plant cells and tissues. The first structural change to take place in the course of pathogenesis is in protective tissues during pathogen penetration, except when inoculation occurs through injury to protective tissue. At penetration most changes occur in the epidermal tissues except for the few pathogens for which the external tissue is the primary tissue of colonization. However, for the majority of plant pathogens it represents only a barrier on the pathway to other cells and tissues. Undamaged periderm tissue is highly resistant to most mechanisms used by pathogens, except for example, for some soil pathogens such as *Armellaria mellea* causal agent of root rot of woody plants.

Structural changes occurring during infection are characteristic of all affected tissues and result from changes in cell walls, membranes, organelles and cytoplasm (see section on structural cytopathological changes). In addition to plant susceptibility, the extent of these changes is influenced by time of infection and extent of tissue development. Structural disorders are more severe if infection takes place in earlier rather than later stages of tissue development. In mature tissues, any disorder is less pronounced, depending upon the causal agent involved. In a broad sense, structural changes are a consequence of all cytopathological disorders.

I. HISTOTROPHIC PARASTISM

Some pathogens in the course of their development have become adapted to parasitizing and living in plant tissues. Such pathogens have been defined as histotrophic pathogens. Pathogen histotropism occurs in all living plant tissues. It is possible to distinguish epidermal pathogens that live exclusively or almost exclusively in epidermal tissue cells. This group, for example, includes the powdery mildew fungi (Erysiphales), *Synchytrium endobioticum* on potato tubers an stolons, *Polymyxa betae* on sugar beet root hairs, as well as others.

Several other more or less specialized pathogens are adapted to living in parenchyma tissue and include some of the photosynthetic parenchyma pathogens such as *Taphrina deformans* in peach leaf mesophyll, the rust fungi in photosynthetic tissue prior to lignification of the prenchyma cells, and others. Members of the Hymenomycetes are examples of a specialized group of pathogenic fungi that destroy wood parenchyma. Other pathogens that degrade parenchyma tissues of potato tubers and root and foligage vegetables, and cause wet rot also may be included among these pathogens. This latter group is represented by several fungi including *Botrytis* and *Fusarium* spp., and bacteria such as *Erwinia carotovora*.

Pathogen histotropism also occurs separately in xylem and phloem. Xylem histotrophic fungal pathogens include, among many others, *Verticillium albo-atrum* in stems of Solanaceous crops and *Pseudopeziza tracheiphila* in vascular bundles of grape leaves. Typical bacterial histotrophes include *Corynebacterium sepedonicum* in potato tubers, *C. michiganese* pv. *michiganense* in tomato plants, *Pseudomonas solanacearum* in Solanaceous plants as well as many others.

Phloem histotrophic pathogens include some phytopathogenic viruses, such as potato leaf roll virus, and all mycoplasmalike organisms that parasitize plants.

Phytophthora cambivora cause of the ink disease of Spanish chestnut, *Castanea sativa*, develops in the root and stem cambiums and can be included among the histotrophic pathogens that parasitize the lateral meristematic tissues. Direct infection of apical meristems also can be caused by fungi in the Peronosporales and Ustilatinales. Unspecialized meristematic tissue pathogens include some of the plant viruses.

J. INDUCED TISSUES

Induced neoplastic tissue formation is a reaction of plants to damage caused by external abiotic or biotic factors. The tissues most frequently induced to form neoplastic tissue are secondary meristem and secondary periderm tissues.

Secondary meristem is formed from somatic cells immediately following damage or infection of such cells. Somatic cells first become hypertrophied and then become hyperplastic resulting in the formation of callus tissue which serves to heal wounded or damaged tissues. Secondary meristem formed around wounds also is defined as traumatic meristem.

In damaged or infected plant parts, the secondary meristem also functions as the secondary phellogen forming secondary or induced periderm tissues. The induced periderm tissue around infection points confines the pathogen involved preventing or hindering its further spread. The formation of such periderm has a broad significance and according to Huber,[1] usually forming around pathogens in roots and leaves, galls caused by rust fungi, nematode injuries, and some tumors induced by viruses.

Induced secondary periderm tissue has been recorded in infected parts of plant organs, for example, in potato tubers infected by potato mosaic virus or around points of infection of *Spongospora subterranea*. Peridermal tissue also is formed around stone cells in pear fruits infected with the stony pit disease.

Induced secondary peridermal tissue also is formed around necrotic spots associated with leaf spots. The development of such tissue is dependent on the characteristics of the pathogen involved and the host reaction. In leaf spots caused by *Alternaria* spp., the pathogens penetrate the cork layer formed in the pathway of the hyphae giving such spots the appearance of concentric rings.[2]

A separate cork layer is formed between the upper and lower epidermal layer and around the necrotic leaf spots of stone fruit trees infected with *Clasterosporium carpophilum*. Such encircled spots fall out causing a shot-hole effect which is characteristic of the disease. Leaf shot-hole in the foliage of fruit trees also is caused by phytopathogenic bacteria such as *Pseudomonas syringae pv. morsprunorum* and *Zanthomonas campestris* pv. *pruni*. Gäumann[3] defined leaf shot-hole as a histogenic defensive reaction in which the pathogen together with the necrotic spot, is eliminated. According to Kiraly[4] leaf shot-hole should be considered a consequence of plant hypersensitivity to the pathogen unrelated to resistance to the disease.

Tissues eliminated from infected plant organs also may be induced in other tissues. The mechanism for the separation of leaves from twigs and branches may be included among such mechanisms.[5] Leaf separation zones are located at the base of the petiole and consists of two layers; the upper by which a leaf blade can be separated from the petiole, and a lower layer, which is considered a protective layer. The upper layer is composed of cells different from the petiole parenchyma cells and is the site of plant part separation. However, the lower layer, exposed after separation, is covered with materials that prevent water loss. On the surface of this lower layer, beneath the scar, cork formation takes place characteristic of injuries or wounds.

The drop of plant organs such as leaves, flowers or fruits of infected plants can be the consequence of various bacterioses, mycoses, or viroses. Causal agents of defoliation include, for example, pathogenic fungi such as *Diplocarpon rosae* on rose plants and *Puccinia pruni-sponosae* on plum trees, as well as *Fusicladium dendriticum*, cause of the drop of infected apple fruits, and *F. pirinum*, cause of the drop of infected young pear fruits. Some strains of tobacco mosaic virus cause defoliation and fruit drop in pepper plants. In Požegača cultivar plum, infected with plum pox virus, fruits drop off as maturation begins.

Separation and drop of plant organs is due to either similar or identical plant mechanisms. According to Kiraly[4] rapid defoliation of infected leaves represents a hypersensitive reaction in which plants sacrifice infected organs to protect others from infection.

The formation of secondary peridermal tissues in necrotic spots and around shot-holes on leaves, as well as infected-organ drop should be considered as defensive plant reactions, particularly when they prevent further spread of infection. The formation of the periderma tissue represents an almost perfect protective plant tissue.

V. PATHOLOGICAL CHANGES OF INDIVIDUAL TISSUES

A. PATHOLOGICAL CHANGES OF PROTECTIVE TISSUES

Protective tissues cover the outer surface of all plant parts and play a fundamental role in protecting plants from unfavorable external factors as well as in maintaining the connection between the interior and exterior environment of the plant.

Protective tissues are of particular importance in the study of plant pathology since they are located in the pathway of pathogen penetration and infection. As a result, pathogens cause various changes both in structure and in function in these tissues. The types and intensity of these changes depends on tissue characteristics, the pathogen involved, and environmental conditions during pathogensis.

Protective plant tissues consist of epidermal and peridermal tissues which differ widely in structure and, therefore, require special descriptions of the pathological changes that occur in them.

B. PATHOLOGICAL CHANGES OF EPIDERMAL TISSUES

Epidermal tissues cover all external surfaces of all herbaceous plants and protect the plant from unfavorable and detrimental external factors. It is constantly exposed to mechanical damage caused by various abiotic factors, mechanical injuries caused by contact between plants, and unfavorable meteorological factors. Frequently, injuries of epidermal tissues are caused by animals, usually insects, by their feeding habits on plants.

From the phytopathological point of view, since epidermal tissues are composed of living cells, it represents a favorable environment for the development of some pathogens. Thus, changes in epidermal tissue caused by pathogens may be classified into one group that causes changes in the course of inoculation and another that causes due to infection. Changes in the course of inoculation pertain to disorders in epidermal cells which occur during fungal penetration in the course of infection, those which develop in the epidermal cells.

C. CHANGES IN THE COURSE OF INOCULATION

Pathogen inoculation or penetration may take place through uninjured epidermal tissues, through natural openings, as well as through natural or induced wounds (Figures 48, 49). The most numerous natural openings are leaf scars appearing on shoots following defoliation. Induced wounds are the result of mechanical damage caused by several abiotic and biotic factors.

The changes that occur during the inoculation process caused by baterical and fungal pathogens usually are the result of specialized mechanisms developed by the pathogens for active penetration.

Pathogenic fungi with plasmodes as their vegetative form, such as in the Plasmodiophorales, primarily use enzymes for penetration and infection of epidermal tissue. While those fungi whose vegetative form is hyphae, penetrate epidermal tissues using the mechanical action of the hyphae in addition to enzymatic action.

According Wood,[5] there are several examples of direct pathogen penetration through undamaged epidermal cells. One example is the mechanical penetration of leaves of *Vicia fabae* by *Botrytis cinerea*.[6] The same mechanism also is used by *Sclerotinia libertiana* for pentration of the leaf tissues of *Phaseolus coccineus* and *Vicia fabae*.[7] Hyphae formed by germinating basidiospores of *Puccinia graminis* mechanically penetrate leaves of *Berberis vulgaris*.[8] In contrast, uredospores of most rust fungi form appressoria over leaf stomata and produce infection hyphae which penetrate and infect mesophyll cells.

Phytophthora infestans infects potato leaf tissue either directly through epidermal cells or through stomata. In either case it forms appressoria.[5] *Plasmopora viticola* infects grape leaves only after zoospores reach stomata, germinate, and form infection hyphae capable of infecting mesophyll cells.

Hyphae formed by germination conidia of *Clasterosporium carpophilum* directly penetrate leaf tissues of stone fruit trees either directly through epidermal tissue or stomata. Penetration through stomata guard cells results in the cells becoming thick-walled and dark in color.[3] The

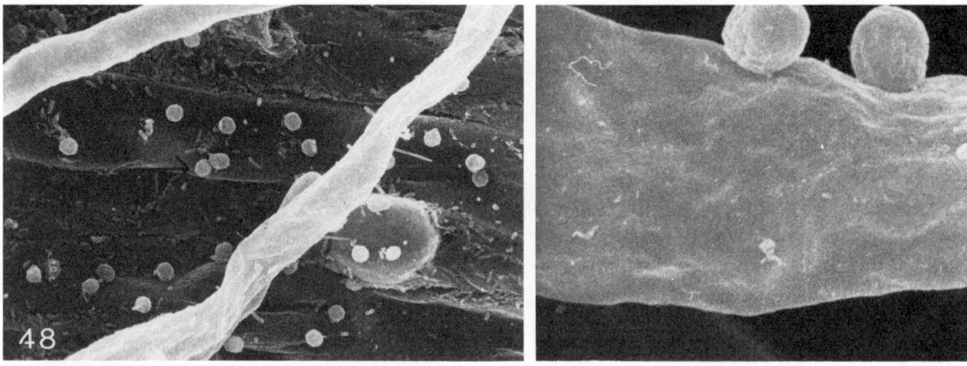

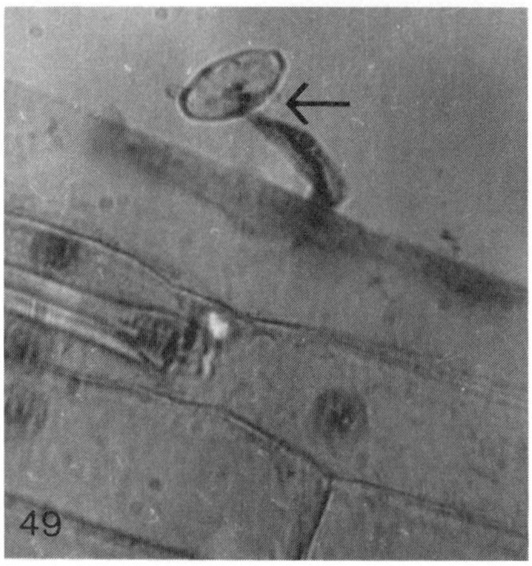

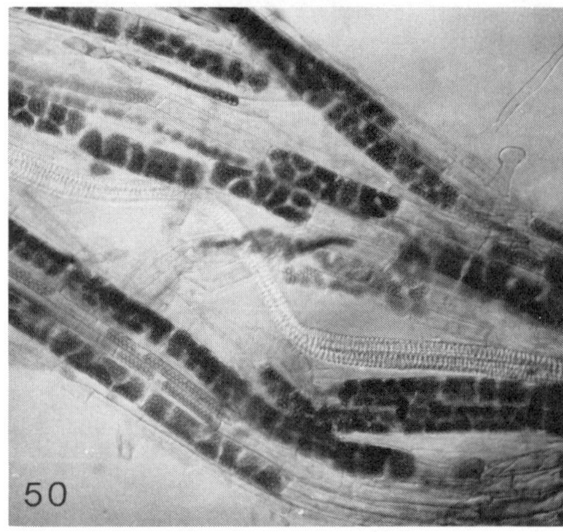

PLATE X. FIGURE 48. Zoospores of *Polymyxa betae* on the surface of epidemal cells of sugar beet root in the process of penetration (arrow) (left), and two zoospores in the process of germination (right). (From Ivanovic, M., Doktaska disertacija, Poljoprivredni fakultat, Beograd (Yugoslavia), 1984.) FIGURE 49. A germinating conidium (arrow) of *Erysiphe graminis tritici* with germ tube and appressorium on an epidermal cell of wheat. FIGURE 50. Necrotic sugar beet epidermal cells with numerous cyctosori of *Polymvxa betae*. (From Ivanović, M., Doltaska disertacija, Poljoprivredni fakultat, Beograd (Yugoslavia), 1984.)

natural wounds occurring on peach shoots following defoliation enable mass infection by conidia of *C. carpophilum*.

In contrast to fungal pathogens, phytopathogenic bacteria penetrate plant tissues almost exclusively though natural openings or wounds in epidermal tissues. In the intercellular spaces of mesophyll and parenchyma tissues they degrade these tissues by enzymatic action and toxin production, depending on the genus and species of the bacterium, its virulence, and host susceptibility.

Inoculation processes as well as structural change occuring as a consequence of these processes, are described in detail elsewhere (see section on structural changes of cell walls).

D. CHANGES DURING THE COURSE OF INFECTION

Such changes may be direct, when occurring as a consequence of parasitism in epidermal tissues, or indirect when changes result from an endophytic pathogen developing in another tissue.

Direct changes in epidermal tissues are caused by a number of specialized pathogenic fungi that attack such tissues. Depending on the mode and place of parasitation, these fungi may be classified as subcuticular, extracellular obligate pathogens, or intracellular obligate pathogens (Figures 50, 51, 54).

Typical representatives of subcuticular pathogens include the causal agents of apple scab,

Fusicladium dendriticum; pear scab, *F. pirinum*; and rose black spot, *Diplocarpon rosae*. Initially, hyphal germ tubes of these fungi penetrate beneath the cuticle and develop in the space between the cuticle and epidermal cell walls. According to Gäumann,[3] hyphae of these pathogens are found in the epidermal cells. However, through osmosis, they absorb sufficient nutrients from these cells for growth and reproduction. Perithecia of these pathogens are formed in the inner destroyed tissues.

According to Tarr,[2] hyphae of *D. rosae* develop primarily between the cuticle and external epidermal cell walls, where they form filamentous colonies with fringed edges. The hyphae then penetrate epidermal cell walls and feed in these cells with haustoria. In the case of severe infections, hyphae probably penetrate mesophyll cells. Infected leaves become chlorotic, die prematurely, and eventually drop off. Conidia formed beneath the cuticle exert mechanical pressure, rupture the cuticle, and are dispersed into the environment. Perithecia are formed in dead leaf tissues. *Venturia inaequalis* has a similar pattern of development on apple leaves.[2]

Graniti[9] included *Spilocaea oleagina,* causal fungus of olive leaf spot, among subcuticular pathogens. This fungus develop beneath the cuticle, forms conidiophores which break through the cuticle either mechanically or through enzymatic action, and releases conidia into the environment.

The fungus, *Taphrina populina,* also a subcuticular pathogen, colonizes space between the cuticle and epidermal cells and, according to Butler and Jones[10] causes epidermal galls in host leaves. Infected cells elongated two to three times and divide once or twice. This process results in an uneven surface of protuberances on the upper side and depressions into these structures on the underside.

Thus, subcuticular pathogens have a high degree of specialization, form haustoria, and do not cause rapid cell death. According to Graniti[9] subcuticular fungi represent transitional forms between extra- and intercellular obligate pathogens.

Extracellular obligate epidermal pathogens include most powdery mildew fungi (Erysiphaceae). These fungi complete their life cycle on epidermal tissues (Figure 51). The mycelia form appressoria with which they attach themselves to epidermal tissues. Haustoria formed from appressoria penetrate epidermal cell protoplasts and absorb nutrients. Haustoria are the only portion of these fungi present in host cells.

Haustoria colonize the protoplasts by exerting pressure on the plasmalemma, which forms a depression in the shape of the haustorium (plasmalemma invagination). Investigations of *Erysiphe graminis* on wheat plants have shown that its branched haustoria are enveloped with a special layer referred to as haustorium encapsulation.[5] According to Ehrlich and Ehrlich[11] the external part of this layer is a membrane which spreads throughout the cell walls. These fungi establish a relationship with the host protoplast which causes functional and structural changes, adapted to their obligate parasitism (see section on protoplast structural changes).

A frequent change accompanying such parasitism is necrosis of individual epidermal cells or cell groups. It occurs particularly in young, susceptible, herbaceous plants. Necrosis usually in the form of irregular spots may be caused by many pathogens including *Sphaerotheca pannosa* on peach trees, *Uncinula necator* on grapevines, and *Erysiphe graminis* f. sp. *tritici* on wheat plants. This type of necrosis is distinguished from other types which occur as a result of other processes and death of tissues infected with endophytic intercellular pathogens.

Intracellular obligate epidermal pathogens include some fungi of the Archimycetes as well as several phytopathogenic viruses. Characteristic representatives of this group belong to the Chytridiales (*Synchytrium endobioticum, S. taraxaci*) and Plasmodiophorales (*Polymyxa betae and P. graminis)*.

S. endobioticum infects epidermal cells in potato root systems, which lack a cuticle, and the eyes and surrounding tissues of tubers. At the beginning of the vegetative period, infection is primarily by zoospores and later by secondary zoospores. Zoospores move in wet soil and upon reaching the epidermal tissue, cease their movement , encyst, and then through a pore formed in the external cell wall, leak their protoplast into the host cell and thus, establish infection. Continuous noninfected cells divide and enlarge to form a rosette. Zoospores formed in these infected cells are released and infect other epidermal cells. Other epidermal cells later become infected with diploid zoospores. These cells divide irregularly and intensely each containing a portion of the pathogen. As a result of hypertrophy and hyperplasia, simple galls are formed which are characteristic of infected epidermal tissues.

Species of *Synchytrium,* induce the formation of the most simple galls.[10] This fungus may infect only a single cell, which enlarges. Hyperplasmic division then occurs in contiguous cells, which envelope the infected cell with a coat of epidermal and subepidermal cells.

Urophlyctis alfalfae infects young buds of alfalfa plants causing wart-like growths on crowns and roots. Zoospores moving through wet soil infect leaf primordia by penetrating epidermal tissues. One, two, or three zoospores may be found in the same epidermal cell.[10] Zoospores continue their development causing changes resulting in the characteristic symptoms of the the disease in epidermal cells. Typical histotrophic epidermal tissue pathogens of wheat and beet root systems are *Polymyxa graminis* and *P. betae*, respectively.

Zoospores of *P. graminis* infect root hair epidermal cells.[12] Primary infection is by zoospores released from cytospores, secondary infection is by zoospores produced in epidermal tissue cells. During infection, the zoospores attach to the outer wall of the epidermal cells and form a penetration tube through which they force their contents into the host cells. Zoosporangia also form penetration tubes inside the host cells releasing zoospores into the surrounding environment. In older plants, the fungus spreads through the root system. *P. graminis* is the vector of the wheat soilborne mosaic virus, and after penetrating host cells, the virus is released into the cells infecting the host.

P. betae infects root hairs and epidermal cells of sugar beets. The infection process and life cycle are similar to those of *P. graminis. P. betae* infects epidermal cells with primary and secondary zoospores, and forms its characteristic structures in infected cells (Figures 50, 54).[12] After extruding their contents through tubular formations into epidermal cells, zoospores retain their coats. *P. betae* infects individual cells but in contrast to *P. graminis* does not spread from cell to cell. Zoospores have been shown to degrade cells walls by enzymatic action and penetrate through a combination mechanical and chemical activities.[13] *P. betae* is the vector of the beet necrotic yellow vein virus, thus when colonizing sugar beet roots, it also infects the roots with the virus.

Other soilborne fungi infect root epidermal cells, but using different mechanisms. For example, various vascular pathogens such as *Verticillium albo-atrum and V. dahliae* penetrate through injured root epidermal cells.[9] *Fusarium oxysporum, F. solani*, and *Helminthosporium gramineum* are capable of penetration through injured roots.[5] Penetration is accomplished with or without appressoria.

Gäumann[3] described necrobiontic mechanisms for *Rhizoctonia solani* on potato tubers and *Cercosporella herpotrichoides* on barley and wheat. On potato tubers and other plant roots, *R. solani* forms superficial layers of hyphae attached to the epidermis. Epidermal cells killed by the fungus are then colonized by the pathogen, which continues its necrobiontic development. This process continues repeatedly in deeper and deeper parenchyma cells and eventually reaches the medullary parenchyma.

Several viruses which are obligate parasites in epidermal cells, highly infectious, and produce high particle content in infected cells induce hypertrophy and hyperplasia in infected as well as other tissue cells. As a consequence of such changes, small galls are formed, for example, in virus-infected plants of *Narcissus* spp.[10] Also, tobacco mosaic virus stimulates in some plants the formation of enations which are composed of the two layers of the epidermis, the palisade, and four layers of the spongy cell tissues.

Root epidermal cells are hosts of some viruses whose vectors are soil fungi, such as wheat mosaic soilborne virus and beet necrotic yellow vein virus. The fungal vectors, *Polymyxa graminis* and *P. betae*, respectively, penetrate the epidermal cells, transmit the virus which becomes systemic in other plant parts and organs.

Indirect changes resulting from endophytic intercellular pathogen development include both structural and functional ones resulting in various symptoms. This type includes mechanical damage caused by the maturing of reproductive apparatus of endophytic pathogens (Figure 55). The damage differs with each pathogen. Reproductive organs such as perithecia, pycnidia, and sporodochia usually protrude through the epidermal cell layer in the form of pustules, such as the acervuli and sori of members of the Albuginaceae and Uredinales. Another group of pathogens include those that produce conidiophores as in the Hyphomycetes and sporangiophores as in the Peronosporaceae, which protrude between epidermal cells or through stomatal openings.

Mechanical damage to epidermal tissues by reproductive organ formation, if numerous, may be highly detrimental to those tissues. For example, the protrusion of hundreds of uredosori of *Puccinia* sp. on cereal stems. A large number of conidiophores protruding through stomatal aperatures can cause death to cells as well as disrupting functions as, for example, by *Plasmopora viticola* on grapevine; *Peronospora parasitica* on cabbage, and *Bremia lactucae* on lettuce. The Ascomycetous fungus, *Rhytisma acerinum* on sycamore causes a splitting of the epidermal tissues parallel with the leaf surfaces within which it produces its vegetative and reproductive stages, resulting in the disorganization and eventually destruction of the leaf tissues.

Many phytobacteria form exudates composed, in part, of bacterial cells, on the surface of infected epidermal tissues. Exudate formation is a characteristic of certain phytobacteria from which a few vascular bacteria and *Agrobacterium tumefaciens* are excluded. Examples of phytobacteria that produce exudates are *Pseudomonas syringae* pv. *phaseolicola* on bean pods (Figure 53), *P. syringae* pv. *morsprunorum* on plum leaves (Figure 53), and *Xanthomonas campestris* pv. *vesicatoria* on tomato leaves.

Endophytic intercellular bacterial and fungal pathogens frequently cause necrosis of infected cells and tissues of herbaceous plant parts in the form of anthracnoses and spots. In such cases, the necrotic processes not only affect the cells and tissues involved, but also epidermal tissues. Endophytic intercellular pathogens that produce toxic byproducts also may alter epidermal tissues remote from the local infections. For example, *Stereum purpureum* separates epidermal from palisade layers in leaves of some forest and fruit trees. The free space then fills with air which reflects sunlight giving a milky or silvery appearance to infected leaves (see section on white coloration).

E. PATHOLOGICAL CHANGES IN PERIDERM TISSUE

Periderm covers the surface of perennial woody stems and underground parts of annual plants. It is a product of secondary meristem known as cork cambium or phellogen. Phellogen is formed by somatic cell meristemization in subepidermal tissues rather than in epidermal tissues and most frequently by meristemization of pericycle cells in roots.

Some periderm tissue characteristics are important in phytopathology. Periderm cells are linked without intercellular spaces. Their walls consist of cork or suberin impermeable to water and gases. These cells die and fill with air. Lenticels, the natural openings in periderm provide a connection between the plant interior and exterior environment. Periderm helps prevent plant desiccation and provides protection from external injuries.

Periderm ranks among the most resistant tissues to plant pathogens. Uninjured periderm is almost immune to penetration by pathogens. Periderm resistance to pathogens is based on its chemical and structural characteristics. The cork is, for the most part, immune or highly resistant to enzyme degradation by phytopathogenic bacteria and fungi. Pathogenic fungi are incapable of penetrating the cells through mechanical action of growing hyphae. Since peridermal cells are dead at maturiy, they cannot be hosts to bio- or necrotrophes, and therefore, cannot be affected by enzymes, hormones or toxins produced by these pathogens.

Fungi with mycelia are incapable of penetrating between periderm cells, since the cells are interlinked in a continuous layer without intercellular spaces. Thus, infection cannot be established and changed due to infection cannot occur in periderm tissue either through inoculation or other infection processes that have been described for epidermal tissues.

To surmount the resistance of periderm, pathogens penetrate through natural openings and wounds, either natural or induced. Some pathogens use both pathways to reach the inner susceptible tissues. Several pathogens, particularly fungal, penetrate the periderm of above- and below-ground organs in this manner. For example, *Nectria galligena*, cause of apple branch canker, penetrates susceptible tissues through lenticels as well as natural or induced wounds. Lewis and Goodman[15] were able to infect 2-year-old Jonathan apple trees by applying a bacterial suspension of *Erwinia amylovora* on lenticels without wounding. Other examples include *Penicillium expansum*, which penetrates lenticels of apple fruit and causes a dry rot; and *P. digitatum*, which invades through wounds causing a citrus fruit rot.

Wound parasites cause many diseases affecting aboveground plant parts of fruit trees, grapevines, and forest trees. This group of pathogens, for example, includes: *Clasterosporium carpophilum* on stone fruit trees, *Sclerotinia* spp. on fruit trees, and *Sphaeropsis malorum* on apple trees. Various fungal agents cause wood rots including: *Fomes fulvus* on stone fruit trees, *Polyporus hispidus* and *P. sulphureus* on apple, pear, and walnut; and *P. squamosus* on pear and walnut.

Certain soil pathogens penetrate belowground plant organs through lenticels and wounds in periderm tissue. For example, *Spongospora subterranea*, *Streptomyces scabies*, as well as the bacterial pathogens, *Erwinia carotovora* and *E. aroidae* invade potato tubers through lenticels and wounds.

Agrobacterium tumefaciens penetrates roots of grapevines and fruit trees through wounds and caused tumor formation (Figure 56). Under certain conditions, this bacterium penetrates roots through lenticels. *Armillaria mellea* penetrates roots of several deciduous and evergreen plant species through wounds in the parenchyma. If this tissue is uninjured, the fungus can penetrate directly through mechanical means or by enzymatic degradation (see section on suberin degradation by pathogens).

Finally, it should be stressed that all the aforementioned tissue changes pertain to primary periderm. Newly formed or secondary periderm surrounding focal points of infection and preventing or delaying further spread of a pathogen may develop in infected plant parts. Such secondary tissue can be defined as induced or supplementary periderm. The mechanism of formation and the protective role of this secondary periderm is similar to that of the primary one.

Induced periderm has been observed in several plants infected with different pathogens implying that the formation represents a general phenomenon of defense in plants. Examples and mechanisms of induced periderm formation are described elsewhere in this publication (see sections on histotransformative changes and induced tissue formation).

F. PATHOLOGICAL CHANGES OF MERISTEMATIC TISSUES

Meristematic tissue is capable of continuous cell production. Cells formed by this tissue becomes transformed into somatic cells forming individual tissues, organs, and whole plants. Primary, originating from embryonic cells, and secondary, originating from meristemized somatic cells, meristem tissue activity can be altered by several biotic and abiotic agents. Any change in this activity disturbs regular tissue formation, which is manifested by abnormal formation of plant parts. In contrast to other tissues, primary and secondary meristematic tissues are found in individual or limited plant parts and represent zones or points of growth. These

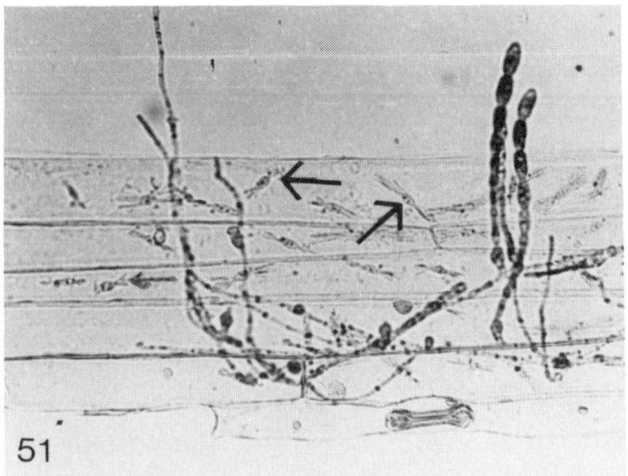

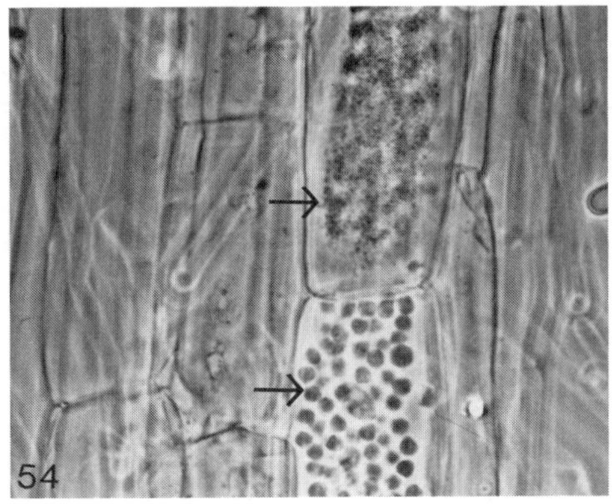

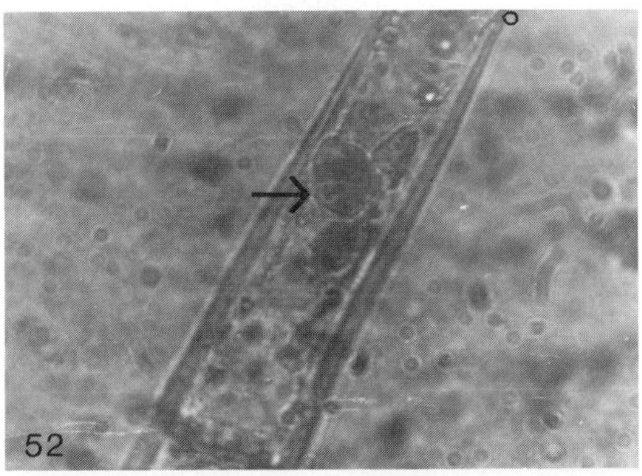

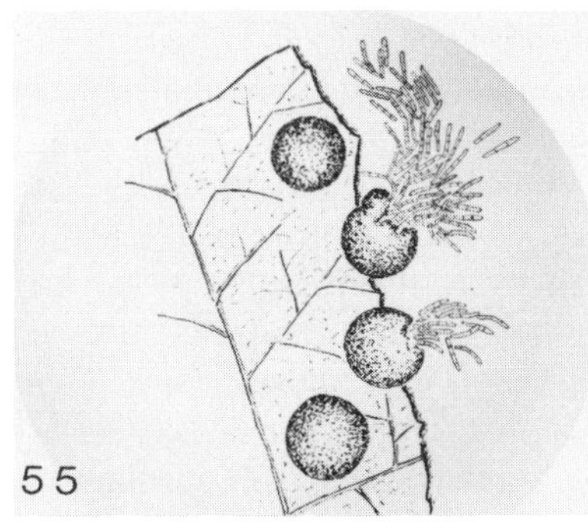

PLATE XI. FIGURE 51. Epidermal leaf cells of wheat infected with *Erysiphe graminis* showing numerous haustoria (arrows) within the cells, epiphytic and two conidiophores with conidia (right). FIGURE 52. Aggregated virus inclusions (arrow) of the tomato mosaic virus in a trichome of an infected tomato leaf. (From Jasnić, S., Doktorska disertacija, Poljoprivredni fakultat, Navi Sad (Yugoslavia), 1984). FIGURE 53. A common bean pod infected by *Pseudomonas syringae* pv. *phaseolicola* showing lesions with bacterial exudate (left), and plum leaf (cv. Pozega) infected with P. syringae pv. *morsprunorum* (right). FIGURE 54. Epidermal cells of sugar beet root showing the initiation of cystosorus formation by (arrows) *Polvmyxa betae*. (From Ivanović, M., Doktoska disertacija, Poljoprivredni fakultet, Beograd (Yugoslavia), 1984). FIGURE 55. Eruption and conidial dispersion of pycnidia of *Septoria helianthi* on a sunflower leaf. FIGURE 56. Large tumor formation on the stem of a grapevine caused by *Agrobacterium tumefaciens*.

tissues are classified as; terminal or apical occurring in shoot and root tips; inserted or intercalary occurring between zones of somatic tissues; and side-lateral occurring in stems and roots. The location of meristem tissues in the plant influences their susceptibility to certain pathogens or groups of pathogens.

A plant's ability to regenerate damaged or lost parts is based on the activity of meristematic tissue, which is formed around mechanical injuries. The tissue thus formed is known as traumatic meristem tissue; and that formed in a similar manner around points of infection is known as pathological meristem tissue.

G. PATHOLOGICAL CHANGES OF APICAL MERISTEM TISSUES

Pathogens may cause changes in the apical meristem of sprouts on stems and shoots, leaf primordia, and that of flower parts and roots.

The apical meristem of stems and shoots may be partially or completely colonized with a variety of plant pathogens, but most frequently by phytopathogenic bacteria and viruses. Several fungi, however, may infect apical meristem tissue, primarily those belonging to the order Peronosporales and Ustilaginales. The meristem tissue of young plants soon after germination are most susceptible to soilborne pathogens.

Members of the Peronosporales that cause direct or indirect damage to apical meristem tissue include primarily those which result in systemic infection. For example, *Peronospora arborescens* may directly infect the apical meristem of poppy plants so that all organs formed from this tissue are systemically infected.[16] Also, *Pseudoperonospora humuli* causes a systemic infection in hop shoots which develop from infected buds.[16] Infected shoots elongate slower and remain shorter than uninfected ones, possibly due to reduced meristematic activity. *Plasmopora halstedii* infects sunflower seedlings during germination and emergence; the mycelium then grows though the stems, reaching the seed by the end of the season. Infected plants are stunted and may be 5 to 6 times shorter than uninfected ones a consequence of structural and functional disorders in meristematic tissues.

Some representatives of the Peronosporales frequently develop on apical shoots in which desiccation occurs followed by death of the tissues, such as *Plasmopora viticola* on grapes and *Peronospora tabacina* on tobacco. Other fungi also can infect meristem tissue and cause rapid plant death, such as *Phoma beta* in the apical meristem of beet stems. The fungus spreads gradually to the youngest and central rosettes of leaves, which desiccate and turn black.[16]

Gäumann[3] pointed out that the meristematic tissues of some plants are highly susceptible to certain soilborne pathogens resulting in seedling infection. This group of fungi includes: *Fusarium* spp., such as *F. culmorum* and *F. graminearum* cause of snow mold of wheat seedlings; *Phytophthora fagi* cause of seedling disease of beech, maple, and other plants; and *Pythium debaryanum* cause of seedling blight of several plants. He further stated that these pathogens primarily attack seedlings, since they require embryonic tissues and favorable environmental conditions for rapid development.

Some viruses cause top necrosis in plants resulting in death of apical meristems. Top necrosis may be caused by various viruses either alone or in combination. Some plants, particularly those belong to the Solanaceae, are particularly susceptible to this type of necrosis (Figure 57). For example, the causal agent for potato top necrosis are the potato viruses X and Y.[17] The necrotic process caused by these viruses usually begins in the stem and spread rapidly into the meristematic zones, where death of cells occurs. The consequence of such processes, particularly in cases of infection by potato virus Y, are wilting and death of leaves. Apical meristem necrosis in potato caused by potato virus X and red clover mottle virus results in a rapid stem and plant decay.[18]

Foliage pathogens represent the most numerous group of plant pathogens that may affect the activity of meristematic cells taking part in leaf formation. Leaves originate from leaf primordia near the base of the vegetative cone. In the course of growth, leaf meristem cells divide continuously until the leaf is completely formed. Infections of under-developed leaves directly influence cell division and the differentiation processes and thus prevent normal leaf formation. Disorders in leaf morphongenesis are more conspicuous in early infections and most damaging on young plants emerging from infected seeds.

Fungi and bacteria that cause leaf spots can be used to illustrate the detrimental influence of these pathogens on leaf cell division. Early infection by these pathogens affects, *inter alia*, cell division, thus causing various cytopathological changes leading to cell death. The result of such changes are underdeveloped and deformed leaves due to uneven growth of individual leaf spots.

Among the numerous fungi that may cause such changes are: *Fusicladium dendriticum* on apple leaves, *F. pirinum* on pear leaves, *Phytophthora infestans* on potato leaves, and *Plasmopora viticola* on grape leaves. Similar symptoms are caused by many phytopathogenic bacteria, such as: *Pseudomonas syringae* pv. *tabaci* on tobacco leaves, *Xanthomonas campestris* pv. *vesicatora* on tomato leaves, and *X. campestris* pv. *malvacearum* on cotton leaves.

Plant viruses that cause systemic infections induce, *inter alia*, a series of changes characteristic of cell division and differentiation in young leaves. Various changes, as evidence by reduced growth and asymmetric leaves, have been observed on tobacco plants infected with tobacco mosaic virus. Also, changes in leaf primorida meristem, caused by growth hormones, prevent normal leaf development.[19]

Uneven development, frequently described as leaf curling, crumpling, and wrinkling, is, *inter alia,* a result of irregular or the absence of cell division at certain points. Such symptoms are caused, for example, by bean common mosaic virus on bean leaves, cucumber mosaic virus on bean and tobacco leaves, and watermelon mosaic virus in watermelon leaves.[21]

A large number of pathogens may interrupt inflorescene and flower formation as well as fruit formation and growth following pollination (Figure 58). Some pathogens, such as members of the Ustilaginales, are adapted for growth and development only in floral parts. Many other pathogens can develop in both flower parts and other plant organs. The Ustilaginales, causal agents of bunt and loose smuts in cereals and other plants, either completely or partially infect apical meristems of floral parts. For example, *Tilletia tritici* and *Ustilago hordei* infect wheat and barley germs, respectively, during seed germination, while others, such as *Ustilago tritici* and *U. nuda* infect wheat and barley ovules, respectively, during plant flowering.

In young plants, infected in either manner, the pathogen mycelia spread systemically and infect parts of the apical meristem. Infection occurs only after the mycelium has penetrated this tissue. When the vegetative cone meristem begins to form inflorescences, the pathogen penetrates all parts at the time of heading including each spikelet. Individual mycelial cells are transformed into chlamydospores in the colonized tissues, and by absorbing host tissue remnants, completely fill the space of the degraded grains and spikelets. The powdery mass of chlamydospores that result is a characteristic symptom of the disease.

Ustilago violaceae, causal agent of anther loose smut, systemically infects some plants of the carnation family (Caryophyllaceae). The pathogen usually infects the embryos of these plants, although it is capable of infecting young tissues and buds of growing shoots. The hyphae develop rapidly accompanying the vegatative cone through to flowering maturation. Finally, the hypae colonize the anthers forming chlamydospores in the pollen sac replacing the pollen grains.[3] In contrast, *U. maydis*, cause of maize smut, occurs in localized infections of various plant parts and, regardless of whether the infection takes place through the epidermis or through wounds, is limited to infected embryonic tissues.[3] The fungus develops intra- and intercellularly within galls and finally, its hyphae are transformed into chylamdospores completely filling the space of the degraded tissues (Figure 2).

On the other hand, *Claviceps purpurea* develops in the ovaries of members of the grass family (Poaceae) preventing embryonic cell development and meristem activity. Infection usually takes place near the base of the unfertilized ovary[20] or by penetration of hyphae through the stigma or style.[10] Instead of grains being developed, the mycelia develop in the infected ovaries forming characteristic sclerotia.

Members of a group of more specialized fungal pathogens, such as *Monilia* spp., develop first in floral parts and later in the fruits. *Monilia cinerea*, for example, causes complete desiccation of infected sour cherry and apricot flowers, in the course of which, meristem tissues also are destroyed, precluding any further flower development. Similarly, grape flowers infected with *Plasmopara viticola* become desiccated resulting in the death of all floral parts. Death of cells resulting from infection of young fruits prevents normal fruit growth and formation; for example, young apple and pear fruits infected with *Fusicladium dendriticum* or *F. pirinum*, respectively, remain underdeveloped and deformed. A similar retardation of tomato fruits occurs when they become infected early with *Pseudomonas syringae* pv. *tomato* or *Xanthomonas campestris* pv. *vesicatoria*. Cherry tree stems infected with little cherry disease, form fruits that are a third to a half the size of those on uninfected ones, which is a result of retardation of cell division and fruit morphogenesis.[21]

Some soilborne phytopathogenic fungi are capable of infecting meristematic tissues in the root growth zone of plants, thus directly affecting cell division. These fungi penetrate through uninjured or injured root parts. According to Talboys,[22] growth and water absorbing root zones are vulnerable points of entry for many fungi. Pathogenic, soilborne fungi, which cause vascular wilts, penetrate parts of young roots, as well as through mechanical injuries or wounds caused by insects or nematodes.[2] According to Tarr,[2] these pathogens spread into the xylem tissues, then into the contiguous parenchyma tissues causing plant death. Examples of diseases of this type are caused by *Fusarium oxysporum* f. sp. *vasinfectum* cause of Fusarium wilt of cotton; and *F. oxysporum* f. sp. *cubense* cause of Panama disease of bananas.

In summary, death of meristem tissues caused by most soilborne pathogens results in apical root rot. This group of pathogens affect a wide range of plants, and includes for example, *Fusarium solani* cause of apical root rot of beans; and *Verticillium albo-atrum* cause of apical root rot of tomato.[10]

H. PATHOLOGICAL CHANGES OF INTERCALARY MERISTEMATIC TISSUES

Intercalary meristem tissue originates from the apical meristem and, in grasses (Poaceae), is located close to the internode base above each node. This tissue, in contrast to others, is less susceptible to direct external infection by pathogens. However, indirect infection may occur by pathogens already established in leaves and stems of susceptible plants. These pathogens spread from infected areas into the meristem (Figure 61). Such indirect infection caused by most pathogenic fungi is less important when the spread of the pathogen is limited or slow.

In contrast, systemic phytopathogenic viruses, due to their general or systemic spread in infected plants (Figure 59), are more likely to infect this tissue. In fact, meristematic tissues are regularly infected with viruses, since they are located in the pathway of transport and spread of virus particles in stem tissues.

A common symptom in grasses attacked by systemic viruses, particularly in the case of early infection, is growth retardation, which is manifested as dwarfing and stunting. Such symptoms indicate a limit or prevention of cell division of meristem tissues, thus, influencing plant growth and development. Experimental data concerned with the effect of systemic virus infection on growth and development of grasses have been reported by several authors.

Investigations carried out in Yugoslavia have shown, for example, that maize dwarf mosaic virus reduced maize plant growth from 4.5 to 18.5%, depending on cultivar susceptibility.[23] Pop and Tusa[24] reported similar symptoms in Romania and found that maize plants infected with the same virus were reduced in growth by 10 to 17% that of uninfected plants. In Romania, dwarfing of plant parts were found above points of infection. Similarly, maize rough dwarf virus caused

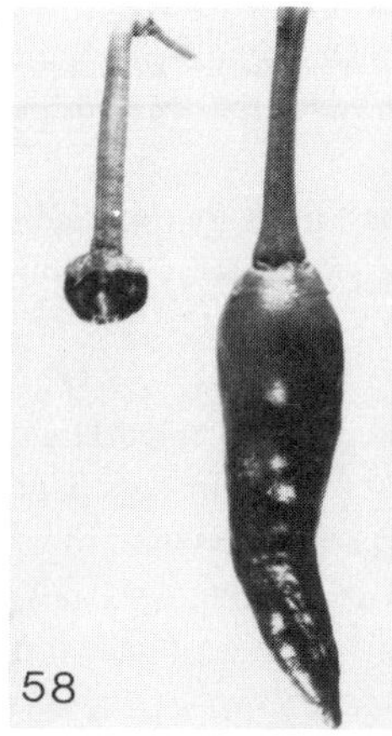
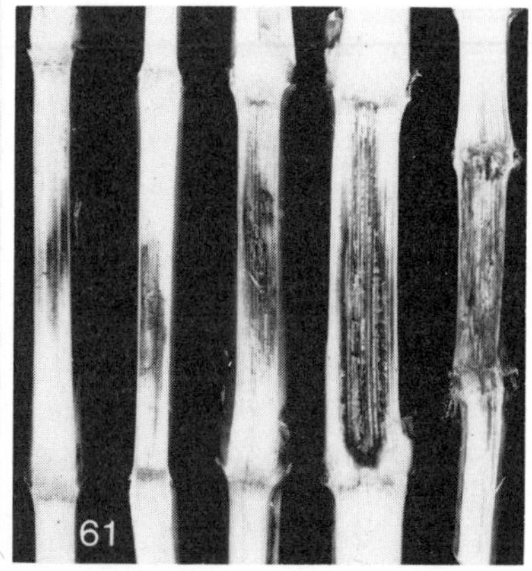

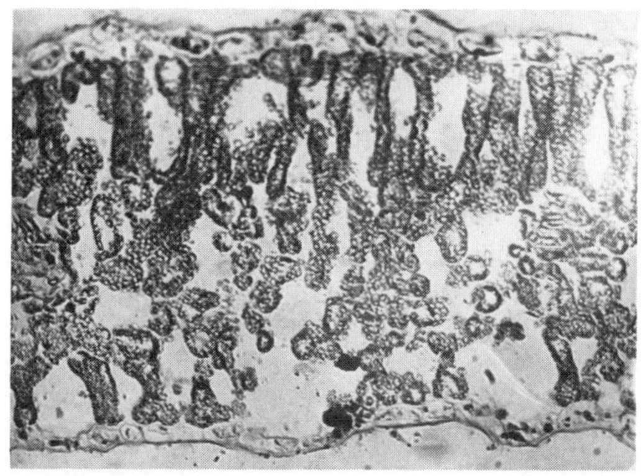
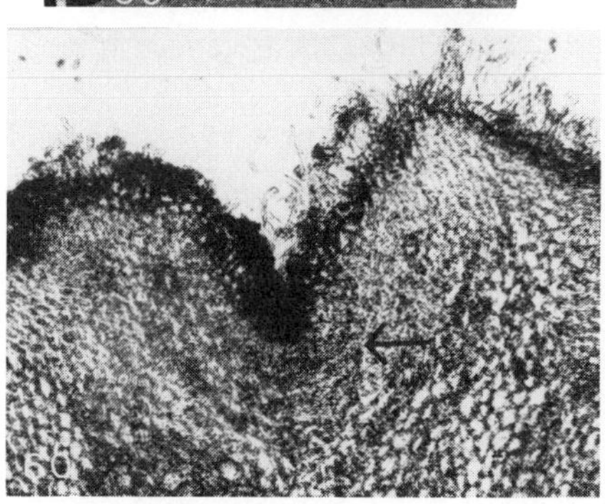
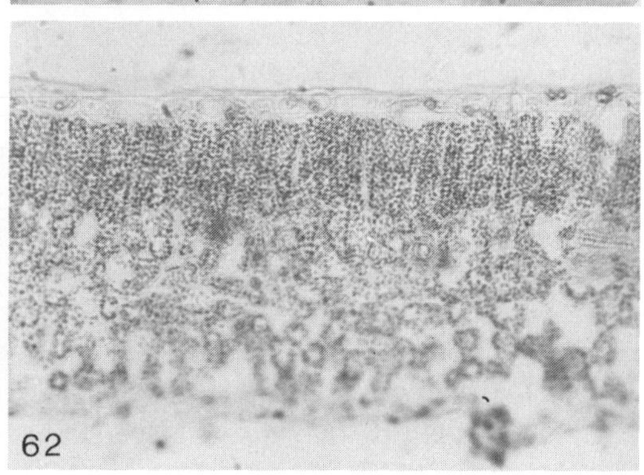

PLATE XII. FIGURE 57. Pepper plant infected with the tobacco mosaic virus showing wilting of the top of the plant due to desiccation and necrosis. FIGURE 58. Pepper fruit (cv. Horgoska slatka) undeveloped because of infection by the cucumber mosaic virus (left) compared with a developed fruit (right). FIGURE 59. Dwarfed wheat plants infected with wheat streak mosaic virus (foreground) compared to uninfected plants (background). (From Tošić, M., in: Šutić, D., *Anatomija i Fiziologija Bolesnik Biljaka,* 1987. With permission.) FIGURE 60. Crossection of a partial portion a root of a Gerbera plant showing undifferentiated and irregularly arranged parenchyma cells (arrow) infected with *Phytophthora crvptogea.* (From Tajcevski, T., Magistarski rad, Polyoprivredni fakultat, Beograd (Yogoslavia), 1985.) FIGURE 61. Vertical sections through maize stems infected with *Gibberella zeae* showing necrosis spreading within internodes. FIGURE 62. Crossections of tobacco leaves infected with the tobacco mosaic virus (above) showing altered and deformed mesophyll cells compared with a uninfected leaf with normal mesophyll cells (below). (From Mikovski, J., Doktorska disertacija, Poljoprivredni fakultat, Beograd (Yugoslavia), 1964.)

internode shortening and growth retardation in maize plants infected early with the virus.[21] Tošić[25] established that stunting of susceptible wheat plants infected with the streak mosaic virus ranged from 15.0 to 45.9%. He also showed that brome mosaic virus reduced the height of infected wheat plants by 13.0 to 17.6%.

Several grass viruses have been named after the characteristic dwarfing and rosette symptoms they cause: Arrhenatherum blue dwarf, barley yellow dwarf, oat sterile dwarf, and rice dwarf.[21]

I. PATHOLOGICAL CHANGES OF LATERAL MERISTEMATIC TISSUES

Pathogens can cause changes in lateral meristems similar to those of apical meristem tissues. The primary symptom of these changes is necrosis, and in some cases, hyperplasia. Secondary meristem formation frequently occurs in infected tissues around points of infection and, depending on the location, may be included among lateral meristem tissues.

The pathogenic fungus, *Phytophthora cambivora,* infects the cambium of chestnut roots and stems, in addition to other tissues.[2,3] This fungus, which normally lives in humose soil, penetrates root and lower stem parts and, when reaching the cambium, develops more extensively. The consequence of cambium infection is cell necrosis and formation of slimy, black exudate in the infected parts.[16]

Some pathogens, on the other hand, infect cambium tissue and stimulate hyperplasia, such as *Plasmodiophora brassicae*, which causes clubroot of crucifers. Hyperplastic root formation occur in the cambium and its initial phloem. The fungus quickly reaches cambium tissues and stimulates hypertrophy and hyperplasia of the affected tissues.[10] The pathogen spreads from the infected stem cambium into the phloem and secondary bark tissues, which may be infected only in this manner.

Hyphae of *Nectria galligena*, causal agent of canker of apple and other woody plants, penetrate the bark through natural openings or wounds and then spread through the stem. During this process it infects the cambium and living cells of medullary rays on its way to the woody cylinder.[10] In the course of pathogenesis this pathogen destroys the infected cambium cells. As a reaction to host infection, healthy cambium cells form hyperplastic excrescences around the point of infection.[16]

Intensive cambium activity occurs in many infected plants resulting in stem thickening. For example, the fungus *Marasmius perniciosus*, cause of witches' broom in cacao, stimulates the increase in stem diameter by two to three times.[5] Stem thickening also occurs as a result of infection by *Peronospora parasitica* on plants of *Capsella bursa-pastoris,* by *Puccinia carthami* on plants of *Carthamus tinctorius*, and by *P. menthae* on peppermint plants.

Agrobacterium tumefaciens, cause of crown gall of fruit trees and other woody plants, survives in soil and penetrates susceptible host through wounds, resulting in the formation of neoplastic tumors in meristematic and parenchyma tissues.[5] Cells contiguous to the damaged ones enlarge and divide giving rise to tumoral excrescences at the points of infection. New vascular elements are formed by the bark cambium in the tumor tissues.

Detrimental effects on lateral meristems also result from infection by some phytopathogenic viruses and viroids. For example, flattening or fasciation occurs in herbacious parts of grape stems infected with the fanleaf virus as a consequence of disturbed cambium activity. Apple stem pitting disease prevents normal cell formation and differentiation in cambium tissues resulting in the development of characteristic cavities.[26] Citrus plants infected by tristeza virus develop necrosis of newly-formed meristem or callus tissues, accelerating the processes of plant decay.[21] Due to irregular meristem activity, the potato spindle tumor viroid causes deformed tubers.

Apple branch flattening is caused by a mycoplasmalike organism which also affects cambium tissue. The flattening of stems is a result of the restriction of branch thickening accompanied by the development of sunken areas and twisting of the stems. Kegler[27] stated that he thought all these disorders in plants actually are a consequence of the disturbing of cambium tissue.

J. PATHOLOGICAL MERISTEM TISSUES

Pathological meristem tissues occur in infected plant parts usually around the points of infection. These tissues can form peridermal cell layers which hinder or prevent further pathogen spread. Depending on the occurrence of the tissue, it can be classified as lateral meristem tissue.

Pathological meristem can perform as a phellogen or cork cambium, if it originates from the transformation of somatic, mostly parenchyma, cells. Regardless of location, the formation of pathological tissues, as supplementary tissues in infected plant parts, represents a defensive plant reaction to pathogen spread.

Pathological meristem and peridermal tissue have been shown to develop in many infected plant organs. The formation of these tissues and their characteristics are discussed in detail in the section concerned with histopathological change.

VI. PATHOLOGICAL CHANGES OF PARENCHYMA TISSUES

Parenchyma tissue consists of botanical traits which makes it unique in the structure and life of plants. Among these traits, from the phytopathological point of view, are (i) the living state of its cells, (ii) its general presence throughout the plant structure, (iii) its highly pronounced changeability which is greater than for any other permanent plant tissue, and (iv) its diversified and abundant activities in the plant.

Being composed of living cells, parenchyma tissues represent a generally favorable environment for the development of pathogens belonging to a variety of groups. This tissue is the host to a great number of plant pathogens which may be defined as parenchyma pathogens. Parenchyma pathogens include most phytopathogenic bacteria, fungi and viruses. In contrast, the smaller group of nonparenchyma pathogens is mainly comprized of specialized pathogens that colonize or invade epidermal or vascular tissues.

Parenchyma tissues are the main constituents of higher plants, which is permeated with other tissues, especially conductive, on the inside and covered with protective tissues on the outside. Due to their omnipresence, parenchyma tissues are more exposed to pathogen influence than any other tissues. Parenchyma tissues represent an unlimited environment for pathogen attack, continuous infection, and uninterrupted pathogen development. Pathological changes of infected parenchyma tissues depend primarily on the type and traits of the pathogen involved.

Parenchyma tissues belong to a group of permanent plant tissues formed in the processes of differentiation and specialization of newly-formed cells from meristematic activity. Distinct from other permanent tissues, parenchyma cells are least differentiated and therefore, histologically, are most similar to meristematic cells. Because of this characteristic, the cells are highly susceptible to histological transformation. Transformation due to processes of reversible or repeated differentiated is a characteristic of living cells of most tissue, but is primarily a trait of parenchyma cells.

Under the influence of abiotic and biotic factors, parenchyma tissue cells may be reversibly transformed into meristem cells (i.e. pathological meristematization of parenchyma cells.). Due to this ability for transormation, parenchyma cells represent the most important source for the formation of new secondary meristem cells and tissues.

In the course of pathogenesis, parenchyma cells may also change histologically and by repeated differentiation, transform into cells of other tissues. For example, parenchyma cells may be transformed into stone cells in a process known as parenchyma sclerenchymization.[18]

Parenchyma tissues perform the greatest number of different activities of vital importance to plants. These activities include, photosynthesis, uptake of nutrients from the environment, reserve food synthesis and storage, secretion and excretion of a variety of compounds, wound healing and others. All these functions are performed by the living parenchyma cell protoplasts.

It follows that any disorders in the vital functions of plants usually occurs in damaged or infected parenchyma tissues. Most physiological and structural cytopathological changes described in plant pathology pertain to those of cells of parenchyma tissues.

Based on morphological and physiological differentiation, parenchyma tissues may be distinguished as chlorenchyma, storage parenchyma, root and stem parenchyma, conductive parenchyma, etc. Different pathogens can cause numerous and various types of pathological changes in these tissues.

A. PATHOLOGICAL CHANGES IN CHLORENCHYMA

Due to its wide dispersal in plants and its structure and function, chlorenchyma has a particular significance in phytopathology. Typically chlorenchyma is located in leaves and is composed of palisade and spongy tissues, which together often are referred to as leaf mesophyll. Mesophyll cells contain chloro- and chromoplasts, the former being photosynthetically active. This activity is carried out primarily in the palisade tissues. The spongy tissue is more important for the exchange of gases, nutrients and water, as well as in the transport of photosynthates.

Chlorenchyma mesophyll may be infected by any of numerous phytopathogenic bacteria, fungi, and viruses. The method of colonization and parasitization depends on its structure and chemical composition as well as characteristics of the pathogen involved.

Several pathogenic fungi colonize the highly developed intercellular spaces of the mesophyll. The following examples can be given; the hyphae of *Taphrina deformans,* for example, colonize the mesophyll by growing exclusively in the intercellular spaces, using nutrient compounds obtained through the wall of contiguous host cells. It is considered a facultative parasite which develops systemically in leaf and young stem tissues. Another facultative parasite is *Pyrenophora graminea,* cause of barley leaf stripe, which also develops systemically in barley leaf mesophyll.[2] *Mycosphaerella fragariae,* cause of strawberry leaf spot, penetrates directly through the cuticle or through stomata and develops intercellularly.[10] Similarly, *Gnomonia erythrostoma,* cause of cherry scorch, penetrates cherry leaves directly where its intercellular mycelium colonizes the spongy mesophyll.

Extracellular obligate parasitic fungi are characterized by their hyphae growing intercellulary in mesophyll tissue and penetrating contiguous cells by haustoria. Haustoria absorb nutrients in the host cells providing for pathogen nutrition and survival. This group of pathogens includes several highly specialized biotrophic fungi in the Albuginaceae, Peronosporaceae, and Uredinales. Biotrophic pathogens, such as *Leveillula* spp., develop mainly in mesophyll cells with their conidiophores protruding through stomatal openings.[9]

Another group of less specialized pathogens, which colonize mesophyll tissues, spread from cell to cell. For example, hyphae of *Botrytis cinerea,* which parasitizes many plants, grow from appressoria, chemically or mechanically degrade cell walls, and reach host cell protoplast. The pathogen spreads in infected tissues causing their death. Hyphae of another fungus, *Clasterosporium carpophilum,* directly penetrate leaf mesophyll. They ramify, grow intra- and intercellulary in all directions, and colonize both palisade and spongy tissues.[28] This pathogen causes cell necrosis resulting in the shot-hole symptom in infected leaves.

Most phytopathogenic bacteria characteristically develop in mesophyll tissues. The only exceptions are a few vascular bacteria such as *Corynebacterium flaccumfaciens* pv. *flaccumfaciens,* and *Pseudomonas solanacearum,* as well as bacteria causing cankers in woody plants, such as *Agrobacterium tumefaciens* and *Pseudomonas syringae* pv. *savastanoi.*

Phytopathogenic bacteria penetrate mesophyll tissues through stomata and wounds and, upon reaching intercellular spaces, enzymatically degrade pectic substances of the middle lamella and later cellulolytic constituents of the cell walls. The development of such bacteria in mesophyll tissues more or less localized around points of infection.

The mesophyll of susceptible hosts can be infected by all or almost all viruses. An exception is among some of the specialized histotrophic viruses, which develop in vascular tissues, such as the citrus tristeza virus, which is found only in phloem.[29] Other phloem viruses are the beet curly top, beet yellow vein, and potato leaf roll viruses.[18]

Phytopathogenic viruses may be introduced directly into mesophyll cells mechanically by contact between plants, with the aid of animal or plant vectors, or by grafting. Mesophyll cells may be infected indirectly by the spread of virus particles from infected cells of one tissue or another. Virus particles spread between mesophyll cells by cytoplasmic flow through plasmodesmata. The diameter of plasmodesmata, for example, in tobacco leaves ranges for 20 to 200 μm, which is sufficiently large enough to allow for the passage of most virus particles.[30] Virus infection of mesophyll tissues of susceptible plants is usually characterized, but not always, as systemic. Thus, it is assumed that all cells of such tissues are infected in the course of pathogensis. Such systemic infection usually results in a mosaic pattern on infected leaves, a common symptom of viral diseases.

Since mesophyll is basically parenchyma tissue, it is susceptible to cyto- and histopathological disorders. Mechanisms for these disorders being interconnected and interdependent differ in the course of pathogenesis (see section of cytopathology). The most frequent symptom of cytopathological change is the development of spots, abnormal coloration, and/or leaf blade or shoot deformation.

Spot formation is another wide spread reaction of infected mesophylls. Such spots usually occur around points of infection. Infected cells change functionally and structurally, lose their vitality, and eventually die. In addition to mesophyll cells, death of other cells takes place within the spots. Spots may differ in shape and size, and, according to movement within the host, systemically in susceptible plants, and localized in hypersensitive plant, differences among spots can be distinguished (see section on cytonecrotic changes).

Mesophyll tissues rank among the most important carries of plant pigments, such a chlorophyll in chloroplasts, other pigments such as carotenoids in the chromoplasts, and flavonoids in the cell sap. Almost all types of infection, except those that are latent, cause disorders in mesophyll coloration. Color changes occur simultaneously with spot formation in many infected mesophylls and therefore, these two changes are most frequently described and used for the identification of many plant diseases.

Dark pigments formed as a result of complex biochemical processes in infected protoplasts also influence color changes. Under certain conditions, white discoloration occurs as a characteristic cytopathological change in the mesophyll (see section on cytochromatic changes).

Many mycoses have been named, based on these two charateristics. A few examples are orange leaf spot of plum, caused by *Polystigma rubrum;* maize ashy leaf stripe or shot-hole disease, caused by *Exserohilum turcicum;* and cherry and peach violet leaf spot, caused by *Clasterosporium carpophilum.* Similarly, several bacterioses have obtained their names from the symptoms produced: bean brown leaf spot, caused by *Xanthomonas campestris* pv. *phaseoli;* tomato black leaf spot, caused by *Pseudomonas syringae* pv. *tomato;* oat halo blight, caused by *P. syringae* pv. *coronafaciens,* etc. Many viruses have obtained the names in a similar manner: tomato ringspot virus, prunus necrotic ringspot virus, etc.

Normal mesophyll formation depends on the number and size of differentiated cells in its structure. However, several pathogens alter the number and size of mesophyll cells resulting in slight to extensive deformations of the tissues. Hypoplasma of mesophyll cells results from inadequate or reduced activity of meristem cells of leaf primorida. Hypoplasmic disorders in meristem tissues frequently influence histogenesis in the mesophyll. It has been shown that a reduction of the number of mesophyll cells may vary, depending on the host, mechanisms of pathogen influence, and the age of the plant at infection. For example, broad bean leaf mesophylls infected with pea enation mosaic are composed of only a single type of tissue consisting primarily of irregular cells.[31] Due to such changes, the diameter of infected leaf blades are reduced by a third to a half.[26]

Some plant viruses either partially or completely prevent the formation of mesophyll cells and tissues. In such cases, infected leaves are formed without certain parts and are referred to as abortive changes. Abortive changes most frequently result in filiform leaves, caused, for example, by cucumber and tomato mosaic viruses. In such cases, parenchyma and other tissues are not formed, so that blades are composed only of the midrib and surrounding mesophyll cells under the epidermis.[18] The grape fanleaf virus also causes abortive changes in leaf mesophyll. Parenchyma tissue is not formed between the midribs of infected grape leaves, and deep notches are formed in the leaves which may extend to the base of the leaf blade. Other leaf parts appear as lobes, all of which gives the disease its common name.

Other studies have shown that some pathogens affect the cell differentiation process in mesophyll tissue, preventing the normal formation of palisade and spongy tissues. Accompanying cell hypotrophy in these tissues is the reduction in number or lack of intercellular spaces.

Disorders in cell differentiation and hypotrophy frequently occur in leaf parts with chlorotic mosaic, which can be caused by several viruses. According to Šutić[18], underdeveloped or immature palisade cells are particularly susceptible to hypotrophy, so that those leaf parts with such cells are thinner than normal and have a rough surface. Matthews[17] also stated that yellow zones on leaves infected with many of the mosaic viruses are thinner than green ones; that mesophyll cells are less differentiated with a smaller number of chloroplasts, and that only a few or no intercellular spaces are formed. Schmidt[26] observed that cells in the yellow zones of sugar beet leaves infected with sugar beet mosaic virus are closely arranged and that these leaf parts are thinner than green ones due to the absence of palisade and spongy tissue differentiation. Bawden[32] also showed that palisade cells in leaves infected with a mosaic virus are shorter than normal with smaller intercellular spaces, and that chlorotic leaf parts usually are thinner than the green ones. For example, potato leaves of the cultivar President simultaneously infected with potato viruses A and X developed palisade cells in the chlorotic leaf areas which were 45 μm long and 113 μm thick, while those in the dark green areas of infected leaves were 57 μm long and 137 μm thick. The palisade cells in uninfected leaves were 72, 52, and 56 μm long, and 148, 140, and 147 μm thick, respectively.

Similar to viruses, several phytopathogenic bacteria and fungi cause leaf growth retardation and malformation in infected plants. Symptoms from the result of these conditions are particularly evident in young leaves infected at the time of mesophyll formation and differentiation. Such symptoms are caused, for example by the fungi *Fusicladium* sp. on apple and pear leaves, and *Phytophthora infestans* on potato and tomato leaves; and by the bacteria *Pseudomonas syringae* pv. *medicagnins* on bean leaves, *P. syringae* pv. *pisi* on pea leaves, *Xanthomonas campestri* pv. *malvacearum* on cotton leaves, and *X campestris* pv. *vesicatoria* on tomato leaves.

Mesophyll and thus, leaf malformations, may occur as a result of palisade and spongy tissue hyperplasia. Some pathogens stimulate hypertrophy and later these tissue are transformed into secondary meristematic cells by processes called "transforming metaplasmic changes" or "parenchyma cell meristemizaion". Then, like meristematic cells, these cells divide and cause localized or general mesophyll thickening. These types of changes are usually associated with fungi which stimulate hormonal activity in infected tissue cells.

The fungus *Taphrina deformans*, cause of peach leaf curl, is an agent of hyperplasmic mesophyll malformation and has been extensively studied.[5,10,28] This fungus develops primarily between epidermal and palisade cell layers, but also invades the mesophyll. According to Wood,[5] it stimulates the division of cells of the spongy tissue, which then lose their characteristic shape and intercellular spaces. Palisade cells also divide and lose their elongated shape thus, becoming similar to cells of the spongy layer. Due to such changes of the spongy tissue, the infected leaves curl downward, giving the characteristic symptoms of the disease. According to some authors, other fungi belonging to the Order Taphrinales cause similar changes in infected plants.

Exobasidium oxycocci, for example, also causes hyperplasmic changes in leaf mesophyll tissue of susceptible plants.[10] This fungal pathogen stimulates the cells in the leaf mesophyll of plants of *Vaccinium macrocarpon* to become uniform, lacking intercellular spaces. Palisade leaf cells of *V.vitis-ideae*, however, change only a slightly under the influence of this pathogen, with leaf thickening resulting from the development of rounded to polygonal-shaped cells lacking intercellular spaces in the spongy tissue.

Mesophyll hyperplasia develops in leaf galls caused by *Gymnosporangium juniperi-virginianae* on apple leaves.[10] This fungal pathogen causes a less pronounced change in palisade cells, however, the spongy tissue cells elongate susbtantially dividing along transverse and oblique walls, resulting in leaf blades two to three times thicker than uninfected ones. Similarly, *Cystopus candidus* causes leaf thickening on plants of *Capsella bursa-pastoris*. The mesophyll cell become large and round, with no difference from palisade and spongy layer cells.[10]

Likewise, *Elsinoe fawcetti*, cause of citrus scab, stimulates changes in citrus leaf mesophylls. Initially it infects leaves on the lower leaf surface with hyperplastic changes first occurring in these spongy parenchyma tissue, then later it infects some of the palisade cells.[10]

In summary, pathological mesophyll changes are heterogenous and therefore difficult to assess in detail. However, in a broad sense, these changes could be generally defined as; an insufficient number of cells for mesophyll formation, incomplete mesophyll cell differentiation and specialization, trophic changes in mesophyll formation, chromatic mesophyll changes, and mesophyll cell pathological meristemization.

B. PATHOLOGICAL CHANGES OF STORAGE PARENCHYMA TISSUE

Storage parenchyma tissue is specialized for the accumulation and storage of organic nutrients and water. The accumulation of organic materials, such as oils, proteins, starch, and sugars, provide for plant nutrition and play an important role in the life fo the plant. Cells of this tissue function only in the living state.

From a phytopathological point of view, it is important to stress that storage parenchyma is widespread in the plant. It can be located in both reproductive and vegetative organs, more or less adapted to the functions of the tissues in which they occur.

Vegetative organs with storage parenchyma include roots of such crops as carrot and parsley, or root with stem parts, such as fodder and sugar beets. Storage parenchyma also is found in transformed shoots, such as bulbs, rhizomes, and tubers. Storage parenchyma also is present in fleshy stone fruits, such as those of apple, peach, pear and plum; and in berry fruits, such as cucumber, melon and tomato. Secondary wood xylem as well as wood ray parenchyma tissue cells, which transport organic nutrients from the phloem to the inner, functional xylem, also play an important role in the storage of reserve nutrients.

A variety of pathogens infect these various types of storage organs and develop in the cells of their tissues. Pathogens cause a variety of pathological changes in storage parenchyma, depending on the susceptibility of the host organ, the type of pathogen involved, and external conditions during pathogenesis.

Pathogen development in storage parenchyma is manifested by a variety of symptoms, such as wet and dry rots, and wood decays, being most specific for these types of tissues.

Wet rot occurs primarily in parenchyma tissue of metamorphytized vegetative root organs as well as roots, stems, and shoots of herbaceous plants. The pathogen influence on these tissues consists mainly of enzymatic degradation of pectic substances in the middle lamella and hemicelluloae degradation in cells walls. With the degradation of the middle lamella, intercellular linkages collapse, tissues disintegrate, and are transformed into moist, pulpy masses of usually dark brown in color, and frequently accompanied by an unpleasant odor. Typical causal agents of wet rot are fungi such as *Botrytis cinerea, Sclerotinia fuckeliana, S. libertians , S. sclerotiorum* , and bacteria, such as *Erwinia carotovora.*

S. libertiana, for example, causes wet rot of carrot and celery roots, potato tubers, and cucumber and gourd fruits; while *E. carotovora* causes a wet rot on potato tubers, carrot, celery and sugar beet roots, and in cucumber, melon, pepper, tomato and watermelon fruits.[33] The fungus *Colletotrichum lagenarium* develops on fruits of the Cucurbitae damaging their parenchyma tissues, which often are colonized by saprophytic bacteria. The development of these bacteria causes a decay or rot which then reduces the market value of the product. Another example is *Pseudomonas syringae* pv. *lachrymans*, which causes a decay of cucumber fruits, transforming them into pulpy mass. Dry rot occurs in the parenchyma storage tissue of pome and stone fruits. It is characterized by drying and wrinkling of infected fruits, which eventually become desiccated, yet maintaining their original form. Such fruits may persist for many years and are referred to as mummied fruit.

Dry rot also is caused by a group of fungi in the *Stromatinia,* with conidial stages in the *Monilia*. The mycelium of these fungi spreads rapidly throught infected parenchyma later forming stroma, the fact for which this suborder is named. Another type of dry rot is caused, under certain conditions by members of *Fusarium* spp., including *F. coeruleum* and *F. solani* on potato tubers.

Wet and dry rots rank among the most frequent disorders in storage parenchyma tissues. The mechanism of decay is described in the section on cytopathological changes.

Storage parenchyma in secondary wood are susceptible to numerous pathogenic fungi. The lignified walls of these cells are composed of cellulose, hemicellulose, and lignin. Pathogenic fungi degrade these compounds with their various enzymes, resulting in different types of wood rots. For example, white rot is caused by several hundred fungi of the Hymenomycetes, which degrade cellulose and lignin. They are well-known producers of cellulose, mannase, xylanase, and lignin-degrading enzymes.[34] *Polyporus versicolor* is a typical example of this type of fungus.[35] Brown rot is caused by fungi of the Holobasidiomycetes, which degrade carbon hydrates and alter lignin, but not using the byproducts for their nutrition.[34,35] These fungi live in the fiber lumen in which they secrete polysaccharide enzymes for degradation of the corresponding cell wall consitiuents.[36] The polysaccharide enzymes involved are cellulose and hemicellulose.

Some fungi belonging to the Ascomycetes cause a wet rot of wood tissue, but these are considered a less detrimental type of rot than those previously mentioned.[37] It occurs under extremely wet conditions. The primary action of these pathogens are cellulose degradation, causing significant lossses. It is generally considered that these fungi degrade only a small portion of lignin.[34] However, Mount[35] thought that the hyphae of these fungi usually are not present in the cell lumen, but in the cylindrical cavities formed in the secondary cell walls. The types and details of wood rot are described in more detail in the section on cytolytic changes.

C. PATHOLOGICAL CHANGES OF STEM AND ROOT PARENCHYMA

Parenchyma tissue, wide spread in roots and stems of plants, is susceptible to numerous pathogens. Symptoms of parenchyma tissue infection differ depending upon the type of pathogen involved, environmental conditions of pathogenesis, and the susceptibility of the parenchyma tissue involved (Figure 60).

Parenchyma tissue in the lower stem, i.e. hypocotyl and roots, of the seedlings of many plant species are susceptible to some soilborne pathogenic fungi. These fungi degrade parenchyma tissue cells through their enzymatic action, particularly with pectolytic enzymes, and other degrading products. The consequences of such action is manifested in the form of seedling damping-off, the most characteristic disease symptom. Fungi belonging to this group include

Fusarium, Pythium, Phytophthora, Rhizoctonia and other genera, which infect young plants prior to cell wall lignification.

Active destroyers of bark, stem, and root parenchyma include polyphagous soil fungi such as *Sclerotinia libertiana* on sunflower and *S. sclerotiorum* on grape berries and shoots. Degradation of infected tissue is manifested in the form of a wet stem and partial root rot. All living tissues in the infected areas are degraded, resulting in irreversible tissue desiccation.

Fungi that cause anthracnose develop deeper in the infected tissues, primarily in fruit and stem parenchyma, where it persists fora relatively long time, causing cell necrosis and death of the infected plant parts. Typical fungi involved are *Ascochyta pisi* on peas and *Colletotrichum lindemuthianum* on beans.

Extracellular obligate parasites, represented by fungi belonging to the Uredinales, predominantly colonize parenchyma tissues of herbaceous plant parts, such as leaves and stems, which consists of cells with nonlignified walls. According to Tarr,[2] the mycelia of these pathogens develop systmeically during aecial stage formation and locally during formation of the uredial stage.

Root and stem parenchyma tissue cells also are susceptible to meristemization (see sections on cytotransformative and histotransformative changes). Thus, such meristematic cells divide rapidly and increase, sometimes substantially, into a parenchyma tissue mass. These enlarged parenchyma tissue masses take on the form of cankers, galls, tumors, etc., usually evident around the points of infection.

Plasmodiophora brassicae pathogenic to roots of various crucifers, infects medullary ray and medulla parenchyma cells, in addition to cambium tissues. Infected cells in which the plasmodia develop, enlarge, then noninfected contiguous parenchyma cells are transformed into secondary meristem cells, which in turn divide uncontrollably and form the typical clubs found on infected crucifer plants.

Agrobacterium tumefaciens penetrates through wounds on crowns and roots of susceptible plants, causing tumors in the meristem and parenchyma tissues. Tumors develop from the enlargement of cells surrounding the wounds the cells continue to divide and eventually form tumors around the points of infection.[5] Olive galls, caused by *Pseudomonas syringae* pv. *savastanoi*, are also induced by meristematic processes. Graniti[9] stated that secondary meristems and undifferentiated tissues usually are responsible for the hyperplastic reaction resulting in the formation of olive cankers.

In summary, parenchyma tissue cells are susceptible to various transformations, which is of particular importance to the survival of plants. Pathological changes are most wide spread and diversified in parenchyma tissue, the fundamental tissue plant structure. Disorders in parenchyma tissues can occur during the course of their formation and differentiation, as well as in differentiated tissues.

VII. PATHOLOGICAL CHANGES IN CONDUCTIVE TISSUES

The role of conductive tissue is to ensure the efficient and unhindered transport of nutrients and water through the plant. In general, conductive tissues consist of the xylem, serving for the transport of inorganic compounds and water; and the phloem for the transport of organic compounds.

Conductive tissues feature some traits deserving of attention by plant pathologists. One such trait is the complex structure of these tissues. Conductive tissues consist of elements, which in part represent their function; conductive parenchyma, mechanical tissue, wood in xylem, bast in phloem, and other transitional cell forms.

Cells of many of these tissues are living and, therefore provide an environment suited for the development of many plant pathogens. Other tissues, however, such as sclerenchyma which make up part of the mechanical tissues die upon reaching maturity, making them less susceptible to pathogens. Conductive tissues are built of unique units, which, according to Zimmerman and McDonald,[38] are present in astronomic number in large plants. These elements are interconnected in such a way as to form an uninterrupted and functional system linking the plant with its environment. Plant pathogens developing in such a system, disorganize its structural units and thus, endanger the system and the plant.

The functions of damaged conductive units may be taken over by undamaged parts, contributing to the preservation of the plant. However, more serious disruption to nutrient transport which cannot be so compensated, may be expected in cases of substantial damage.

Meristems which give rise to xylem and phloem are the procambium and cambium ring. Procambium cells give rise to primary, and cambium ring cells to secondary conductive tissues. Depending on their origin, both primary and secondary tissues are considered primary conductive tissues.

Some pathogens, such as *Agrobacterium tumefaciens* and *Plasmodiophora brassicae,* stimulate the formation of new conductive tissues in infected plant pathological or induced conductive tissues. They are formed by differentiation of cells originating from the induced meristem in infected plant parts.

A. PATHOLOGICAL CHANGES OF XYLEM TISSUES

Xylem tissues provide for mineral and water transport form the soil to all plant parts. It consists of living cells, except for the fully differentiated tracheae, and therefore provides a suitable environment for colonization by several plant pathogens. This complex structure is composed of conductive elements, such as tracheae and tracheids, vascular parenchyma, wood ray parenchyma, wood fibers, as well as other cells representing transitional forms between tracheids and sclerotic cells.

A major portion of xylem consists of conductive elements, conductive parenchyma, including parenchyma cells of wood rays, and are the tissues most often colonized by plant pathogens. Pathological changes in these tissues result in xylem necrosis and death followed by a failure in its functions (Figure 63).

Causal agents of xylem disease include pathogens belonging to several groups, including bacteria, fungi, mycoplasma, rickettsias, and viruses (see the section on intracellular pathogens).

Pathogenic fungi represent the most numerous, and possibly the most destructive group of all xylem pathogens. Some of them develop in the xylem, as well as in other tissue and thus, are considered as poor vascular pathogens. Included among these less specialized pathogens are *Clasterosporium carpophilum* found on the leaves of stone fruit trees, *Pseudopedziza tracheiphila* on grape leaves, and *Pythium debaryanum* cause of damping-off of many herbaceious and woody plants.

Another group includes soil fungi which have specialized for invasion of xylem, particularly its vascular elements. Xylem elements, in contrast to phloem elements, have a comparatively open system connected, through its parts, to the soil environment. Certain soilborne fungi have become adapted to living in xylem and are referred to a tracheiphyllic pathogens.[9] These pathogens leave the soil and enter into the conductive elements, then colonize the xylem parenchyma tissues, eventually causing their death. These specialized vascular pathogens include *Fusarium oxysorum*, *F. vasinfectum*, *Verticillium albo-artum*, and *V. dahliae*. Examples among the woody plant pathogens include *Ceratocystis fagacearum* on oak trees and *C. ulmi* on elm trees.

Phytopathogenic bacteria that specific vascular pathogens include *Erwinia stewartii* in maize plants, *Corynebacterium michiganense* pv. *michiganese* in tomato plants, *C. sepedonicum* in potato plants, *Pseudomonas solanacearum* on potato and other hosts (Figure 64), and *Xanthomonas campestris* pv. *campestris* in crucifers.

A small number of viruses has been reported to invade xylem, probably because it is a less suitalble environment for virus multiplication.[18] The xylem serves more as a means of transport

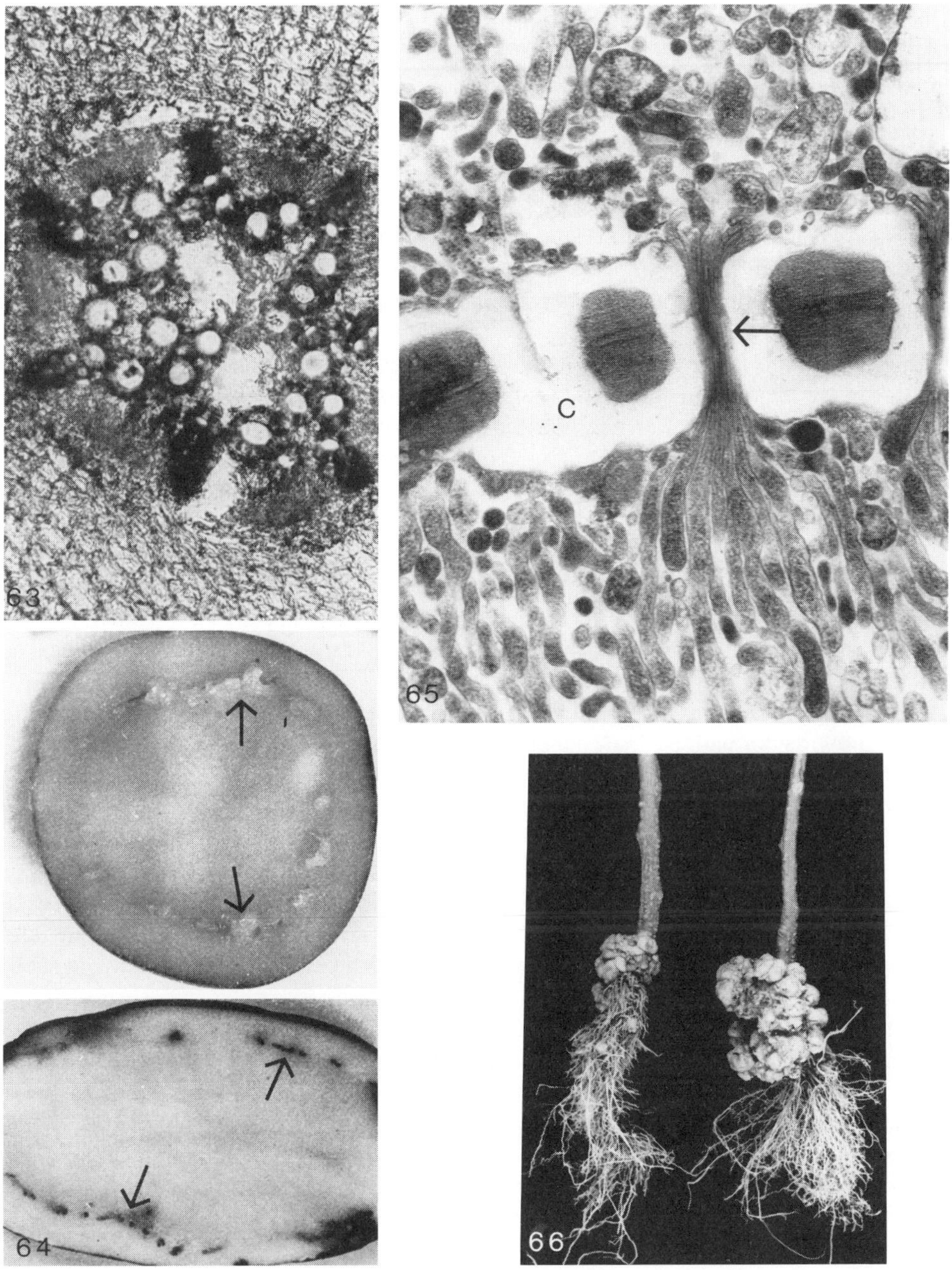

PLATE XIII. FIGURE 63. Crossection of a root of a Gerbera plant infected with *Phytophthora cryptogea* showing destruction of xylem (From: Trajcevski, T., M. Sc. thesis, Belgrade, 1985) FIGURE 64. Potato tubers infected with *Pseudomonas solanacearum* showing whitish bacterial exudate from the tuber vessels (arrows) (above) and the death of xylem elements and dark discoloration (arrows) (below). FIGURE 65. Vertical section through a tomato stem infected with the stolbur mycoplasma showing the passage of mycoplasma through the pore of a sieve plate (arrow) and callose formation (C) in the sieve tube. (From Marwitz, R., Šutić, D., and Aleksić, Ź., in: Šutic, D., *Anatomija i Fiziogija Bolesnik Biljaka*, 1987.) FIGURE 66. Development of tumors on the roots of tomato plants inoculated with *Agrobacterium tumefaciens*. (From Arsenijević, M., Bokterioze biljoka, Poljoprivredni fakultet, Novi Sad (Yugoslavia), 1975.)

in the plant than as a source of inoculum. Since they occur in the transport system, virus particles can appear in guttation droplets on leaves.[39] Chambers and Francki[40] found characteristic particles of the lettuce necrotic yellow virus in xylem cells of young lettuce leaves. Graniti[9] stated that plant viruses replicate in immature conductive tissues, where they persist during the differentiation process.

Vascular pathogens exert their influence on the conductive elements and vascular parenchyma cells with the aid of enzymes and toxins. The enzymes degrade cell structure and the toxins cause cell protoplast death. The degradation and necrotic processes occur more rapidly in young cells in the process of differentiation than in old, differentiated cells. The pathogens less specialized as vascular colonizers cause a more rapid breakdown of vascular tissues than the more specialized ones.

Early infection interrupts the formation of tracheal elements and conductive parenchyma resulting in irregular xylem development. This in turn limits nutritional and water transport. Due to the resulting nutrient deficiency, these cells lag in growth resulting in hypotrophy and a reduction in size resulting in atrophy. Such disorders result in reduced growth of the infected plant which remain dwarfed and stunted. An example of such a pathogen is the causal fungus, *Verticillium albo-atrum.*

Localized xylem necrosis caused by viruses has been observed on broad bean plants infected with the bean southern mosaic virus[41] or tomato ringspot virus.[42] Also, conductive parenchyma and even conductive tracheal necrosis was caused by some potato viruses.[10] Since viruses are not truly vascular pathogens, xylem necrosis caused by viruses must be considered as a result of a general necrosis of associated living tissues.

Hyperplasia and other transformative changes leading to excessive xylem tissue growth and, in some cases, the formative of new conductive tissues, also occurs in the course of pathological xylem histogenesis. Visible symtoms of such changes are the malformation of vascular bundles and other internal structures of the infected plant parts. Such changes have been found in hyperplastic tissues caused by phytopathogenic bacteria, fungi, and viruses.

In clubs formed on crucifer roots in response to infection by *Plasmodiophora brassicae*, hyperplastic parenchyma develops, which is evident by the increase in the number of medullary rays, and the vascular ring separates into isolated, remote bundles.[5,10] Due to the uncontrolled development of medullary rays, these bundles may be further malformed. New tracheal tissues also may be formed in these medullary rays.

In the thickened parts of cocoa plants with witches' broom, caused by *Marasmius perniciosus*, there is an increase in the number of cells in the xylem as a consequence of excessive cambium cell division. Tracheae also are affected and those portions of xylem nearest the cambium contain a larger number of parenchyma cells than those more remote.[10]

In tumors caused by *Agrobacterium tumefaciens*, transformation of some cell groups into conductive elements, such as tracheae and tracheids, also occurs (Figure 66). According to Wood[5] these conductive elements may be numerous and wide-spread in the tumors, but insufficiently organized and slightly or not at all linked with the host plant vascular system. Gäumann[28] reported that these neoplastic vascular bundles are formed from cambium and bark parenchyma.

Hyperplasmic activities occur in some plants infected with certain viruses. For example, cacao branch thickening, caused by the cacao swollen shoot virus, is a result of excessive xylem tissue formation.[43] Gibbs and Harrison[43] reported that a similar hyperplastic reaction occurred in sugar beets, tobacco, and tomato plants infected with the beet curly top virus.

The occurrence of foreign matter or structures in conductive elements of diseased plants is an interesting pathological phenomena associated with vascular pathogens. Such structures include endovascular trabeculae or intracellular strands, typloses, and resins. These foreign materials occur in the lumina of conductive tissues. Endovascular trabeculae are lignified strands, which can vary in number, and usually small enough to pass through the conductive elements. It has been assumed that they are modified plasma excrescences originating from the immediately contiguous conductive parenchyma cells. Endovascular trabeculae have been reported in the conductive elements of grape plants infected with the grapevine fanleaf virus and their presence used as a means of diagnosing the disease.

Tyloses are vesiculated formations formed exclusively in tracheae and tracheids. They originate from parenchyma cells which envaginate through pores into the contiguous vascular elements. Esau[44] stated that both vascular parenchyma and medullary ray parenchyma take part in tyloses formation. The nucleus and portions of the protoplast of the parenchyma cells usually move into the tyloses,[22,44] which also may contain starch grains, crystals of calcium carbonate, tannins, resins, and other compounds.[45] Formation of tyloses is a nonspecific process usually occurring in infected or injured conductive elements. Tyloses also are formed in plant parts showing signs of senescence, such as the infected pith cells of wood plants.[46] Several pathogenic fungi and viruses stimulate the formation of tyloses. For example, tyloses have been observed in chestnut plants infected with *Endothia parasitica;*[28] and in small grains infected with the barley yellow dwarf and other leaf yellowing viruses.[43]

Resinous substances also may occur in injured conductive tissues. Resins are produced by contiguous parenchyma cells and discharged through vascular pores and into the vascular lumina. Although primarily made up of saccharides in which glucuronic acid is bound to pentose and hexoses, the chemical composition of these substances is changeable.[22] The resinous substances may also contain oxidized phenolic compounds which give them a dark discoloration, making them easy to see and use for diagnostic purposes.

The mechanism(s) for resin and tyloses formation is not fully understood. One explanation is that ethylene is released from infected or injured cells and stimulates their formation; other workers believe that the stimulatory factor may be the penetration of air (embolism) into the infected or injured conductive elements.[22] Regardless of the mechanisms involved, it must be consisdered as a protective response, since the clogging of infected or injured conductive elements prevents the uncontrolled loss of nutrients and water from those tissue.

Changes in diseased or damaged xylem, whether by necrosis, structural changes, or the development of new structures, hinders or prevents the normal activites of this tissue. Disorders in the xylem functions are manifested by external symptoms, such as reduced plant growth, vascular necrosis along stems and veins, wilting, chlorosis of plant parts, leaf epinasty, and defoliation. Similar to other diseased or damaged tissues, xylem elements may be part of hyperplastic reactions resulting in the formation of neoplastic excrescences.

B. PATHOLOGICAL CHANGES IN PHLOEM TISSUES

The phloem, like xylem, consists of different tissue cells. The conductive elements are sieve tubes, which are composed of longitudinally linked sieve cells. According to Zimmerman and McDonough[38] sieve tubes tubes resemble xylem tracheids. Other tissues associated with phloem, such as parenchyma phloem in the bark rays, and mechanical tissues, are similar to parallel structutres in the xylem.

In contrast to xylem, phloem tissues consist of living cells, thus offering an environment for the development of various pathogens. However, sieve tubes are a closed sytem in the plant. As a result, obligate parasites such as mycmoplasmas and viruses have adapted to living in such a system. A few phytopathogenic viruses have a developed a degree of specialization for development in the phloem. Examples of some of these viruses that develop exclusively or predominantly in phloem tissues include, the beet curly top, citrus tristeza and potato leaf roll viruses, and probably the beet necrotic yellow vein virus. Viruses belonging to this group replicated in phloem cells causing necrosis in the course of pathogenesis. This process results in disorders in the transport of organic matter through the plant the accumulation of nutrients in the leaves. Curling and stunting of leaves is a characteristic symptom of such dysfunctions.

Internodes of infected plants have reduced growth rates and infected plants appear dwarfed or stunted. Eventually, the consequences of these disorders and dysfunctions is plant death.

Some less specialized viruses develop in the phloem as well as in other plant tissues, such as the beet yellow virus and potato Y virus (potato streak virus). These viruses disturb or prevent phloem tissue formation, resulting in symtoms on infected plants such as leaf yellowing and vein necrosis.

Plant mycoplasmas belong to a group of plant pathogens that specialize in colonizing phloem tissues, but also are capable of developing in sieve tubes and parenchyma tissues. Thus, the detrimental effects of mycoplasma infection occur in all tissues associated with phloem elements. Example of diseases caused by plant mycoplasmas are stolbur on pepper, tomato, and other plants; strawberry petal greening; and pear decline. Mycoplasma multiply in sieve tubes, thus, reduce sieve permeablility, which leads to dysfunctions in the phloem tissues (Figure 65).

Sieve tubes, as well as phloem parenchyma are relatively thin-walled, nonlignified cells which are susceptible to the necrotic processes which frequently occur in infection plants. Such infected cells are characterized by hyperplasia, which results in proliferation of cells in infected phloem. The symptoms caused by these phenomenon as well as the presence of mycoplasma bodies, virus particles or other such pathogen structures in phloem are so characteristic that they can be used in the diagnosis of infection.

Viruses specialized for development in phloem tissues typically cause phloem necrosis, such as the potato leaf roll (PLRV) and potato streak (PVY) viruses. The PLRV causes necrosis which is limited to phloem tissue and gradually spreads in infected plants.[17] According to Bawden,[32] PVY causes phloem necrosis in leaves, usually spreading along the veins in the collenchyma tissue without affecting the xylem conducting elements. Top necrosis is a frequent symptom caused by viruses on potato plants infected with a variety of viruses. Necrosis usually progresses up the plant beginning in the inner phloem, then encompasses the contiguous parenchyma, and then spreads into the top of the stem. This necrotic development is accompanied by the formation of phellogen, which forms a corky ring layer around the points of infection , which prevents further spread of the disease. The beet curly top virus causes a spreading phloem necrosis, which may also affect the pericycle of infected plants.[32]

The citrus tristeza virus causes a necrosis at grafting sites of sweet orange and other susceptible rootstocks, such a grapefruit and common orange. In this situation, sieve tubes are blocked and medullary rays cells show a slower than normal growth rate in the course of the necrotic processes.[21] These disorders prevent the transport of nutrients from leaves to the root system, resulting in a rapid decline of the roots, decay, and eventual wilting of the entire plant.

In hypersensitive plants, general necrosis of the phloem as well as other living tissues occurs in the form of localized spots. This type of necrosis results in spots that can be caused by many pathogenic bacteria and fungi on leaves and other plant parts.

In apple trees, for example, infected with apple stem pitting virus, changes in the vascular bundles take place in addition to the necrotic reaction. These changes are characterized by pitting in the branches and stems, due to insufficient cambium activity. Normal cells are no longer differentiated and phloem cuneately spreads into the xylem. The consequences of such changes are necrosis of the sieve tubes, damage to the conductive elements, and a general reduction in the number of conducting vessels.[47]

Cells of phloem tissue are also capable of hypertorphy and hyperplasia, reactions resulting in the formation of tumors, which differ in shape and size depending on the host and host susceptiblity. For example, beet curly top virus first causes hypertrophy and later death of parenchyma cells nearest to infected sieve tubes. However, more remote cells divide, give rise to a hyperplastic mass of altered sieve tubes, which lack sieve plates and sometimes bear only compaion cells.[10] The cambium continues of form new vascular elements, but they are primarily composed of incomplete sieve tubes with companion cells and parenchyma. This parenchyma together with the pericycle, enlarges and divides forming neoplastic tissue resembling callus tissue. This neoplastic tissue protrudes on the veins of the lower leaf surfaces in the form of small warts. The sieve tubes and companion cells associated with these outgrowths die rapidly.

Rasa and Esau[48] studied the effect of beet curly top virus on phloem changes in infected tomato plants. Schmidt[26] described these changes in a number of interconnected stages. The first sign of such changes was an intense protoplast coloration in the cells close to the differentiated sieve tubes, which were sometimes enlarged (primary hypertrophy). Later, these cells are subject to necrotic processes (primary necrosis). Prior and just following necrosis, hyperplastic division of contiguous cells (primary hyperplasis) takes place. Abnormal sieve tubes lacking companion cells and parenchyma are formed during this type of cell division. Hyperplastic tissues formed in this manner also are subject to necrosis (secondary necrosis) followed by death of the cells. Cells of tissues contiguous to the necrotic cells also divide (secondary hyperplasia). All of these pathological processes result in significant changes in the phloem and other infected plant tissues.

Tumors also are produced in plants infected with some of the reoviruses as a result of meristematic activity of the phloem parenchyma cells. For example, tumors caused by the clover wound tumor virus are the result of intensive hyperplastic activity of phloem parenchyma cells in susceptible plants.[43,49] Tumors formed on sugar cane infected with Fiji disease also are a result of meristematic division of phloem parenchyma cells, which give rise to sieve tubes and tracheids.[10]

As discussed earlier, the presence of mycoplasma bodies and virus particles in sieve tubes and the cells of other phloem tissues are important in the diagnosis of phloem-related disease (see section on intracellular pathogens).

Cell inclusions also occur in phloem and other tissues infected by various viruses and can be diagnostic. Inclusions bodies occur only in virus-infected cells, their occurrence following the spread of the virus in the plant.[43] For example, the amorphous inclusions in sugar beet plant infected with the beet yellow virus occur primarily in phloem tissues, whereas in plants infected with the clover wound tumor virus, the inclusion bodies are present only in the pseudophloem (hyperplastic phloem) (see section on intracellular pathogens).

The formation and deposition of callose sometimes takes place in diseased phloem tissues of plant infected with mycoplasmas and viruses. Change in sieve tubes, phloem parenchyma, and companion cells in mycoplasma-infected plants usually are accompanied by callose deposition.[50] The deposition of callose often results in increased fluorescence of the infected plant parts.[51] Callose fluorescence sometimes is used to determine the presence of tissue damage caused by mycoplasma.

Callose deposition can also take place in cell walls of virus-infected cells and sometimes, in the cell walls of noninfected contiguous cell walls. Callose formation has been observed in sieve tube pores of virus-infected plants, clogging the pores, which then cease to function.[52] Such callose deposits can be stained in infected phloem cells using the Igel Lange Test, particularly for the detection of callose in potato tubers infected with the potato leaf roll virus.[43,53] Since callose deposit formation is dependent on a variety of factors and may be caused by a variety of agents, this method is not always reliable for detecting virus infection in plants.

Callose deposit may also occur in the walls of noninfected cells surrounding a localized virus infection, thus preventing further virus spread.[17] Opel[54] stated that the formation of callose and calcium pectate around necrotic cells and tissues probably blocks the connections of plasmodesmata between infected and noninfected cells. Thus, callose formation and deposition in phloem and other tissues represents a defensive reaction by the plant which hinders or prevents the spread of viruses from the original points of infection.

In conclusion, plant mycoplasmas and viruses, as the most frequent phloem invaders and pathogens, cause various pathological changes in this tissue. Particularly important among these changes are necrosis, structural and hyperplastic changes, and the development of neoplastic formations in the phloem tissues. All these changes disturb or prevent phloem functions. As a result of the disruption in the supply of organic nutrients to the plant, there is a delay or retardation in the development of plant organs. Because of the lack of transport of organic materials from leaves, they become thickened, firm and brittle as well as curled and rolled. Chromatic changes also occur, particularly a general chlorosis on leaves and other herbaceous parts. Diseased or damaged phloem tissues are prone to hyperplasis resulting in the development of local tumorlike excrescence or organ proliferations.

References

1. **Huber, M. D.**, Disturbed mineral nutrition, in *Plant Disease An Advanced Treatise III*, Horsfall, J.G. and Cowling, B.E., Eds., Academic Press, Inc., New York, NY, 1978, 163
2. **Tarr, J. A. S.**, *The Principles of Plant Pathology*, MacMillan Press, London, 1972.
3. **Gäumann, E.**, *Pflanzliche Infektionslehre*, Verlag Birhauser, Basel, 1946.
4. **Király, Z.**, Defenses triggered by the invader: Hypersensitivity, in *Plant Disease An Advanced Treatise* V, Horsfall, J.G. and Cowling, B. E., Eds., Academic Press, Inc., New York, NY, 1980, 201.
5. **Wood, R. K. S.**, *Physiological Plant Pathology*, Blackwell Scientific Publications, Ltd., London, 1967.
6. **Blackman, V. H. and Welsford, E. J.**, Studies in the physiology of parasitism. II. Infection by *Botrytis cinerea*. Annals Botany 30, 389, 1916.
7. **Boyle, C.**, Studies in the physiology of parasitism. VI. Infection by *Sclerotinia libertiana*. Annals Botany 35, 337, 1921.
8. **Waterhouse, W. L.**, Studies in the physiology of parasitism VII. Infection of *Berberis vulgaris* by sporidia of *Puccinia graminis*. Annals Botany 35, 557, 1921.
9. **Graniti, A.**, Tissue and organ specificity in plant diseases, in *Specificity in Plant Diseases*, Wood, R.K.S. and Graniti, A., Eds., Plenum Press, New York and London, 1975.
10. **Butler, J. E. and Jones, G. S.**, *Plant Pathology*, London Co., Ltd., England, 1949.
11. **Ehrlich, H. G. and Ehrlich, M. A.**, Electron microscopy of the sheat surrounding the haustorium of *Erysiphe graminis*. Phytopathology 53, 1378, 1963.
12. **Rao, A. S.**, Biology of *Polymyxa graminis* in relation to soil-borne wheat mosaic virus, *Phytopathology* 58, 1516, 1968.
13. **Ivanović, M.**, *Polymyxa betae* Kesk., parazit šećerne repe i vektor virusa nekrotičnog žutila nerava repe. Doctorska disseratija. Poljoprivedni Fakultet, Universitet u Beogradu, Zemun, 1984.
14. **D'Ambra, V. and Mutto, S.**, The ultrastructure of *Polymyxa beta* zoospore exit-tube differentiation. *Canadian Journal of Botany* 55, 831, 1977.
15. **Lewis, S. and Goodman, N. R.**, Mode of penetration and movement of fire-blight bacteria in apple leaf and stem tissue. *Phytopathology* 55, 719, 1965.
16. **Josifović, M.**, *Poljoprivredna Fitopatologija*, II Izdanje, Naučna knjia, Belgrade, 1956.
17. **Matthews, F. E. R.**, *Plant Virology*, Academic Press, Inc., New York & London, 1970.
18. **Šutić, D.**, *Biljni Virusi*, Nolit, Belgrade, 1980.
19. **Tepfer, S. S. and Chessin, M.**, Effects of tobacco mosaic virus on early leaf development in tobacco, *American Journal of Botany*, 46, 496, 1959.
20. **Campbell, W. P.**, Infection of barley by *Claviceps purpurea*, *Canadian Journal of Botany*, 36, 615, 1958.
21. **Šutić, D.**, *Viroze Bijaka*, Nolit, Belgrade, 1983.
22. **Talboys, W. P.**, Dysfunction of the water system, in *Plant Disease An Advanced Treatise* III, Horsfall, J.G. and Cowling, B.E., Eds., Academic Press, Inc., New York, NY, 1978, 141.
23. **Tošić, M. and Misovic, M.**, Uticaj viroznog mozaika na porast i prinos nekih sorti i hibrida kukuruza, *Zaštita Bilja* 1971, 93 and 173.
24. **Pop, I. and Tusa, C.**, Influence of maize mosaic on the growth and yield of maize hybrids, in *Viruses of Plants, Proceedings of the International Conference on Plant Viruses, July, 1965*, Wageningen, 1966, 170.
25. **Tošić, M.**, Virus diseases of wheat in Serbia. II. Some changes in wheat plants infected with wheat streak mosaic virus (WSMV) and brome grass mosaic virus (BMV). *Phytopathologische Zeitschrift*, 71, 327, 1971.
26. **Schmidt, B. H.**, Pathologische Zytopathologie und Anatomie, in *Pflanzliche Virologie*. I., Klinkowski, M., ed., Akademie Verlag, Berlin, 1980, 84.
27. **Kegler, H.**, Untersuchungen über Virosen des Kernobstes IV. Die Flachästigkeit des Apfels, *Phytopathologische Zeitschrift*, 50, 297, 1964.

28. **Gäumann, E.**, Über Abwehrreaktionen bei Pflanzenkrankheiten, *Experimentia* 12, 411, 1946.
29. **Price, W. C.**, Flexuous rods in phloem cells of lime plants infected with citrus triteza virus, *Virology* 29, 285, 1966.
30. **Shalla, T. A.**, Relations of tobacco mosaic virus and barley stripe mosaic virus to their host cells as revelaed by ultrathin tissue sectioning for the electron microscope, *Virology* 7, 193, 1959.
31. **Ulrich, J. and Quantz, L.**, Morphologische und anatomische Untersuchungen über pathologische Veränderungen durch das scharfe Adernmosaik (pea enation mosaic virus) an *Vicia faba* und *Pisum sativum* und durch eine Virose an *Trifolium repens, Phytopathologische Zeitschrift*, 51, 1, 1964.
32. **Bawden, G. F.**, *Plant Viruses and Virus Diseases*, Waltham, MA, 1956.
33. **Arsenijević, M.**, Bakterioze biljaka, *Poljoprivredni Fakulter*, Novi Sad, Yugoslavia, 1973.
34. **Kirk, T. K.**, Effects of microorganisms on lignin, *Annual Review of Phytopathology*, 9, 185, 1971.
35. **Mount, S. M.**, Tissue is disintegrated, in *Plant Disease An Advanced Treatise*, III, Horsfall, J. G. and Cowling, B. E., Eds., Academic press, Inc., New York, NY, 1978, 279.
36. **Cowling, E. B. and Brown, W.**, Structural features of cellulosic materials in relations to enzymatic hydrolysis, in *Cellulases and Their Application; Advances in Chemistry* 95, Gould, R. R., et al., Eds., American Chemical Society, Washington, DC, 1969, 152.
37. **Duncan, C. G.**, Wood-attacking capacities and physiology of soft-rot fungi, *United States Forest Service Report* 2173, Washington, DC, 1960, 1.
38. **Zimmermann, H. M. and McDonough, J.**, Dysfunction in the flow of food, in *Plant Disease An Advanced Treatise*, III, Horsfall, J. B., and Cowling, B. E., Eds., Academic Press, Inc., New York, NY, 1978, 117.
39. **Wetter, C.**, *Einführung in die Virologie,* Universitat des Saarlandes, Saarbrucken, 1968/69.
40. **Chambers, T. C. and Francki, R. I. B.**, Localization and recovery of lettuce necrotic yellows virus from xylem tissues of *Nicotiana glutinosa, Virology* 29, 673, 1966.
41. **Mitchell, J. W., Preston, H. W., and Beal, M. J.**, Stem inoculation with southern bean mosaic virus a promising method for use in screening chemicals for antiviral activity, *Phytopathology,* 46, 479, 1956.
42. **Smith, F. H. and McWhorter, P. F.**, Anatomical effects of tomato ringspot virus in *Vicia faba, American Journal of Botany,* 44, 470, 1957.
43. **Gibbs, A. and Harrison, B.**, *Plant Virology - The Principles*, Edward Arnold, Ltd., London, 1976.
44. **Esau, K.**, *Plant Anatomy*, John Wiley and Sons, Inc., New York, NY, 1969.
45. **Cincović, T., Čanak, M., and Kojić, M.**, *Botanika* 4, Izdanje, Naučna Knjiga, Belgrade, Yugoslavia, 1980.
46. **Küster, E.**, *Pathologische Pflanzenanatomie* 3., Verlag von Gustav Fischer, Jena, 1925.
47. **Hilborn, M. I., Hayland, F., and McCrum, C. R.**, Pathological anatomy of apple trees affected by the stem-pitting virus, *Phytopathology* 55, 34, 1965.
48. **Rasa, E. A. and Esau, K.**, Anatomic effects of curly top and aster yellows viruses on tomato, *Hilgardia* 30, 469, 1961.
49. **Lee, C. L. and Black, L. M.**, Anatomical studies of *Trifolium incarnatum* infected by the wound tumor virus, *American Journal of Botany* 42, 160, 1955.
50. **Maramorosch, K.**, Plant mycoplasma diseases, in *Physiological Plant Pathology*, Heitefuss, R. and Williams, H. P., Eds., Springer-Verlag, Berlin, 1976, 150.
51. **Hiruki, C. and Dijkstra, J.**, Light and electron microscopy of *Vinca* plants infected with mycoplasma-like bodies of the sandal spike disease. *Second International Congress Plant Pathology Abstracts*, 640, 1973.
52. **Eschrich, W.**, Sealing systems in phloem, in *Transport in Plants. I. Encyclopedia of Plant Physiology, New Series I*, Zimmerman, M. H., and Milburn, J. A., Eds., Springer-Verlag, Berlin, 1975, 39.
53. **Schuster, G. and Buhan, O.**, Zur Präzisierung des Kallosetestes für den Nachweis der Blattrollkrankheit bei Kartoffeln, *Z. Landwirtsch, Versuchs-u. Untersuchungswesen* 4, 37, 1968.
54. **Opel, H.**, Allgemeine Pathophysiologie, in *Pflanzliche Virologie* I, Klinkowski, M., Ed., Academie Verlag, Berlin, 1980, 212.

Chapter 3

ORGANOPATHOLOGY

Organpathology is the study of diseased plant organs, which are characterized by various functional and structural changes, and defined by the general term "organopathological changes". Such changes are the result of disorders in the cells and tissues that make up the organ structure. Organopathological changes are characterized by symptoms which usually are in the form of organ malformation. Some of the distortions or malformations are specific enough to serve for diagnosis, while others are less typical and diagnostically unreliable.

I. DISEASED PLANT ORGANS

Generally, plant organs are considered complex multicellular specialized associations composed of several tissues. Thus, all disorders occurring in them are the result of changes taking place in the cells and tissues involved. Infection of plant organs cannot be separated from those of its cells and tissues.

Diseases of plant organs, and their cells and tissues, are caused by a variety of abiotic and biotic agents. Plant organs differ in their susceptibility to the various causal agents or groups of agents.

From a symptomatic and topographic point of view, two types of changes are evident in diseased plant organs: external and internal. Internal pathological changes prevent or interrupt organ formation, which is manifested in appearance and shape of the organ. Organ malformations are easily diagnosed by external symptoms and are the most familiar symptoms in plant pathology.

Changes in the shape of diseased organs represent a general reaction to infection and can also be defined as morphological change or teratomas (malformations). As a common phenomenon, organ malformation is known as teratosis (morphosis) and the particular study of such malformations is called teratology.

Plant organ malformation rather than complete organ destruction is a result of a process of coevolution between the plant and pathogen(s) involved. Highly susceptible individual plants and highly virulent pathogens would be self-eliminating early in the evolutionary process. Such coevolution allows the ultimate acculmulation of resistance and fitness genes in the host and various parasites. Plants that have developed on this principle are capable of surviving and propagating their kind under a range of adverse environmental conditions including plant organ malformations.

One of the means of countering the loss of portions of plant organs is through the development of a larger external structure, such as in the canopy and roots. A larger than basically required external leaf surface ensures an adequate supply of solar energy and carbon dioxide; a larger than required root surface enables for the intake of minerals and water in spite of losses due to disease.

Diseased foliage, generally, is characterized by the loss of all or portions of the leaf surface. Leaf surface reduction can be caused directly by pathogens developing on leaves, or indirectly by pathogens affecting others organs, such as roots.

Examples of direct leaf damage include leaf spots caused by many bacteria and fungi (see section on cytonecrotic changes). Infections taking place early in the phenologic stages of development usually prevent normal leaf growth such as by: *Phytophthora infestans* on potato, *Plasmopora viticola* on grape, *Pseudomonas syringae* pv. *tabaci* on tobacco, and *Xanthomonas campestris* pv. *vesicatoria* on tomato. In some cases, leaf blades have deep, irregular lobes, such as those produced on grapevines infected with the grape fanleaf virus; and in other cases, the leaf blade is reduced to threadlike elongations (filiform), such as those produced on tobacco plants infected with the cucumber or tobacco mosaic viruses.

Under severe disease conditions, some pathogens can cause complete defoliation, such as the fungi *Polystigma rubrum* and *Puccinia pruni-spinosae* on plum trees.

An example of indirect damage to foliage is damage to roots. Many pathogens reduce and in some cases completely destroy the active cells and tissues on root surfaces. This group includes soilborne pathogenic fungi causing complete or partial root necrosis of herbaceous plants, such as *Fusarium* and *Verticillum* spp.; and of woody plants, *Armillaria mellea* and *Rosellinia necatrix*.

Even with larger leaf and root surfaces, many pathogens continue to threaten the normal development of the external structures of plants. By damaging susceptible organs, pathogens can completely or partially disturb the physiologic functions in plants. For example, early defoliation of grape by *Oidium tuckeri* and *P. viticola* and of plum trees by *P. rubrum* and *P. pruni-spinosa,* deprive the infected plant of the normal functions performed by the leaves. Another example, is the desiccation of flowers of apricot and sour cherry by *Monilia cinerea* which prevents flower and fruit development.

In plants with diseases caused by mycoplasmalike organisms, such as stolbur of tomato and strawberry green petal, the greening of floral corollas substantially changes the function of the affected flowers. Some members of the Ustilaginales, such as *Tilletia tritici* and *Ustilago tritici* on wheat, and *U. nuda* on barley alter the fruits or seeds transforming them into a mass of chlamydospores, thus preventing the function of these organs.

Similary, by causing degradation or necrosis of root and lower stem parts, pathogens reduce plant consistency. For example, lodging of herbaceous plants, such as members of the grass family, causes losses up to 60% or more,[1] with an average range of 5 to 20%[2] in any year. Fungal causal agents frequently associated with cereal plant lodging are *Cercosporella herpotrichoides, Fusarium moniliforme,* and *Ophiobolus graminis. Diplodia zeae* increases the susceptibility of maize plants to lodging; and *Piricularia oryzae* degrades the peduncle of rice inflorescences causing the heads to droop. *Kabatiella caulivora* degrades red clover stems resulting in the breakage and wilting of plant parts above the points of infection.[3] Some soilborne fungi such as *F. moniliforme, Phytophthora* spp., *Pythium* spp., *Rhizoctonia solani,* and *Sclerotinia* spp., enzymatically degrade lower stem and root parts causing seedling damping-off and lodging of a wide range of plants.

Causal agents of wood rot in perennial woody plants also reduce the consistency of stems, which then become more susceptible to damage caused by storms and other nonparasitic agents (see the section of cytoloytic changes). Cowling[2] estimated that over 2000 fungal species can cause wood rot in living stems, some of which are highly specialized, such as *Polyporus betulinus* restricted to *Betula* spp.; and those which are less so, such as *Fomes igniarius* and *F. pini* which infect a wide range of woody plants.[4] Wood rotting organisms cause greater damage in forests throughout the world than any other group of disease agents.

Viruses that cause stem pitting in fruit trees reduce the mechanical consistency of the wood of infected trees. This group includes the apple stem pitting diseases and the tomato ringspot virus, which causes stem pitting on apricot, cherry, peach, and sour cherry. Stem and bark disorganization on infected plants occurs near the stem base and main root, resulting in increased susceptibility to breakage below the soil line.[2]

Cells of apple branches infected with the apple rubbery wood mycoplasma have a greater cellulose than lignin content and affected branches have a reduced consistency and increased flexability. In addition, it was shown that the lignin in infected cells differed from that in noninfected plants.[5]

Plant organs develop and become mutually linked within plants through a principle of the most economical utilization and management of "construction material". This principle is expressed, for example, as the ratio between size and shape, of surface to volume, and by the internal distribution of tissues within individual organs. Plant pathogens disturb or destroy the application of this principle in the destruction or in the abnormal development of diseased organs.

All pathological changes in cell, tissue, or organ structure are a result of either a lack or loss, or uneconomical excess of the "construction materials." Local and general necrosis is accompanied by a corresponding loss of materials excluded from normal plant activity. For example, in filiform tobacco leaves caused by either the cucumber or tobacco mosaic viruses, construction materials are absent, resulting in the abnormal leaf development. In the development of neoplastic formations, such as in cabbage clubroot, caused by *Plasmodiophora brassicae* and in crown gall, caused by *Agrobacterium tumefaciens,* the "construction materials" are used in the uncontrolled formation of club and gall formation (Figure 66). As a result, although the infection may be localized, the entire plant lags in growth and reproductive ability. Similarly, in plants showing shoot proliferation, such as witches' broom, caused by *Taphrina cerasi,* and apple proliferation, caused by a mycoplasma, a large number of shoots are produced from dormant buds causing an uneconomical consumption of materials for their formation.

Thus, diseased organs, as part of the entire plant, disturb all the fundamental activities of that plant. All pathological changes in one location can cause physiological and structural changes on plant parts in other locations.

II. INFECTION OF ORGANS

Organogenesis is the formation of organs in the course of ontogenetic (individual) plant development. The internal structure and shape of various organs are formed through organogenetic processes. In their final stages of development, each organ assumes a specific shape, reflecting their development, and therefore, organogenesis is identified with morphogenesis.

Organogenesis in multicellular organisms is a result of the processes of cell differentiation (cytodifferentiation) and tissue organization (intercellular linkage). The process begins with the fertilized egg cell which divides rapidly forming the cellular material which is morphologically and physiologically differentiated, then organized into tissues and organs. Organ development is the result of a series of consecutive processes: cell division, cell enlargement, cell and tissue specialization and differentiation, and finally, organ formation.

Plant growth and the general appearance of a plant are based on organ development which in turn depends on numerous factors in and outside of the plant. According to Kozlowski[6] the growth of multicellular plants is the final result of a series of coordinate physiological processes depending on genetic factors, and various biochemical and physiological processes which are mutually linked by regulatory processes controlling photosynthesis, respiration, translocation, etc.

Organs may be infected at any time during the course of their development with the consequences differing depending on the stage of development. Early infection at the time of cell division and differentiation may prevent organ formation. For example, *Claviceps purpurea* infects the ovaries of rye plants directly or indirectly.[7] The hyphae colonize the ovule and then develops on the surface of the ovary forming a multitude of spores. Early ovary infection prevents grain formation resulting in the development of sclerotia of the pathogen. The internal hyphae of fungi *Ustilago nuda* in barley and *U. tritici* in wheat grains prevent normal seed differentiation by absorbing the cell constituents and using them for the formation of a large number of chlamydospores. Also, tomato aspermy virus prevents seed formation in infected tomato fruits.

The pathology of plant organs generally is studied in either of two stages, incomplete or mature organ development. In plant organs that have not matured, active cell division and differentiation are taking place. These processes are most active in initial organ formation with the intensity gradually decreasing in the course of growth and maturation. Infections taking place during these early stages of organ development usually cause the most conspicuous disorders in organogenesis since they directly influence cell activity in tissues and organ formation. Organs infected during this time have a malformed appearance and differ significantly in size and shape from healthy organs. Two examples are apple fruit and leaf malformation caused by *Fusicladium dendriticum,* and pepper fruit and leaf malformation caused by the cucumber mosaic virus.

Generally, organs infected during the final growth stage ususally are characterized by a reaction indicating tolerance to infection. Depending on organ susceptibility and environmental conditions, pathogens may develop locally or systemically. Disease symptoms may not be pronounced or even unexpressed. If infection takes place in the later stages of organ development, there may be no significant influence on organ shape and size. For example, tomato leaf spot may not develop or develop poorly in plants infected late with *Xanthomonas campestris* pv. *vesicatoria* or mild instead or severe mosaic symtoms may develop on leaves of tobacco and tomato plants infected late with the tobacco mosaic virus.

In the course organogenesis, individual plant parts may display temporary or permanent susceptibility to some pathogens. Temporary susceptibility as a rule, is determined by the chemical composition of a given host cell, and the ability of various pathogens to enzymatically or mechanically destroy its constituents. In the course of cytohistological cell differentiation, the chemical composition of cells is altered depending on the structure and function of the tissue to which they belong. Alterations in the chemical composition of the cell wall can change its ability to provide protection against the penetration of a large number of pathogens.

Changes in the chemical composition of a cell also can result in the alteration of their susceptibility to individual pathogens. The chemical composition of seedling plants, for example, is susceptible to enzymatic activity of several soilborne fungi causing damping-off of young plant. Bark tissues in these plants have not formed through lignification and suberization and they have only a thin external cuticle cell layer. Because of the lack of such protection, seedlings are susceptible to pathogens which produce enzymes that degrade the cellulytic and pectic constituents of the cell walls. Fungi that produce only pectolytic enzymes include *Pythium* spp., which cause damping-off of seedlings; those that produce either cellulolytic enzymes or both, such as *Fusarium moniliforme* and *Rhizoctonia solani,* degrade stem and root tissues of more mature plants.[8] The susceptibility of plants to damping-off organisms decreases and ceases with plant growth and the lignification of peripheral cells.

During epidermal tissue cell cythohistological differentiation, a cuticle layer composed of cutin is formed on their outer surface. The cuticle is resistant to the action of various pathogens. However, as long as a cuticle layer is not formed, the seedling is susceptible to pathogens. This susceptibility is transitory since it persists only until the cuticle is completely formed. For example, hyphae of the fungus *Alternaria tomato* readily penetrate the cuticle of young tomato fruits, but when the cuticle on these fruits becomes thicker, infection is more difficult and in fruits approaching maturity, infection is impossible.[9]

Temporary host cell susceptibility is also dependent on the chemical composition of the cell protoplast. This composition may influence pathogen development providing a favorable environment, or conversely, it may act as a parasiticidal substance. *Conothirium diplodiella*, for example, infects only damaged grapes which contain no less than 0.5% sugar; the development of pathogen mycelia then increases with increases in sugar content.[10]

Many plant parts are permanently susceptible to some pathogens, depending on the genetic characteristics of the host, and not depending decisively on the other factors such as the chemical composition of the cells. Organs possessing permanent susceptibility, depending on their stage of development, may differ little by the mode and intensity of their reaction to infection. For example, all herbaceous organs or grapevines are permanently susceptible to certain fungal pathogens, such as *Oidium tuckeri* and *Plasmopora viticola*. The herbaceous organs of tobacco are threatened throughout their life by a large number of pathogenic fungi, such as *Erysiphe cichoracearum* and *Peronospora tabacina*; the bacterium *Pseudomonas syringae* pv. *tabaci*; and the cucumber mosaic and tobacco mosaic viruses.

Similar to the relationship between host tissues and pathogens, the relationship between plant organs and pathogens is a dynamic living system which has become established in the course of evolution. In these processes, pathogens have adapted to colonizing plant organs as living environments. Some have more or less specialized as pathogens of specific organs.

The least specialized pathogen group comprises those infecting only herbaceous or only woody plant organs. An exception is mycoplasmas and viruses which causes a general systemic infection and usually are present in living cells of all plant parts.

The group of pathogens that infected herbaceous plant organs comprises the largest number of pathogens. These are, for example, pathogenic fungi found in the fungal groups of Erysiphales, Peronosporales, and Uredinales which develop in a variety of herbaceous plant parts. Pathogens of individual wood/organs, include *Armillaria mellea* and *Rosellinia necatrix* as well as several wood destroyers, such as members of the Hymenomycetes. Highly specialized pathogens which partially or completely develop only in one plant organ are defined as being organotropic. Organotropic pathogens include, for example, the fungus *Polystigma rubrum* and *Puccinia pruni-spinosae* both of which infect only plum leaves. An even more specialized organotrophic pathogen is *Taphrina pruni* which infects only plum fruits and rarely any other herbaceous organs of the plant. Another example is the mycelium of *Claviceps purpurea* which infects on by rye ovaries. Other organtrophic pathogens of woody plants include *Nectria cinnabarina*, *N. ditissima*, and *N. galligena*, which colonize woody stems and branches but not roots of various trees.

In conclusion, plant organs represent a favorable environment for different types of plant pathogens. In the process of adapting to this environment, these pathogens have further developed their parasitic activity. Developing in and on plant organs, pathogens cause various alterations of these organs. Several of these changes are conspicuous, other are not and may be unknown, so that it is impossible to enumerate all of them. However, for didatic reasons, these changes may be grouped into several types based on the mechanisms of occurrence and symptomology.

III. TYPES OF ORGANOTROPHIC CHANGES INDUCED BY PATHOGENS

Organopathological changes pertain to structural and funtional disorders in individual organs. Structural changes are determined by disorders in the organization and structure of plant organs. These changes can range from slight, without a great influence on organ function, to severe when organs are distorted or do not form at all, or die and are dropped from the plants.

A. ORGANOPLASTIC CHANGES

Organoplasmatic changes arise from disorders in cell division and differntiation in the course of organ formation. They are a result of cytoplasmatic and histoplasmatic organ changes (see sections on cytoplasmatic ahd histoplasmatic changes). Cytoplasmatic organ changes usually occur in organs infected during the initial stages of organogenesis.

The consequences of cytoplasmic organ disorders in the course of organogenesis may be manifested in one of three ways, (i) prevention of formation, (ii) dwarfing or stunting, or (iii) enlargement of the affected organs.

Organ formation may be prevented in instances where infection affects meristematic tissue cells prior to or at the beginning of their division activity. Such tissues may be subject to necrotic processes which deny them of their functional activity. An example is the infection of rye ovules by *Claviceps purpurea,* which prevents ovule formation. Other examples are the failure of endocarp, fruit stone, formation in plum because of early infection by *Taphrina pruni;* lack of seed formation in tomato fruits because the presence of the tomato aspermy virus causing pollen abortion;[11] and abnormal pollen in barley plants infected with the barley stripe mosaic virus which disturbs fertilization resulting in sterile spikes.[12] If *Cieba petandra*, a West African forest tree, is infected with the cacao swollen shoot virus, if forms branches lacking spines.[13]

Organ dwarfing is the most frequent symptom of organ infection. The extent of dwarfing may differ depending on the stage of tissue differentiation at which infection takes place. Organs on plants infected early are likely to show the most severe dwarfing and stunting.

Infection of sugar beets with the beet necrotic yellow vein virus, which causes rhizomania, results in the production of roots that may be five or more times smaller than those on noninfected beets. Potato tubers produced from seed potatoes infected with any one of several potato viruses can be so stunted that their yield is uneconomical.

Stem dwarfing is usally manifested as internode shortening which give an infected plant a rosette appearance. The symptoms on young sugar beet plant infected with *Peronospora schachtii* for example, are called top rosette. Grapevine fanleaf causes pronounced shortening of the internodes in grapevine shoots. Similar changes occur on peaches infected with the peach rosette mosaic virus, on maize infected with rough dwarf virus, and on barley infected with the barley yellow dwarf virus.

Inhibition of fruit growth is also a common symptom of disturbed cell plasmatic activity during organogenesis. For example, early infection by *Fusicladium dendriticum* causes apple fruit dwarfing, while infection by *F. pirinum* causes fruit dwarfing in pear. Cherry fruits on trees with little cherry disease are much smaller than those on unifected trees. Cucumber mosaic virus causes fruit dwarfing as well as sterility in pepper.

Plant organ enlargement results from hypertrophy and hyperplasia of cells. In some cases organ enlargement is slight and hardly discernable, while in other cases infected organs enlarge to such an extent that the growth may be many times that of uninfected tissues.

Organ enlargement is often accompanied by thickening of the tissue involved. For example, fungi belonging to the genus *Taphrina* cause thichening of infected leaves in several plant species; peach leaves infected with *T. deformans* are thicker than normal due to excessive cell proliferation in the spongy and palisade cell layers. Leaf thickening also occurs in plants of *Capsella bursa-pastoris* infected with *Albugo candida*. *Peronospora parasitica* causes stem thickening on the same plant, and stem thickening also occurs on plants of *Carthamus tinctorius* infected with *Puccinia carthami*, as well as on peppermint plants infected with *P. menthae*.[14] The perennial mycelium of *Gymnosporangium sabinae* stimulates the production of swellings on stems and branches of *Juniperus sabinae*.[9]

Abnormal elongation of plant organs is also an organoplasmatic phenomenon. For example, *Gibberella fujikuroi* (*Fusarium moniliforme*) causes internode elongation on rice stems.[14] Rice internodes become very elongated resulting in giantism.[15] Another example of internode elongation is caused by *Uromyces pisi* on plants of *Euphorbia cyparissias*.[14] Also, potato spindle tuber viroid causes abnormal elongation of potato tubers.[16]

Pathogen-stimulated enlargement may affect only parts of an organ as well as the entire plant organ. For example, lettuce big vein virus causes a thickening of veins, particularly the midribs, on leaves of of infected plants. In anemone plants infected with the anemone alloiophylly virus, leaves are broader and much thicker than uninfected ones.[16]

In summary, organoplasmatic changes are expressed by a variety of symptoms on many types of plant organs, and can be caused by a variety of pathogenic agents.

IV. MORPHOLOGICAL ORGAN CHANGES

Morphological organ changes pertain to the alterations in the shape of damaged or diseased plant organs. The shape of individual organs is genetically controlled and are recognized as a taxonomic trait. The eventual shape of a plant organ takes place during the processes of organogenesis, which is associated with morphogenesis. Any change in shape from the normal usually is the result of alterations occurring during organ formation.

Changes in the shape of an organ are easily recognized and can be used as a reliable symptom of damage or disease. Symptoms may occur either on parts of an organ or involve the entire plant. Changes in the shape of whole organs or an entire plant may be expected in the case of systemic infection, or those taking place early in ontogenesis. However, the extent to which individual organs will be altered will depend on the mechanisms involved. Many of the pathogens involved in altering the shape of plant organs influence the amount and composition of hormones and other growth substances in host cells.

Often the changes in the shape of plant organs are characteristic of the pathogen or pathogens involved and can be used to a certain extent for diagnostic purposes. It is impossible to describe and enumerate all the various types and possibilities of changes in the shape of diseases plant organs. They are discussed in more detail in the section on pathological changes of individual plant organs.

Morphological changes in plant organs are the most common changes that occur in disease plant organs. The causal agents involved belong to a variety of pathogen groups. Those diseases caused by mycoplasmas and viruses, many of which are graft transmitted, in the past were attributed to genetically controlled mechanisms, because of the continuity of symptom production.

V. ASYMMETRICAL ORGAN CHANGES

In general asymmetrical changes in diseased plants are a result of alterations in the symmetry of plant organs and structure. The overall structure of plants in an unobstructive environment is symmetrical with a more or less even distribution of the organic mass around an imaginary axis, which gives plants and their organs a symmtrical and proportioned appearance. This symmetry is used in taxonomic definitions of some plant species.

As a genetically-influenced structural principle, symmetry is an expression of the processes of organogenesis including cell division and differentiation, and tissue formation and differentiation, which results in a proportional linkage of tissues and organs. External as well as internal structural symmetry is linked to organ and whole plant organization. As a characteristic trait, symmetry can be altered simultaneously with changes in the plant shape. Morphological organ changes induced by numerous abiotic factors commonly cause changes in organ and plant symmetry.

Changes in organ symmetry may be more or less profound depending on the time of infection. The least changes occurs in organs that become infected after maturation. In such cases external organ symmetry may not be visibly altered. However, in other cases, the changes may result in the complete loss of structural symmetry with infected organs displaying an asymmetrical appearance.

Asymmetrical organs result when infection occurs early in organ development, which results in the death of cells or groups of cells during cell division. Organ development is altered or ceases at those points of infection where cell death takes place, resulting in organ malformation and loss of symmetry. Examples of this type of asymmetrical organ development are plum fruits infected with *Taphrina pruni* and the bending and twisting of pea fruits infected with *Ascochyta pisi*. Leaf epinasty is another example of an alteration in organ symmetry (see section on morphological organ changes).

Similar asymmetrical organ development occurs in plants infected with plant viruses and mycoplasmas. Pepper plants, for example, infected with the cucumber mosaic virus produce asymmetrical leaves and fruits; and the internodes of grapevine shoots infected with the grapevine fanleaf virus deviate from their normal symmetry resulting in a zigzag shape. Rice roots infected with the rice dwarf virus are slow growing and tend to grow horizontally along the soil surface. The apple flat limb mycoplasma causes, in addition to limb flattening, a twisting of infected stems and branches. Apple shoots and branches infected with the apple rubbery wood mycoplasma curve downward, outside of their normal symmetry.

Complete loss of symmetry is also chanracteristic of diseases resulting in abortive disorders of infected plants. In such disorders, parts of individual organs do not form, which results in malformed organs with a complete loss of symmetry. Examples of such malformed organs are the filiform leaves of tomato plants infected with the cucumber mosaic virus and lobed leaves of grapevines infected with the fanleaf virus.

Loss of symmetry also occurs in plants infected with the causal agents that result in tissue proliferation. These pathogens stimulate hyperplasia in which new pathological growths are formed altering the shape and symmetry of affected organs. Examples of these types of changes include the formation of asymmetrical roots in cabbage and other crucifers infected with *Plasmodiophora brassicae* and the formation of asymmetrical roots and stems of grapevines infected with *Agrobacterium tumefaciens.*

VI. ARRANGEMENT CHANGES

Changes in the arrangement of plant organs pertain to the alteration in the distribution or organs within the plant structure. Normally plant organs are arranged in such a pattern as to enable the best possible performance of their function in the plant. Leaves of healthy plants, for example, are distributed in a regular manner along stems and shoots to reduce mutual shading and ensure maximum utilixation of solar energy; and root systems are arranged in soil to absorb sufficient nutrients and water required by the plant The arrangement of organs is genetically controlled and therefore is of use in taxonomic studies.

Certain pathogens influence the arrangement and distribution of plant organs in several ways. One means is to prevent organ formation or development early in their formation. *Claviceps purpurea*, for example, prevents rye grain formation resulting in an irregular arrangement of grains on the spike. The almond bud failure disease causes both death of almond buds on symptomatic stems and prevents bud development along asymptomatic shoots.[16] Morvan[17] described a pear virus disease which causes bud drop in infected stems. Some buds dropped at the beginning of winter; those that remain develop into thin shoots in the spring, which desiccated, or gave rise to shoots in the form of a rosette.

Changes in the arrrangement of leaves in the plant canopy occur with the early defoliation of infected plants. *Polystigma rubrum* and *Puccinia prunispinosae*, for example, cause premature drop of foliage on severely infected plum trees. *Marsonia juglandis* also causes a premature defoliation on walnut trees resulting in a severe altering of canopy during the vegetative growth period.[10]

When rosetting occurs due to internode shortening, such as that caused by the peach rosette mosaic and grapevine fanleaf viruses, the regularity of leaf and shoot distribution is completely lost.

The occurrence of witches' brooms on cherry trees infected with *Taphrina cerasi*, or pepper plants infected with the cucumber mosaic virus, or apple plants infected with apple proliferation mycoplasma, is accompanied by an abnormal or asymmetrical arrangement of the abnormal shoots.

An asymmetrical arrangement in flowers and flower parts may result from infection by plant pathogens. For example, *Peronospora radii* causes an inverse arrangement of flowers on parts infected inflorescence of *Matricaria inodora.*[14] Also, the petals of zinnia flowers infected with the cucumber mosaic virus are irregularly distributed along the flower head.[16]

Various root pathogens alter the root systems of infected plants by either reducing or increasing the number of rootlets or root hairs. For example, *Agrobacterium tumefaciencs* causes neoplastic excrescences on roots of a number of plants which results in a significant reduction in the number of rootlets. In contrast, a large number of irregularly distributed rootlets are formed on sugar beet roots of plants infected with the beet necrotic yellow vein virus, which results in a condition called rhizomania or hairy root.

VII. CORRELATIVE ORGAN CHANGES

Correlative organ changes refers to disorders that occur in the interdependence within plant organs and their function. According to Küster[18] this interdependence may be mutal or unilateral and is manifested by changes in one part of an organ or its function with resulting in a simultaneous change in other parts or function of a plant organ. This interdependence of form and function of a plant organ is the principle on which all plant parts depend. Such a correlation is present during cytogenesis, the interdependence between the nucleus and cytoplasm, for example; and histogenesis and organogenesis, examples of interdependence among contiguous and remote cells.

The regulators of interdependence in plant organs are plant growth hormones which provide for the corrdination and linkages in morphogenesis and during metabolic processes in the course of ontogenesis and individual plant development, and whose activity is under genetic control.[19]

In describing the mechanism of plant growth, Kozlowski[6] stressed the importance of the following hormones: auxins and gibberellins, which stimulate cell distribution; cytokinins, which enhance cell division; and ethylene, which induces root and stem growth and fruit maturation; as well as inhibitory hormones, such as abscisic acid, which decreases cell size.

It has been shown that pathogens can alter the quantity and activity of all cell hormones. Such changes result in the disturbance or loss of a given interdependence in organism development. For example, interruption in auxin activity in apical shoots where lateral bud formation takes place, results in uncontrolled shoot branching and formation, giving rise to a witches' broom-type growth.

Pathological changes in hormone quantity and functions have a pronounced influence on the development of plants and their organs giving rise to either enlarged or stunted form (see section on cytohormal changes).

Defoliation caused by leaf pathogens and insects as well as stem girdling caused by canker- or lesion-causing agents reduce basipetal hormone transport resulting in decreased root growth.[6]

In the absence of the influence of interdependence factors, irregular tissues may be formed. For example, inverse vascular bundles are formed where the phloem is internal and the xylem is external in tumors caused by *Agrobacterium tumefaciens* (Figure 66) and clubs caused by *Plasmoiophora brassicae* (see section on pathological changes of vascular tissues).

The loss or disturbance of interdependence due to tissue damage may be reestablished when the lost parts are formed again in a natural manner. The establishment of such interdependence is characteristic of the regenerative processes associated with traumatic wounds.[18] By analogy, these processes may be reestablished in the healing of wounds caused by necrobiontic pathogens such as *Nectria galligena* on apple branches and stems.

Data reported by Wilson[20] indicated that morphological changes in plant organs (teratomas) could be classified into three groups, those under the control of the host plant, the pathogen, or outside influences.

A limited number of morphological organ changes characterized by intense organ growth are under the control of the host plant, and include the symptoms of fasciation, rasp leaf, phyllody, and hairy root. Causal agents of these types of changes include certain insects or pathogens.

In contrast, galls caused by certain nematodes, nodules induced by rhizobacteria, and galls induced by certain insects are under the control of the causal agent. The occurrence of these abnormal growths are, in part, determined by the needs of their causal agents.

According to Wilson,[20] outside control of these changes is due to the introduction of a genome from the pathogen into the host plant genome. Such genomes occur in plants infected with

Agrobacterium tumefaciens or the clover wound tumor virus. *A. tumefaciens* transfers the bacterial plasmid DNA into the plant cell resulting in an increase of that DNA in the tumor cell.[21] The presence of the bacterial plasmid results in the loss of normal cell division in the tumor cells. Under the influence of the new genome, tumor excrescences are developed automonously and their growth in infected organ tissue is unlimited.

In summary, interdependent disorders in the development of plant organs are complex and influenced primarily by the host, the pathogen or both resulting in the formation of new, pathological growth in infected plants.

VIII. TRANSFORMATIVE ORGAN CHANGES

Some plant organs are characterized by the ability to be transformed into other organs and such an occurrence is defined as an organ transformation. The ability for organ transformation is a feature of plant life and is characteristic for plant cells and tissues, as well as organs. Somatic cells, for example, may be transformed into meristematic cells (meristemizaion), into stone cells (sclerenchymization), or other tissues (redifferentiation) (see section on histotransformative tissue).

The transformation of organs into other organs represents a complex alteration in the plant. Glišić[19] referred to the transformation of vegetative tissues into floral buds as a process in which a flowering hormone, florigen, plays an important role. This hormone alters the the genetic activity in the vegetation buds on which the transformation into flower buds is based.

Transformative organ changes also may occur in plants infected with various plant pathogens, such as the greening of flower parts on infected plants. In such cases diseased flower parts turn green, and significantly alter their structure and formation. Causal agents of such changes are mycoplasmas causing strawberry petal and tomato anther and pistil greening (see section on cytochromatic changes).

Phyllody represents a characteristic transformative organ change characterized by the transformation of sexual organs into leaves.[22] Phyllody is caused by several fungi, including those that cause downy mildew, rust, and loose smut.[14]

On millet plants infected with the downy mildew fungus, *Sclerospora graminicola*, grains and sometimes anthers are transformed into short leaflike shoots. On crucifer plants infected with *Albugo candida*, the anthers are transformed into organs resembling leaves or carpels. Leaflike formations also replace reproductive organs in inflorescences of maize plants infected with *Ustilago maydis* or *Sphacelotheca reiliana*.

Phyllody also has been described on *Carthamus tinctorius*[23] and sesame plants caused by mycoplasmas.

The cause of phyllody is not fully understood. Bos[25] assumed that the repression of vegetative characters occurred in shoots resulting in their transformation into flowers. However, if this repression caused by a pathogen is controlled, vegetative tendencies resume their influence and the leaves are formed once again. It is presumed that hormones direct this entire process.[22]

IX. TRANSVESTIC ORGAN CHANGES — TRANSVESTISM

Transvestic organ changes are characterized by the development of flower parts of one sex into the other in diclinous flowers of some plants. Mathre[22] defined such a change as transvestism.

Transvestism is characteristic of flowers. Since flowers are shortened shoots displaying limited growth and are derived from leaves metamorphosed or transformed in to floral parts, they represent an entity in which transvestic changes may be expected. Transvestism can be caused by flower pathogens which presumably disturb the balance of hormonal substances responsible for the normal, controlled development of flower parts. Transvestic processes may stimulate the development of anthers in female flowers and of rudimentary ovaries in male flowers. Examples of causal agents of transvetic changes in flowers are found in the order Ustilaginales and include the following genera; *Sorosporium*, *Tilletia*, and *Ustilago* (see section on pathological flower changes - transvestism).

X. CHROMATIC ORGAN CHANGES

Coloration is a general characteristic of all plant organs, which is important in the preservation of plant life. Changes resulting in decoloration or discoloration in plant organs is defined as a chromatic organ change.

The color of a plant organ depends in part on structure, function, and location on the plant. It is dependent on the coloration of the cells and tissues taking part in the organ structure.

Carriers of various pigments are found both in living and dead plant cells. Living cells carry pigments in plastids, such as chloroplasts and chromoplasts, and cell sap carries anthocyanins and flavonoids. Dead peridermal cells, for example, carry various pigments in the bark peridermal tissues of roots and stems. Coloration of these cells and tissues is a result of their chemical composition as well as the reflection of light from air layers within their structure.

Decoloration of plant pigments may take place in and around points of infection in plants as a consequence of induced biochemical processes. For example, dark pigments may be formed by the binding of oxidized products of phenolic compounds with cell proteins.

From a phytopathological point of view, the varying presence and relationship among individual pigments in the structure of plant organs should be mentioned. Green pigment is the most wide-spread color in plants and is characteristic of most organs with surfaces exposed to light. It is the color of most living herbaceous organs, such as green flower parts, fruits, leaves and stems. The second most widely occurring group of pigments are the carotenoids, antocyanins and flavonoids, which appear in flower parts, ripening fruits, and dying leaves.

Color changes are the most frequent and wide-spread changes in plant life and are used in applied and theoretical studies in the plant sciences. Color changes in infected plants are used to represent disease symptoms, which are used to describe and name plant diseases.

Changes in green organ coloration, either decoloration or greening, have been described in many plants infected with one or more pathogenic organisms. Decoloration of green organs may be local, or partially or wholly systemic. Local decoloration is caused by a variety of fungi, such as *Phytophthora infestans, Plasmopara viticola,* and *Puccinia* spp., and bacteria such as *Pseudomonas syringae* pv. *tabaci, Xanthomonas campestris* pv. *malvacearum* and *X. campestris* pv. *vesicatoria.* Partial systemic decoloration is caused by, for example, *Peronospora arborescens* on poppy plants and by *P. tabacina* on tobacco plants. Those agents that cause a systemic decoloration include those that cause systemic infection or the pathogenesis involves the action of toxins remote from the point of infection, such as *Fusarium* spp., *P. syringae* pv. *syringae* and *Vericillium* spp.

As causal agents of systemic infection, plant viruses cause systemic discoloration in the form of mosaics, mosaic spots and mottling, as well as general chlorosis and yellowing. Causal agents of mosaic-type diseases include, for example, alfalfa mosaic, bean common mosaic, beet mosaic and tobacco mosaic viruses. Ringspots are caused by several viruses, for example, Prunus necrotic ringspot, red clover clorotic ringspot, and red clover chlorotic ringspot, and tomato ringspot viruses. A general decoloration is caused by the following viruses, beet yellows, potato leaf roll, and rice yellows viruses.

In contrast, pathological greening of flower parts occurs in some plants and has been described in several plants infected with mycoplasmas such as the raspberry stunt and strawberry green petal mycoplasmas. Degradation of green pigments and disruptions in their synthesis results in the appearance of other pigments, such as the anthocyanins and carotenoid pigments, a condition described as characteristic of the following diseases; almond red leaf spot, caused by *Polystigma ochraceum*, barley yellow rust caused by *Puccinia glumarum*, plum orange leaf spot caused by *Polystigma rubrum*, and wheat yellow rust caused by *P. glumarum*. On peach leaves infected with *Taphrina deformans*, a redish anthocyanin coloration is conspicuous.

A localized decoloration resulting in white areas is characteristic of some diseases caused by bacteria, such as halo blight of oats caused by *Pseudomonas syringae* pv. *coronafaciens* and other diseases caused by *Xanthomonas campestria* pv. *gardneri*. White decoloration also occurs in various plant organs of plants infected with one of the following viruses; alfalfa mosaic, which causes white leaf spot and white stripe on pepper; and white streak on narcissus, and white leaf on rice (see section on cytochromatic and histochromatic changes).

XI. PATHOLOGICAL NEW GROWTHS

Pathological excrescences or neoformations occur on a variety of plant organs differing in shape and structure. The new growths are caused by a variety of plant pathogens, including nematodes, and other entities, such as insects. Histoplasmic and hypertrophic changes, alone or in combination, take part in the formation of these new growths.

Depending on the mode of formation and development, new growths differ in their complexity and can be classified into three main groups. Some neoplastic growths may overlap any two of them.

The first group includes those of determinant neoplastic growth which usually occur and spread in and around points of damage or infection, i.e. new growth around traumatic wounds. Representatives of this type of pathological new growth are enations, galls, nodules, and warts. Examples of causal agents are; for enations, tobacco mosaic virus; for galls, *Agrobacterium tumefaciens;* for nodules, *Bradyrhizobium* the nitrogen-fixing rhizobacterium; and for warts, *Pseudomonas syringae* pv. *savastanoi* on olive trees and *P. syringae* pv. *syringae* on cherry branches. Most of these represent more or less localized infections. Some determinant warts are caused by some diseases, such as those on peach fruits caused by the peach wart disease. Other members of this group cause enations on lower leaf surfaces in some plants, including the cherry rasp leaf, grapevine enation, and pea enation mosaic viruses.

The second group is made up of neoplastic growths of an indeterminate nature including cankers and tumors. Such growths are caused by pathogens spreading from cell to cell in infected tissues which are constantly enlarging during the course of disease spread. This group includes pathogens that cause malformed excrescences such as *Plasmodiophora brassicae* on crucifers, *A. tumefaciens* on woody plants, and clover wound tumor virus on *Melilotus* sp. and other plants.

Representative of the third group caused the development of new organs as neoplastic formations. An example of this type of new growth is witches' broom which is caused by *Taphrina cerasi* on cherry trees, the apple proliferation mycoplasma on apple trees, and proliferation on *Convolvulus arvensis* caused by a mycoplasma. Beet necrotic yellow vein virus causes rhizomania on beets resulting in a proliferation of new root hairs. Small tuber formation is caused on other potato tubers on plants infected with *Phytophthora infestans*[26] (see sections on cytoplasmic, cytotrophic, histoplasmic, and histotrophic changes).

XII. DISEASED ORGAN DROP

Organ drop is a frequent sign of plant disease. Some authors include diseased organ drop among the most frequent symptoms of disease along with chlorosis, epinasty, necrosis, and wilting.

Organs are susceptible to organ dropping during the entire course of organogenesis, although dropping is most prounounced if infection takes place early in developmental stages. Organ drop occurs as a consequence of a series of interactive reactions in infected plants and may be better defined in terms of elimination or removal of plant organs.

During natural organ drop, such as leaves on deciduous plants, secondary tissues are formed for organ separation from the plant. Similar types of tissues are formed in a similar manner for the elimination of damaged or diseased organs. Plants do not distinguish between abiotic and biotic factors as causal agents and therefore react to either in a similar manner.

Diseased organ drop is not an accidental phenomenon, but represents a directed plant activity accomplished by certain mechanisms. It is of fundamental biological importance, with the elimination of diseased organs the plant sacrifices certain of its components in order to preserve its enterity.

There are little or no reliable data as to the exact processed involved in leaf or organ fall. Some authors believe it is caused by disturbances in the balance of hormonal substances in diseased plants.[2] Other believe that it is due to enzymatic activity relating to the degradation of cell wall constituents which takes place in the layers of the separating tissues.[27] Zimmerman and McDonough[28] observed that some plant organs die or drop off due to disruptions in the nutrient supply in phloem tissue. Kiraly[29] believes defoliation occurs as a hypersensitive reaction of the plant tissue to the pathogen.

For example, certain pepper cultivars hypersensitive to tobacco mosaic virus, react to inoculation with necrotic spots on the leaves, which fall off, so that the plant remains uninfected. Similarly, certain bean cultivars inoculated with common bean mosaic react in a similar manner, thus insuring protection from the virus. Plant resistance achieved by the elimination of infected organs, such as leaves, has been defined as an “amputative resistance.”[29]

Campbell et al.[30] distinguished two strategies by which plants protect themselves from their enemies, one by keeping invaders away from the attack areas, and the second, to provide protective barriers should an invader reach the plant surface. A third strategy should be included and that is the elimination or removal of diseased parts with the aim of hindering or preventing the spread of invaders.

Organ drop may be caused by abiotic or biotic factors. Biotic factors include various pests, such as insects, and numerous plant pathogens (see sections on changes in individual organs).

Organ drop does not necessarily occur within a growing season or less. Sometimes infected organs remain on the plant from one season to the next or even longer. In such cases, organ drop is probably not under the control of the plant, but that of the pathogens involved. For example, dead and dried leaves infected with *Gnomonia Crythrostoma* do not fall off, but remain in the plant crown until the following spring.[10] According to Wood,[14] this fungus penetrates the leaf petiole thus preventing the formation of the separation layer at the leaf scar and dropping of the leaf until spring.

Many dried or mummified stone fruits infected with *Monilia* spp. remain hanging on tree branches for more than one season or even through several seasons. Although a complete explanation for this phenomenon is not known, it would appear that the fungus prevents the processes of organ drop, thus providing a means of long survival.

XIII. PATHOLOGICAL CHANGES IN INDIVIDUAL PLANT ORGANS

Disease symptoms are a result of various pathological changes taking place in cells, tissues, and organs. These numerous, various and changeable symptoms are a visible sign of the occurrence of disease in plant organs. Some of the symptoms are diagnostic when pertaining to specific cases; while others are nonspecific, occurring in a number of cases and therefore diagnostically unreliable. Although, due to their diversity, it is impossible to analyze all the possible symptoms, it is necessary to mention those characteristic symptoms occurring in individual plant organs.

A. PATHOLOGICAL CHANGES IN ROOTS

1. Form

Changes in the form of roots usually appear as plasmatic and trophic responses of infected cells and tissues. They are caused by a variety of pathogenic agents as well as other microorganisms and animals.

For example, *Plasmodiophora brassicae* causes tumors or clubs on roots of several crucifers and *Agrobacterium tumefaciens* causes galls on roots and crowns of fruit trees, grapevines, and other woody, plants. Root and other organ malformations are also caused by the cacao swollen shoot and clover wound tumor viruses.

Changes in the normal shape of roots are caused on legumes by *Bradyrhizobium* which stimulates the formation of nodular excrescences. In some cases virus infections in legumes interfer with nodule formation.[31] For example, nodule formation was reduced 80% in soybean roots of plants infected with either the soybean mosaic or bean pea mottle viruses, or both.[32] Root knots or galls caused by nematodes are induced by the hormone auxin (Figure 67).[33]

2. Necrosis

Necrotic processes have been recorded in roots of some plants but occur primarily on root hairs (Figure 71). For example, necrosis occurs on maize roots infected with the maize rough dwarf virus on tobacco roots of plants infected with the potato streak mosaic virus.[34] The tobacco necrosis virus causes local necrotic reaction on roots of tobacco.[31] The grapevine fanleaf virus causes root hair necrosis in grapes particularly at the apical regions. Necrosis also occurs on roots and root hairs of sugar beet plants infected with the necrotic yellow vein virus[16] as well as on pepper roots infected with the stolbur mycoplasma. In the case of stolbur disease, necrotic processes begin on the apical portions of the root hairs, then spreads inward, eventually affecting the entire root system.[35]

3. Rot

Root rot on vegetable and other herbaceous plants can be caused by a number of pathogenic fungi including *Didymella lycopersici, Phytophthora infestans* (Figure 68), *Pythium debaryanum,* and *Rhizoctonia solani* (Figure 69).[35] These pathogens also cause damping-off of young host seedlings. Root rot on some plants is caused by soilborne vascular pathogens which penetrate the apical portions of young roots or root hairs, or through mechanical injuries and wounds. After colonization these pathogens enter the xylem and later parenchyma tissues causing death of cells, tissues and organs.[37] Included in this group of pathogens are *Fusarium oxysporum* f. sp. *cubense,* cause of Panama disease of banana, and *F. oxysporum* f. sp. *vasinfecturm,* cause of cotton wilt.

Some soilborne fungi cause decay of root hairs in fruit trees and other woody plants. For example, *Rosellinia necatrix* causes root hair rot on apricot, cherry, grapevine, peach, and plum trees. *Armillaria mellea* causes severe losses due to rot of root hairs on root of several woody plants including grapevine.[10]

4. Number

Pathological changes are characterized by either a reduction or an increase in the number of root hairs on a root system. Reduced numbers of root hairs and their eventual death is caused by many of the soilborne fungi discussed previously. A significant number of root hairs are lost on roots of plants infected with *Agrobacterium tumefaciens.* Adventitious roots of maize plants infected with the maize rough dwarf virus are shorter than normal and have fewer numbers of secondary root hairs. Grapevine roots infected with grapevine fanleaf virus are less developed;

root hairs of grapevine cuttings from diseased grapevines are less numerous and elongated than those from uninfected vines. Barley yellow dwarf virus caused a reduction in roots number on barley plants.[16]

In contrast, the number of roots and roots hairs can be above normal on some diseased plants. For example, adventitious roots formation on tobacco plants infected with *Pseudomonas solanacearum* is probably due to hyperauxiny in the diseased plants.[15] Hairy root occurs in sugar beets infected with the beet necrotic yellow vein virus which is due to the increased number of root hairs.

5. Color

Root color changes are not common since there is an absence of a diversity in root coloration. However, some diseases result in chromatic changes in roots. For example, the vascular pathogens in the genera *Fusarium* and *Verticillium* cause a darkening of infected roots and necrotic xylem. Several of the causal agents of root and root hair rot on herbaceous plants cause a general darkening of infected plant parts (see Chapter 2). Potato streak virus causes darkening of root tissues in infected tobacco plants.[36] Dark necrotic areas develop on sugar beet root hairs caused by beet necrotic yellow vein virus.

XIV. PATHOLOGICAL CHANGES IN STEMS

A. FORM

Some characteristic disorders occur in the stems of diseased plants but are not as diverse as are the changes in diseased leaves, flowers and fruits (Figure 70).

Localized stem malformations are manifested as random swellings or thickenings and other types of pathological excrescences. Some examples are the branch and stem swellings caused by *Gymnosporangium* spp. on *Juniperus* spp. Also, such symptoms are characteristic for cacao plants infected with the cacao swollen shoot virus. Excresences on maize stems can be caused by *Ustilago maydis*. Other examples are those formed on branches and stems of apple caused by *Nectria galligena*, on olive caused by *Pseudomonas syringae* ssp. *savastanoi*, and on grapevines caused by *Agrobacterium tumefaciens*.

More general or other than localized stem malformtions are characterized as abnormal node development, general stem thickening or flattening, and loss of branch turgor, such as on apple and pear infected by the rubbery wood pathogen.

Internode and stem elongation is symptomatic on rice plants infected with *Gibberella fujikuroi* and on *Euphorbia cyparissias* infected with *Uromyses pisi*.

Internode and stem shortening is also a frequent symptom on diseased plants. For example, this symptom occurs on sugar beets infected with *Peronospora schachtii*, on peppers infected with *Verticillium albo-atrum*, on peach plants infected with the peach rosette mosaic virus, on barley plants infected with the barley yellow dwarf virus, and on oats infected with the oat sterile dwarf virus (Figure 72).

Stem thickening is characteristic of plants of *Capsella bursa-pastoris* infected with *Peronospora parasitica* and on peppermint plants infected with *Puccinia menthae*.

Stem flattening, also described as stem fasciation, occurs both in herbaceous and woody plants. Various forms of stem flattening have been observed in apple branches and stem infected with the apple flat limb mycoplasma. Flat trunk in grapevine is another example, but its causal agent is not known.

Rhodococcus fascians causes numerous thickened shoots on *Lathyrus odoratus*.[14] *Convolvulus arvensis* plants infected with the stolbur mycoplasma produce numerous shoots giving it a bushy appearance.[35] Proliferation of lateral shoots is common on cherry plants infected with *Taphrina cerasi*.

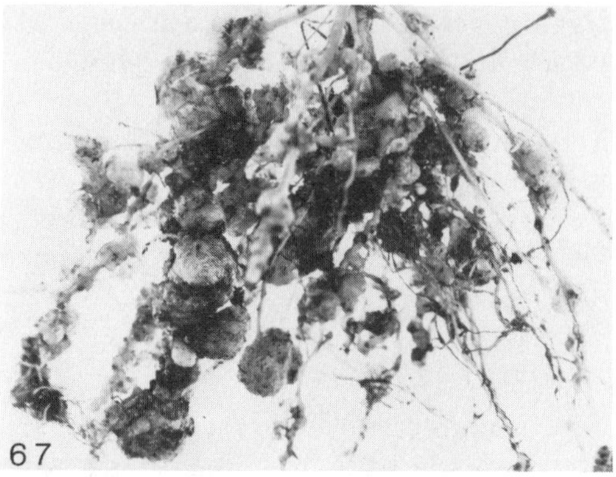

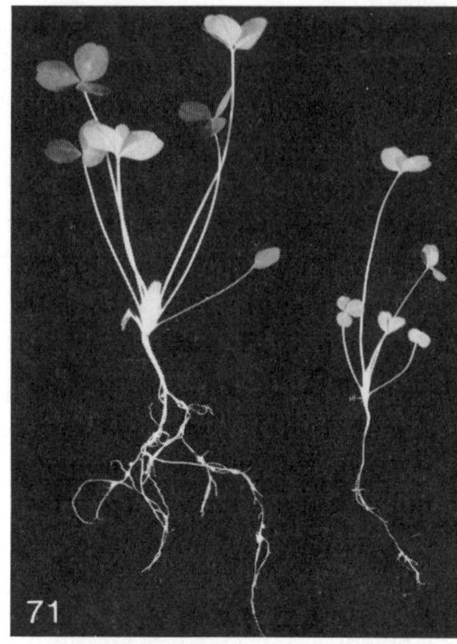

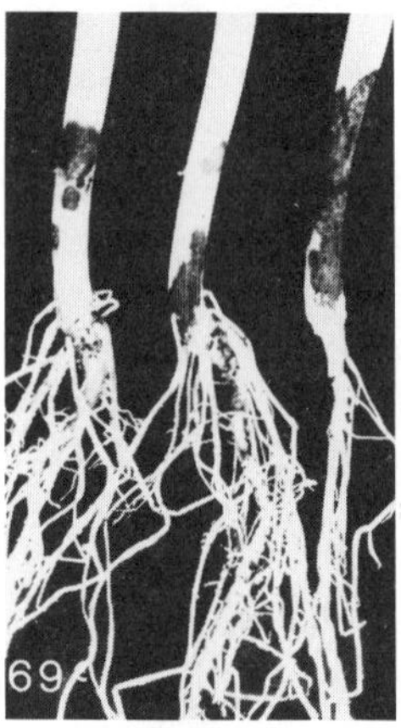

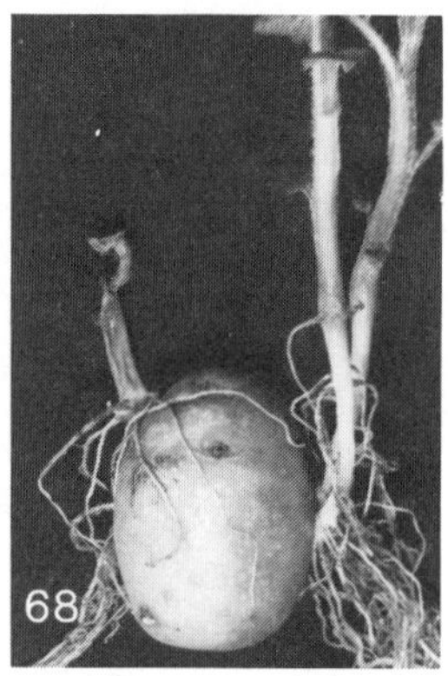

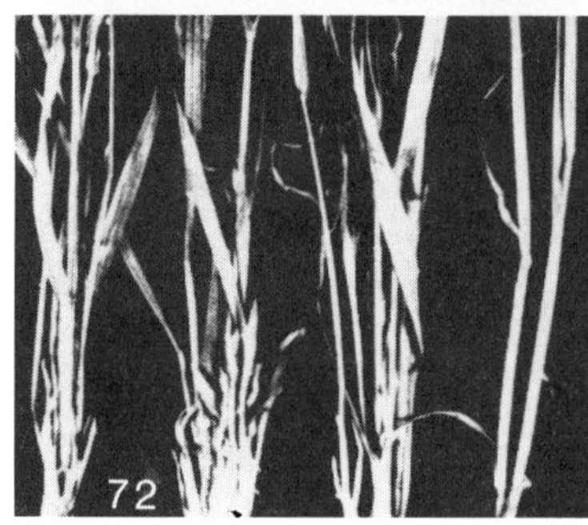

PLATE XIV. FIGURE 67. Development of knots on soybean roots caused by the root-knot nematode, *Meloidogyne* sp. (From Sinclair, J.B. and Backman, P.A., APS Press, Inc., St. Paul, MN. With permission.) FIGURE 68. Sprouts on a potato tuber showing reduced number of roots of one infected with *Phytophthora infestans* (left) compared with uninfected one (right). FIGURE 69. Young plants of common bean showing lesions on the roots, crowns and hypocotyl caused by *Rhizoctonia solani*. FIGURE 70. A portion of grapevines with (left) and without leaves (right) showing shortened internodes caused by infection with the grapevine fanleaf virus. FIGURE 71. Red clover seedlings from a seed infected with the bean yellow mosaic virus (right) and a uninfected seed (left). FIGURE 72. Three oat plants infected with the oat sterile dwarf virus showing stunting and tillering (left) compared with one uninfected plant (right). (From Lindsten K., *Annales Royal Agricultural College,* Sweden 27, 137, 1961.)

B. NECROSIS

All living cells of herbaceous plants are susceptible to necrosis, either localized or systemic (Figures 73,74). Local necrosis, for example, is caused by leafspotting agents such as *Phoma macdonaldi* on sunflower, *Didymella applanata* on raspberry, *Ascochyta pisi* on pea plants, and *Colletotrichum lindemuthianum* on bean plants as well as stem pathogens such as *Phytophthora megasperma* f. sp. *glycinea* on soybeans (Figure 73) and *Phomoposis viticola* on grape (Figure 74). Local necrosis, for example, is caused by some phytopathogenic bacteria, such as *Pseudomonas syringae* pv. *tomato* on tomato, *Xanthomonas campestris* pv. *malvacearum* on cotton, and *X. campestris* pv. *vesicatoria* on tomato.

Systemic stem necrosis can result from the death of phloem tissues in potato plants infected with the potato leaf roll virus. Potato viruses X and Y cause necrotic, longitudinal spots or streaks along stems of some potato cultivars.

The necrotic streak strain of the tobacco mosaic virus causes stem necrosis on the pepper cultivar Niška Šipka. Dark to black spots occur along stems of tobacco plants infected with the tobacco rattle virus with the necrotic tissue spreading into the center of the stem. Necrosis also develops in the apical portions of rose bushes infected with the rose wilt disease, which spreads along shoots resulting in their death and the eventual wilting of the entire plant.

C. RESINOSIS

Resinosis is associated with different abiotic and biotic factors. Abiotic factors include a wide range of various mechanical injuries and wounds as well as nutrient deficiencies. Resinosis is also associated with a large number of pathogenic bacteria, fungi and viruses. For example, *Pseudomonas syringae* pv. *morsprunorum* causes resinosis on cherry. *Clasterosporium carpophilum* causes resinosis on twigs of several types of fruit trees. The citrus psorosis disease complex causes resinosus on the bark of various citrus cultivars. Other fungi that cause the formation of resins on forest and fruit trees are some of the Basidiomycetous fungi such as *Armillaria mellea* on citrus, *Prunus* spp., walnut, and some forest trees; *Fomes pomaceus* on cherry and plum trees, and *Stereum purpureum* on susceptible woody plants.

D. LODGING

Many fungi through their enzyme and toxin production degrade the lower stems of herbaceous as well as woody plant seedlings resulting in lodging or damping-off. Some of these pathogens also cause root rots (see section on root rots). Lodging of mature stems commonly occurs on several cereal plant species. Wheat lodging, for example, is caused by *Cercosporella herpotrichoides*. Inadequate nutrition as well as infection with other pathogens, such as members of *Puccinia* spp., can predispose plants to lodging by other fungi.

E. BREAKAGE

Stem breakage can be caused by a variety of abiotic and biotic factors, alone or in combination. Stems weakened by a pathogen can predispose plants to breakage due to high winds and heavy precipitation (Figure 76). Stem breakage in herbaceous plant can result from stem girdling by a pathogen, such as in the case of sunflowers infected with *Phomopsis helianthi*, and breakage on potato plants due to infection by *Phytophthora infestans*. *Xanthomonas campestris* pv. *malvacearum* causes spots on the stems of cotton plants making them more susceptible to breakage at the infection points.[10]

Fungi in the order Hymenomycetales include a large number of wood rotting pathogens. Among them, *Polyporus sulfureus,* which degrades woody tissues of apple, cherry, pear, and walnut; and *Fomes fulvus* which degrades woody tissues on stone fruit plants. Wood degradation by these fungi plays an important part in reducing the mechanical strength of branches and trunks of trees making them more susceptible to breakage due to high winds or the weight of heavy snows.

F. COLOR

Changes in color take place in many plant organs besides leaves, including stems of herbaceous plants during the process of suberization. Various pathogens also can cause discoloration on young herbaceous stems, such as *Pleospora papaveracea*, which causes black spot on infected poppy stems. Dark brown colors develop on pea stems infected with *Ascochyta pisi* and on bean stems infected with *Colletotrichum lindemuthianum*. *Didymella applanata* causes violet spots on shoots and stems of raspberry bushes. Dark brown spots occur along stems and particularly at the base of young pear trees infected with *Sphaeropsis malorum*. Raspberry plants infected with the black raspberry streak virus develop irregular bluish to violet elongated spots on shoots and stems.[38]

XV. PATHOLOGICAL CHANGES IN LEAVES

A. FORM

Changes in the shape of leaves due to infection by a variety of pathogens are some of the most described and early recognized of plant disease symptoms. A typical symptom is an altered leaf size (Figures 75,77,78). Leaves infected during their formation usually are smaller than normal. Underdeveloped and malformed leaves, for example, are conspicuous on grapevines infected with *Plasmopora viticola* and *Uncinula necator*. The development of leaf spots and anthracnose lesions on leaves in their early development can result in suppressed leaf growth such as on sugar beets infected with *Cercospora beticola*, on alfalfa plants infected with *Pseudopeziza medicaginis*, and on pea plants infected with *Ascochyta pisi*. Early bacterial infection by *Pseudomonas syringae* pv. *glycinea* on soybean plants and by *P. syringae* pv. *lachrymans* on cucumber plants results in similar symptom development. Leaves of various plants infected with many of the mosaic viruses usually are thinner than normal in the yellow areas due to the decrease in mesophyll cell development and the reduction or absence of intercellular spaces.[31]

In contrast, some pathogens cause leaf enlargement or thickening as on plants of *Capsella bursa-pastoris* infected with *Cystopus candidus* and peach leaves infected with *Taphrina deformans*. The anemone alloiophylly virus causes the enlargement of entire leaves in anemone.[16]

Another group of pathogens cause a change in the shape of leaves, such as leaf crumpling, curling, and wrinkling. For example, *Taphrina deformans* causes leaf curling on peaches, and *T. bullata* leaf curling on pears. The strawberry leaf crinkle virus causes leaf curling on strawberry plants. In addition to other symptoms, the bean common mosaic virus causes leaf curling and wrinkling on bean leaves.

Another common alteration in the form of leaves caused by pathogens is leaf rolling. The potato leaf roll virus causes this symptom on potatoes and the cherry leaf roll and grapevine leaf roll disease causes similar symptoms on their respective hosts.

There are other changes in leaf form caused by plant pathogens. In cherry trees the cherry twisted leaf virus causes a downward cupping of leaves along the midribs. Epinasty is a symptom of a similar nature except that the downward leaf curling occurs along petioles of infected leaves. In the apple cultivar Spy 227, epinasty is caused by the apple spy epinasty and decline disease. The Prunus ringspot virus causes epinasty as well as leaf roll on *Prunus* spp.[39] Epinasty also occurs on tomato plants infected with species of *Fusarium* and *Verticillium*.

B. NUMBER

The number of leaves per plant, which is important in view of the role of the canopy in photosynthesis, may be reduced by abiotic and biotic factors. Abiotic factors include late spring

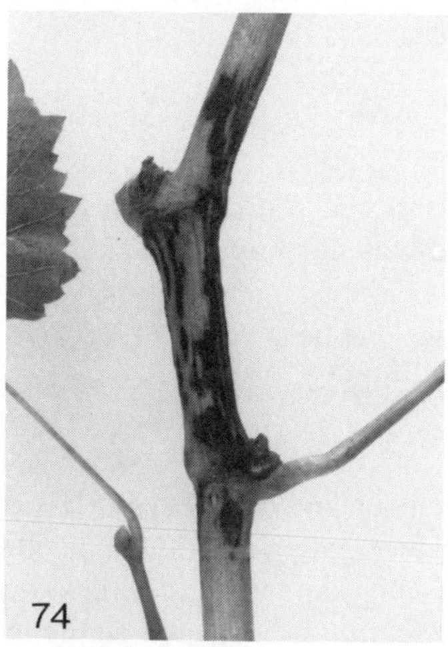

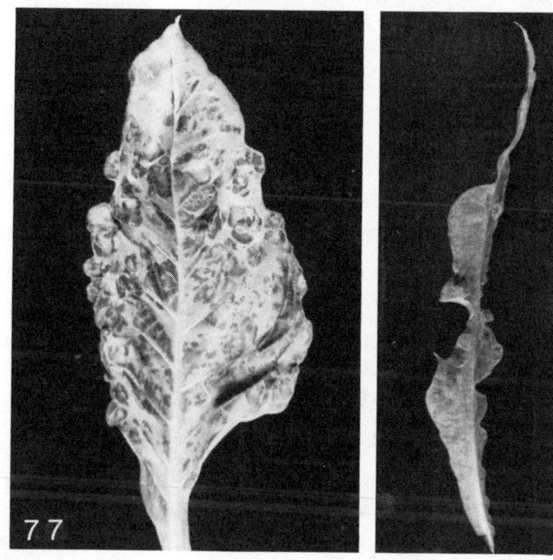

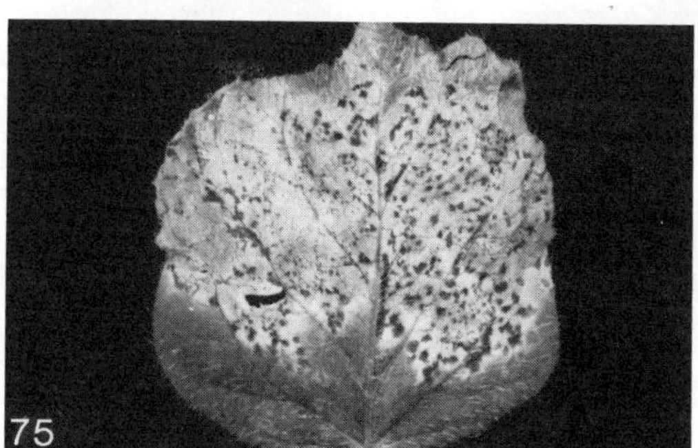

PLATE XV. FIGURE 73. Soybean plant with wilt symptoms showing lower stem decay caused by *Phytophthora megasperma* f. sp. *glycinea*. (From Sinclair, J. B. and Backman, P. A., APS Press, Inc., St. Paul, MN, 1989. With permission.) FIGURE 74. Dark, necrotic spots on the internodes of a grapevine caused by *Phomopsis viticola*. (From Mihajlovic , D., Doktorska disetacija, Poljoprivredni fakultat, Beograd (Yugoslavia), 1982.) FIGURE 75. Deformed soybean leaflet with local spots caused by *Pseudomonas syringae* pv. *glycinea*. FIGURE 76. Lodging of maize plants due to infection by *Gibberella zeae*. FIGURE 77. Tobacco leaves (cv. Virginia) showing symptoms caused by the cucumber mosaic virus. FIGURE 78. Peach leaves showing symptoms caused by inoculation with the plum pox virus compared with a uninfected leaf (left). (From Festic, H., Magistarski rad, Poljoprivredni fakultat, Beograd (Yugoslavia), 1965.)

frosts or the misapplication of a pesticide, both of which may result in complete destruction of buds and leaves.

Leaves also rank high among plant organs for having a favorable environment for the development of a large number of pathogens. Some of these are exclusively leaf pathogens, such as *Polystigma rubrum* and *Puccinia pruni-sponosae* on plum leaves. A frequent consequence of such infection is leaf fall, resulting in a significant reduction in the number of leaves per plant. In some cases, complete defoliation occurs, such a during epidemics caused by these two fungi on plum and *Taphrina deformans* on peach. Other potentially dangerous fungal pathogens that could cause a massive leaf fall are: *Coccomyces hiemalis* on cherry trees, *Diplocarpon rosae* on rose plants, *Plasmopara viticola* on grapevines, and *Pseudopeziza medicaginis* on alfalfa plants.

Another group of pathogens that cause leaf fall, depending on the host susceptibility and environmental conditions include the fungi; *Gnomonia juglandis* on walnut, *Septoria lycopersici* on tomato, and *Stigmatea mespili* on quince; and among the bacterial pathogens, *Xanthomonas campestris* pv. *juglandis* on walnut and *X. campestris* pv. *malvacearum* on cotton.

The reduction in the number of leaves on diseased plants represents the most detrimental of pathological occurences. Some pathogens represent a permanent threat to plants by their potential to cause complete or massive leaf fall.

C. ABORTION

Abortive or loss of certain parts leaf formation may be manifested in several ways including fanleaf, and filiform and willow leaves. Grapevines infected with the grape fanleaf virus give rise to leaves with irregularly-shaped lobes; filiform leaves are formed on tobacco plants infected with the cucumber mosaic or tobacco mosaic viruses; and peach leaves infected with the strawberry latent ringspot virus are narrow resembling willow leaves.

D. NEW GROWTHS

Growths differing in appearance and structure may occur on infected leaves. The most frequent type of growth is the protuberances associated with the uredo- and teleutopustules on leaves of plants infected with one of the various rust fungi. Such protuberances develop on the lower leaf surface of pears bearing the aecidia of *Gymosporangium sabinae* and on the lower leaf surfaces of apples caused by *G. tremelloides*. *Taphrina populina* causes epidermal galls on popular tree leaves, which are the result of the fungus developing between the cuticle and epidermis, stimulating cell division and elongation, which give rise to the new growths.[40] Smut galls differing in size and shape are formed along the midribs of maize leaves as a result of infection by *Ustilago maydis*.

Several viruses cause the formation of enations on the lower leaf surfaces of susceptible plants. Some of these growths are formed along the veins, such as those associated with maize rough dwarf and oat blue dwarf viruses. Others cause enations on the leaf tissues between veins and formed fringed areas on leaves, such as the cherry rasp leaf virus. A typical causal agent of enations is the pea enation mosaic virus which causes new growth around leaf veins. Characteristic enations also occur on leaves of *Prunus* sp. plants infected with Prunus ringspot virus.[39] Clover wound tumor virus causes growth in damage leaf parts as well as small internal tumors in phloem tissues of leaves.[31]

E. NECROSIS

Necrosis most frequently occurs in leaves which are, due to their compositon and structure, plant organs most susceptible to necrotic processes. Necrosis may be localized, when interveinal leaf parts are involved, and systemic when it develops in conductive tissues.

Local necrosis may be caused by a number of pathogenic agents. Typical causal agents are the necrotrophic pathogens which destroy cells through the action of enzymes and toxins. These agents usually cause leaf spots, such as *Alternaria solani* on potato and tomato plants, *Ascochyta pisi* on pea plants, *Colletotrichum lindemuthianum* on bean plants, and *Septoria helianthi* on sunflower plants. Bacterial pathogens that cause necrotic leaf spots include *Pseudomonas syringae* pv. *cannabina* on hemp plants, *P. syringae pv. tabaci* on tobacco plants, and *Xanthomonas campestris* pv. *juglandis* on walnut trees.

In hypersensitive plants, certain viruses cause local lesions, such as the necrotic strain of the plum pox virus on leaves of *Chenopodium foetidum* and the tobacco mosaic virus on leaves of *Nicotiana glutinosa*. Other viruses cause necrotic spots of different shape and sizes, such as the lettuce necrosis virus on lettuce and the Prunus necrotic ringspot virus on *Prunus* spp.

Frequently, the dead tissues in the center of necrotic spots fall out, giving a shot-hole appearance to affected leaves. This symptom can be caused by such fungi as *Clasterosporium carpophilum* and *Phyllosticta prunicola* both pathogens on the leaves of stone fruits (Figure 79). Shot-holing also occurs on plum leaves infected with *Pseudomonas syringae* pv. *morsprunorum*.[41] Typical shot-holing occurs on sweet and sour cherry trees infected with the Prunus necrotic ringspot virus. Tobacco rattle virus also causes shot-holing on tobacco leaves.

Leaf necrosis may develop as a result of systemic infection by *Xanthomonas campestris* pv. *campestris* on cabbage. After penetration through the hydathodes of leaves, the bacterium causes a brown to blackish discoloration of the leaf veins, followed by a decay of infected leaves. Another example is the mycelium of *Pseudopeziza tracheiphila,* which initially penetrates intracellularly into grape leaves, then enters the vascular system in which it develops profusely resulting in the filling of the vessels with a dark-colored material. Infected vessels thus become clogged which is followed by the death of affected leaves.

Necrosis of vascular tissues, particularly that of the phloem, is of particular importance in leaf tissue. Vascular necrosis in leaves is conspicuous due to the color changes in affected tissues. Such discoloration is used for diagnosis, such as on potato leaves infected with the potato streak (Y) virus and beet leaves infected with the beet necrotic yellow vein virus.

F. CONSISTENCY

Changes in the consistency of leaves of plants infected with certain viruses are a result of the loss of assimilates and not a consequence of altered leaf turgor.

Increases in leaf consistency and leaf thickening is usually associated with viruses that cause phloem necrosis in leaves, which prevents the normal translocation of organic matter from the leaf. Due to the accumulation of these compounds, leaves can become erect, fleshy and brittle with a tendency to break or crack. The blades of affected leaves sometimes are leathery and bronze or chlorotic. Such symptoms are associated with infections by the beet yellows virus on sugarbeets, the cherry leaf roll virus on cherry, and the potato leaf roll virus on potato. The barley yellow dwarf virus causes such symptoms on barley. Tobacco leaves infected with the tobacco rattle virus become so brittle that infected leaves produce a rustling sound when disturbed. Such a symptom can be diagnostic.

G. ROT

Decay or rot ususally occurs in large, fleshy leaves rich in organic matter, such as those of leafy vegetable crops, cabbage heads and onion bulbs. Under conditions of high moisture and temperature, and reduced aeration, leaves of grapevines, shrubs and trees may be susceptible to rot. Some examples of fungi that cause leaf rot are *Botrytis cinerea, Fusarium* spp., *Rhizopus* spp., and *Sclerotinia sclerotiorum.* The most common bacterium to cause leaf rot is *Erwinia carotovora.* This bacterial leaf rot is characterized by the formation of a slimy matrix often accompanied by an unpleasant odor and usually colonizes damaged tissues.

H. COLOR

Leaves of most plants have the highest pigment content of any other organ and because of the diversity in leaf color, the frequency of occurrence of chromatic changes is one of the highest (see section on chromatic, cytochromatic and histochromatic changes).

In general the patterns in color changes due to bacteria and fungi are strickingly different than those caused by mycoplasmas and viruses. Most bacteria and fungi cause localized areas of discoloration, while those caused by mycoplasmas and viruses are systemic in susceptible plants.

Bacteria and fungi usually cause chromatic changes in necrotic and contiguous tissues, evidenced around lesions and spots. The contiguous cells are usually influenced by pathogen toxic materials resulting in uncontrolled discoloration and eventually cell or tissue death. The colors produced around lesions and spots vary widely with some being typical of certain pathogens to be diagnostic. Examples of diagnostic symptoms caused by bacteria are bean halo blight caused by *Pseudomonas syringae* pv. *phaseolicola,* oat halo blight caused by *P. syringae* pv. *coronafaciens,* and tomato speck caused by *P. syringae* pv. *tomato.*[42,52] Examples of those caused by fungi are plum orange leaf spot caused by *Polystigma rubrum,* potato and tomato early blight caused by *Alternaria solani,* and violet leaf spot of stone fruits caused by *Clasterosporium carpophilum.*

Chromatic changes caused by vascular bacteria and fungi may by more widespread in plant than those causing localized infections, such as in *Pseudopeziza tacheiphila,* which causes a leaf redness on grapevines, and *Xanthomonas campestris* pv. *campestris,* which causes vein blackening on crucifers.

Mycoplasmas and viruses cause systemic chromatic changes in susceptible plants usually appearing as a uniform leaf chlorosis, yellows, reddish color, or other changes. The most frequently observed system is that of the mosaic pattern with alternating chlorotic and green leaf parts, caused by a number of plant viruses, such as the alfalfa mosaic, peach mosaic potato and tomato mosaic, and tobacco mosaic viruses.

Leaf mottling is a type of mosaic with unevenly colored leaf parts and can be caused by lettuce mosaic virus on lettuce, squash mosaic virus on squash, and cucumber mosaic virus on tobacco, the latter resulting in tobacco white leaf mottling.

Some viruses cause systemic leaf symptoms as chlorotic spots, such as on raspberry caused by the raspberry leaf spot virus, or rings, such as on cherry and red clover caused by the cherry chloratic ringspot virus and the red clover ringspot virus, respectively. The plum pox virus causes a variety of symptoms on plum leaves including mosaic, semi- and irregular rings, and leaf spots. Pronounced decoloration occurs along the tissues surrounding the veins, which also represents a type of mosaic, on plum and walnut leaves of virus infected plants.

Vein clearing, or the decoloration of tissue along the veins of leaves infected with a virus, is a characteristic mosaic pattern. This type of symptom is caused by plum pox virus on peach, and the virus that causes chlorotic vein clearing on salvia. In some cases leaf veins become very chlorotic, and the leaf surface appears netlike. Such a symptom, caused by the chlorotic net virus, appears on infected raspberry plants.

In monocotyldonous plants with parallel veins, mosaic patterns in leaves frequently appear a short or long streaks or stripes limited to the tissues between the veins. This type of symptom occurs on wheat leaves of plants infected with the wheat streak mosaic virus and on barley plants infected with the barley strip mosaic virus.

XVI. PATHOLOGICAL CHANGES IN FLOWERS

A. FORM

Flowers are produced in a huge array of shapes, sizes, and colors, due to their fundamental role in sexual reproduction in plants. Distinct from other plant organs, flowers consist of unique tissues and therefore the reaction of these tissues to pathogens are manifested in different manners than other parts of the plant. Changes in the morphology of infected flowers may be similar whether a single flower or an inflorescence of flowers is involved.

Symptoms are manifested depending on the flower structure, the flower part involved, the pathogen, and the environmental conditions. Sepals usually reaction to various pathogens in a manner similar to leaves because of their location and similarity to leaves. Other flower parts are more delicate and usually die soon after infection. In general, pathogenic bacteria and fungi cause such a rapid loss in infected flower parts that it is often difficult to study the morphological changes that take place during disease development. Rapid flower death is caused by a number of pathogens, including *Phytophthora capsici* on peppers and *Plasmopora viticola* on grapes; and by *Pseudomonas syringae* pv. *tomato* and *Xanthomonas campestris* pv. *vesicatoria* on tomato.

Josifović[10] stated that *Peronospora brassicae* causes peduncle and flower hypertrophy in susceptible crucifer plants, and observed that poppy flowers infected with *Peronospora arborescens* either failed to appear or did not open if they did appear. Peach blossoms infected with *Taphrina deformans* are deformed, become desiccated, die and drop off. Sunflower inflorescences infected by *Phomopsis helianthi* are smaller than normal. *Peronospora radii* causes the distribution of florets in the inflorescent clusters of *Matricaria indora* to be inversed.[14]

Flower malformations occur in plants infected with mycoplasmas and viruses, which do not cause rapid flower decay as noted by infections by bacteria and fungi. As a result, it has been easier to monitor and study morphological changes on flowers infected with one of these pathogens. A number of deformations have been described because of the striking appearance of affected flowers.[16] Severe deformations are caused on carnation flowers infected with the carnation vein mottle virus. In Castle Gilly flower plants infected with any of several viruses, and especially the cabbage black ringspot virus, the flowers are dwarfed, dense, and have small petals. Morphological changes occur on the individual flowers of inflorescence clusters of plants in the Asteraceae. For example, flower heads on aster plants infected with the cucumber mosaic virus are small with altered petals. The sepals of zinnia flowers infected with cucumber mosaic virus, are asymmetrical and irregularly distributed in the flower head. Chrysanthemum plants infected with the chrysanthemum virus B are smaller than normal and those infected with the tomato aspermy virus are deformed and produce curled petals of an uneven length.

Mycoplasmas cause the enlargement of flower parts as well as a fusion of the sepals on tomatoes infected by the stolbur mycoplasma. The sepals of strawberry plants infected with the strawberry green petal mycoplasma become enlarged.

B. NUMBER

The number of fertilized flowers on a plant plays an important role in its reproduction, and may be altered by a variety of abiotic and biotic factors. Abiotic factors include a wide range of possibilities and includes various environmental ones and the misuse of pesticides. A reduction in the number of flowers can also be caused by various plant pathogens, which cause desiccation, death and drop of flowers or inflorescences.

Characteristic biotic agents that cause desiccation are *Cercospora beticola* on sugarbeets; *Fusicladium* spp. in apples and pears; *Monilia* spp. in stone fruits, particulary on apricot and sour cherry trees; *Physalospora cydoniae* on apples and pears; *Plasmopara viticola* on grapevine; and *Pseudoperonospora humuli* on hops. Massive floral desiccation can be caused by *Erwinia amylovora* on apples and pears and by *Pseudomona syringae* pv. *syringae* on pears.

Viruses also disturb the formation of flowers on infected plants. For example, bean common mosaic virus causes, *inter alia,* flower drop on beans. In broad bean plants infected with the bean leaf roll virus, all the flowers fall off and are not replaced. Following the first signs of chlorosis in pepper plants infected with the stolbur mycoplasma, the plants cease to produce flowers; and flowers seldom form on peach trees infected with the peach rosette virus.[43]

Specialized flower pathogens, such as members of the Ustilaginales, inhibit flower development and fruit differentiation (Figure 80)(see section on pathological changes in fruits).

In summary, the reduction in the number of flowers ranks among the most detrimental of plant infection, since any reduction in fertile flowers has a direct impact on fruit and seed production.

C. ABORTION

Abortive changes includes not only the absence of individual flower parts but also the production of sterile flowers or flowers with irregularly developed parts. A variety of plant pathogens may cause such disorders. For example, sugarbeet inflorescences systemically infected with *Peronospora schachtii* are malformed and sterile and those infected with *Cercospora beticola* are sterile.[10] Flowers of crucifer plants infected with *Albugo candida* also are sterile.[10] Sands and McIntyre [44] observed that *Pseudomonas syringae* pv. *syringae* causes flower abortion on infected pear trees. Sterile flowers also frequently occur on pepper plants infected with the cucumber mosaic virus and on pepper and tomato plants infected with the stolbur mycoplasma (Figure 81).

In some cases, abortion of flowers takes place during pollen formation, such as on tomato plants infected with the tomato aspermy virus.[11] Pollen abortion has been recorded on barley flowers of plants infected with the barley stripe mosaic virus.[12] The tomato ringspot virus causes male sterility in flowers of plants in the genus *Geranium*.[45] Sour cherry pollen from trees infected with the yellow virus is less efficient than that from uninfected plants.[38]

The plum pox virus causes larger number of spherical than normally triangular or slightly rounded pollen grains on the plum cultivar Požegača.[46] Such pollen grains showed a reduction in germination of 1.5 to 5 times that of normal grains.[46] Abortive changes can also occur in ovaries and during ovule genesis in infected plants.[22]

D. NECROSIS

The rate of flower necrosis depends on the pathogen or pathogens involved. Pathogens that stimulate enzyme or toxin production usually cause a more rapid necrosis resulting in desiccation than those that do not. If *Plasmopora viticola* infects grape flowers before they open, the corolla cap dries up and turns black. *Monilia* spp. can infect flower blossoms of various fruit trees immediately upon bud break, which results in desiccation and death of petals and anthers. *Venturia* spp. infect the flower buds of apples and pears immediately following bud break with the first symptoms of the disease being necrotic spots on the sepals, receptacles and peduncles. *Phytophthora capsici* causes flower necrosis of pepper plants beginning at the base and eventually causing necrosis of the entire flower.

An example of a bacterial infection is *Erwinia amylovora*, which causes a complete necrosis of apple and pear flowers. The bacteria enter through nectary glands of the flowers, the sepals and the style of the flowers. As a result of both enzymes and toxins, this is followed by a rapid death of all flower parts.

Necrosis of flower parts is also associated with virus infections. For example, the arabis mosaic virus causes ring necrosis of petunia petals, while some strains of the tobacco mosaic virus causes necrotic spots and stripes on tulip flowers.[36] According to Holmes[38] lettuce mosaic virus may occasionally destroy the first flower heads formed on this plant.

E. TRANSVESTISM

Transvestism is a process characterized by the development of sex parts opposite to what would be expected in unisexual flowers; anthers may develop into ovaries and visa versa. It is caused by some of the loose smut fungi, such as *Ustilago violacea*. This pathogen develops on the dioecious flower of plants belonging to the family Caryophyllaceae. It develops systemically in infected plants, which transforms anthers into masses of chlamydospores.[14,37] In diseased plants of *Melandrium rubrum* and *M. album*, the female flowers develop rudimentary anthers. However, pollen is not formed, but are transformed into chlamydospores. Ovaries may also become sterile in infected plants. Similar changes take place in plants of *Muscari comosum* infected with *Ustilago vaillantii;* the upper florets in inflorescences may form anthers containing chlamydospores.

Other pathogens stimulate the development of rudimentary ovaries, which may also contain chlamydospores. Examples of this abboration are found in plants of *Buchloe dactyloides* infected with *Tilletia buchloeana;* and on plants of *Andropogon furcatus* infected with *Sorosporium everhartii*.[14,37] The exact mechanisms of transvestism is not fully understood, but Mathre[22] suggested that such disorders are a result of a distrubed hormone balance.

F. COLOR

Flowers are the most varied-colored organs on plants either as a result of pigments, such as the anthocyanins, or white, which is the lack of any pigment and reflection of all light rays.[47]

Color changes may be caused by a variety of factors. However, the rapid decay of infected flowers is usually accompanied by a darkening or blackening of the tissues resulting in no variation in coloration during the process. Such rapid discoloration are caused, for example, by *Monilia* spp. on flowers of stone fruits and by *Xanthomonas campestris* pv. *vesicatoria* on tomato flowers.

In plants infected with mycroplasmas or viruses, flower discoloration is slow and immediate death does not occur. Under these conditions various chromatic changes take place, frequently spectacular in appearance. Often the natural coloration of petals on infected flowers does not develop or disappears and is replaced by other colors in the form of irregular spots, mottles, stripes, or other features. Color changes in petals frequently are due to the disappearance of the anthocyanin pigments, leading to the expression of otherwise masked pigments. According to Mathews[31] virus infections usually affect only anthocyanin production in vacuoles without affecting the chromoplast pigments. Some examples are the whitish to pink irregularly-shaped spots and mottles occur on peach flower petal infected with the peach mosaic virus.[48] Parish et al.[49] described a disease of cherry trees called cherry flower anomaly which was characterized by the occurrence of narrow green stripes on the petals. Sometimes entire petals became green giving them a leaflike appearance. In tulip flowers infected with the tulip color-breaking virus, the natural colors are replaced by other colors in irregular stripes on the petals.

Such changes are particularly noticeable in the brightly-colored red and violet cultivars. Gilly flower petals contain anthocyanin, cyanine in the vacuoles, and yellow in the plastides of vacuoles.[50] When infected with the turnip mosaic virus, all the colors except yellow disappear.[31] Other examples are the bean yellow mosaic and cucumber mosaic viruses which cause gray and yellowish spots to occur on unicolored gladiolus flowers. Also, yellow striped spots are produced on unicolored flowers of cannas infected with the canna mosaic virus. A variety of colored spots appear on otherwise pink or red colored petals of carnations blossoms infected with the carnation mottle virus.

In contrast to other pathogens, many mycoplasmas cause virescence on colored flower parts. Such greening is characteristic of those that cause yellow-type diseases, such as aster yellows, strawberry green petal, and tomato stolbur (see sections on cytochromatic, histochromatic, and chromatic organ changes).

XVII. PATHOGLOGICAL CHANGES IN FRUITS

A. FORM

Similar to flowers and leaves, various changes occur on fruits as a result of disease. Generally, the intensity of morphological changes varies with the pathogen involved, host susceptibility, and the time or stage of development when infection takes place.

Partial malformation of fruits result from localized infections, such as the infection of apple fruits by *Fusicladium dendrictum* and pear fruits by *F. pirinum*. Many of the anthracnose pathogens cause abnormal fruit development particularly if infection takes place early in fruit development. Some examples are on bean pods infected with *Colletotrichum lindemuthianum*, on melon and watermelon fruits infected with *C. lagenarium*, and on pea pods infected by *Ascochyta pisi*. Similar malformations occur in early infections of bean pods by *Pseudomonas*

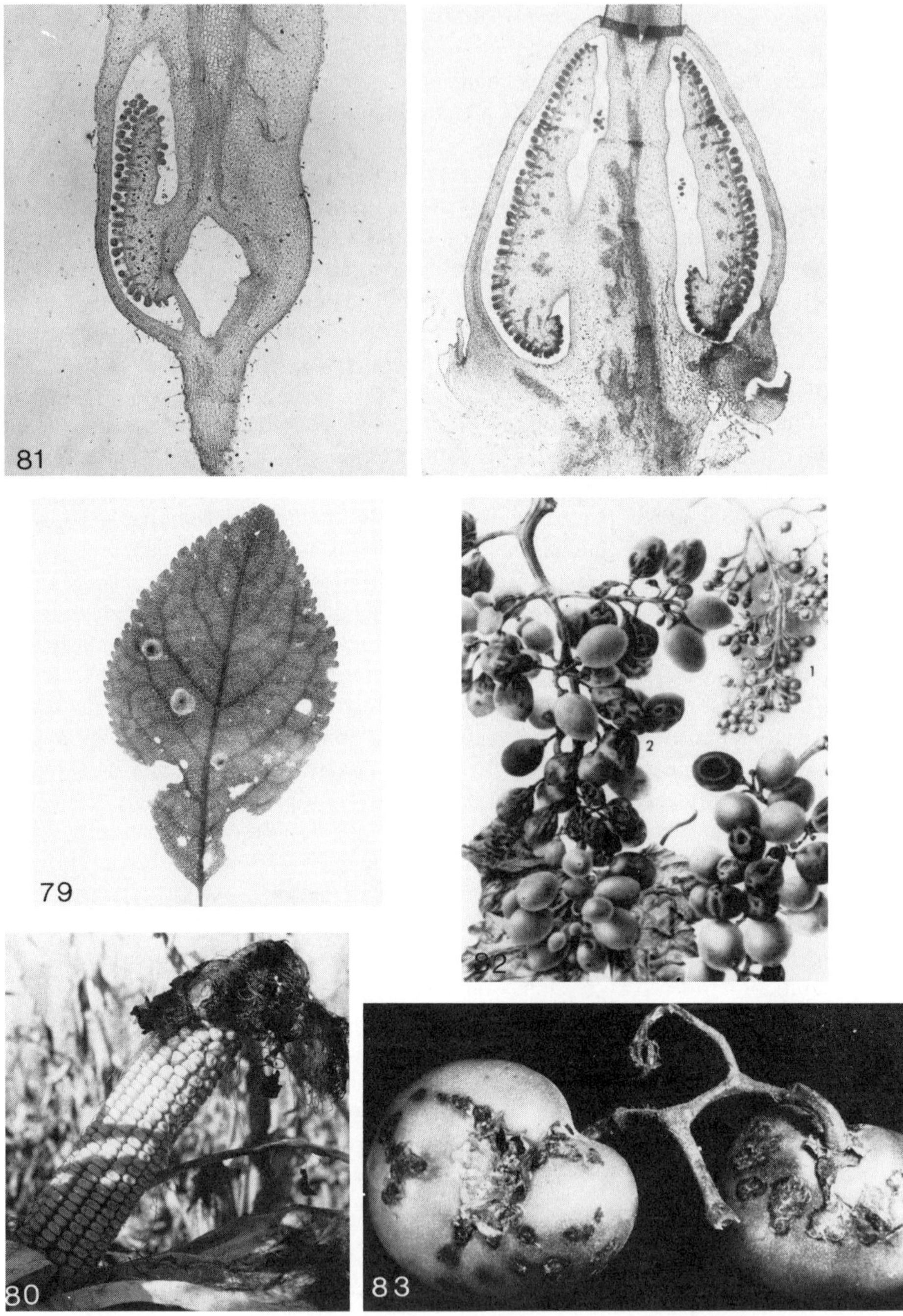

PLATE XVI. FIGURE 79. Necrotic spots on a plum leaf caused by *Clasterosporium carpophilum*. FIGURE 80. A maize ear showing black tumor formation in tissues infected by *Ustilago maydis*. FIGURE 81. Vertical sections through embryos of tobacco (cv. Prilip) showing deformed ovary (left) caused by the stolbur mycoplasm compared with a uninfected one (right). (From Mickovski, J. Doktorska disertocija, Poljoprivredni fakultat, Beograd (Yugoslavia), 1964.) FIGURE 82. Wrinkled and desiccated young fruits of grape (2) caused by *Plasmopora viticola* compared to uninfected older fruits (1). FIGURE 83. Scabby and malformed tomato fruits caused by *Xanthomonas campestris* pv. *vesicatoria*.

syringae pv. *phaseolicola* and of tomato fruits by *Pseudomonas syringae* pv. *tomato* and *X. campestris* pv. *vesicatoria* (Figure 83).

Extensive rather than localized changes on fruits are caused when infection takes place through the peduncle, such as in the case of plum fruits infected with *Taphrina pruni* resulting is "bladder plums".

Changes involving the whole fruit are caused by many plant viruses, such as the apple green crinkle disease which causes protuberances on apple fruits, and the pear stony pit disease which causes extreme malformations on pear fruits.

The cucumber mosaic virus causes dwarfing and malformation of cucumber and pepper fruits infected with the virus. The plum pox virus causes small, crinkled fruits with various -shaped depressions on fruits of the plum cultivar Požecača (Figure 84).

B. NUMBER

The number of fruits produced by a plant can be reduced by a variety of environmental factors, one of them being low temperatures. Freezing temperatures may destroy many if not all the flowers preventing any fruit set or cool temperatures accompanied by persistent rains can result in a reduced pollination. A reduction in the number of fruits regularly occurs on plants infected with a variety of pathogens. The level of reduction depends upon the pathogen or pathogens involved, the susceptibility of the flowers or fruits, and environmental conditions at the time of infection. Fruit reductions by pathogens can occur at flowering or after fruit set. Such damage is usually associated with diseases caused by bacteria and fungi.

Those pathogenic fungi that cause complete fruit destruction include *Monilia* spp. which infect flowers and fruits of pome and stone fruit trees. Members of this genus can destroy all the blossoms on apricot and cherry trees resulting in no fruit set. Other representative of this group are *Oidium tuckeri* and *Plasmopora viticola* which infect grapevine flowers and fruits (Figure 82). Similar detrimental effects are also caused by *Fusicladium dendriticum* in apples and *F. pirinum* on pears. The fungus, *Gnomonia juglandis,* and bacterium, *Xanthomonas campestris* pv. *juglandis,* infect young walnuts, which die and drop off. The number of wheat grains can be greatly reduced by *Tilletia tritici* and *Ustilago tritici* and on barley by *U. nuda.*

Various viruses and mycoplamas in many cases abortion of flowers, or sterility in flowers resulting in limited fruit production. For example, the barley strip mosaic virus causes flower sterility on barley, reducing yields by half. Infections by the maize rough dwarf virus can cause whole ear as well as grain atrophy, particularly in the apical portions. Inflorescences frequently are not formed on rice plants infected with the rice dwarf virus. In soybean plants infected with the soybean mosaic virus, there is a reduced number of pods and many of those formed are seedless. Due to the malformation of flowers and pollen sterility, peppers from plants infected with the cucumber mosaic virus are either not formed or significantly reduced in size below that of uninfected plants (Figure 85).

Fruit reduction can occur in various fruit trees infected with a variety of viruses, such as the prune dwarf strain of the sour cherry yellows virus which causes pollen sterlility and thus reduced fruit set on cherry. The plum pox virus causes early fruit drop of up to 40% on some cultivars of plum. Fig trees infected with the fig mosaic disease, drop most of its fruits early in the growing season. The grapevine fanleaf virus causes a reduction in the number of berries (Figure 88).

The fertility of plants infected with mycoplasmas can be significantly reduced, such as on tomatoes infected with the stolbur mycoplasma. The same mycoplasma reduces the number of flowers on peppers and those that do form unusable fruits.

C. NEW GROWTH

The formation of neoplastic growth on fruits is less frequent than on other plant organs. Usually any new growth on fruits is comparatively simple and results from hypertrophy and hyperplasia in the epicarp tissues. Hypertrophy usually occurs at points of infection by such fungi as *Clasterosporium carpophilum* which causes neoplastic growths on apricot fruits.[10] The hypertrophied tissues turn reddish brown with age and become fissured, resulting in the development of a typical fruit scab. This fungus also causes less pronounced hypertrophic reactions in the epicarp tissues of peach and plum fruits.

New growths due to hyperplasia occur on some fruit trees infected with some viruslike diseases. For example, peach wart disease causes wartlike excresences in the form of whitish wrinkles or light brown to bright red rims on young peach fruits (Figure 86). Such warty fruits also may occur as a natural characteristic on some peach cultivars or due to damage caused by insects. Slight protuberances, colorless at first then turning dark and necrotic, occur on apricot fruits infected with the apricot ring pox disease. These protuberances coalesce and form reddish violet rings or semi-rings of scabby tissue.[16] These scabs resemble those caused by *Clasterosporium carpophilum*. Wartlike excrescences with necrotic spots are formed on watermelon fruits infected with the tobacco ringspot virus.[38] Similar symptoms develop on cucumber fruits infected with the cucumber mosaic virus along with fruit malformation. Infection of Turk's cap (*Cucurbita maxima*) fruit by the same virus results in the formation of similar wartlike growths.

Although the pathological growths on fruits are less complicated and frequent than on other plant organs, they still can be used for diagnosis and can decrease the value of affected products.

D. SUBSTITUTION

In plant pathology substitution refers to the replacement of an entire fruit by a pathogen. Some examples are *Claviceps purpurea,* cause of ergot of rye, *Tilletia* spp., cause of wheat bunt, and *Ustilago* spp., cause of loose smut. The mycelium of *C. purpurea* is restricted to the grain it infects forming a sclerotium which develops in place of the grain. The mycelium of *Tilletia* spp. develop in the fruits or grains of rye and wheat, replacing the grain with chlamydopsores. Fungi of the genus *Ustilago* completely destroy the individual grains it colonizes. These and other pathogens that result in fruit substitution rank among the most detrimental and significantly reduce crop yeilds.

E. NECROSIS

Necrosis in fruits is characterized by a more or less rapid death of cells or groups of cells either on the surface of the fruit involved, or internally in the deeper pericarp layers.

External necrosis often occurs on young fruits still protected with only epidermal tissues and can affect either or both the epidermal and pericarp cells. External necrosis is usually manifested in and around points of infection in the form of anthracnose lesions, scabs, spots or similar types of symptoms. Their causal agents, in general, are bacteria and fungi, and occasionally, viruses. External scabs and spots are typified by *Fusicladium dendriticum* on apple fruits (Figure 87) and *F. pirinum* on pear fruits, *Colletotrichum lindemuthianum* on bean pods, *C. lagenarium* on melon and watermelon fruits, and *Ascochyta pisi* on pea pods.

Certain bacteria also cause characteristic external necrosis on fruits, such as the necrotic spotting on cotton bolls caused by *Xanthomonas campestris* pv. *malvacearum,* which results in dwarfing and dropping of infected bolls. *Pseudomonas syringae* pv. *phaseolicola* causes necrotic spots on bean pods; and *Xanthomonas campestris* pv. *vesicatora* and *Pseudomonas syringae* pv. *tomato* cause a scab-like necrosis on tomato fruits.

Superficial fruit necrosis can be caused by certain viruslike diseases, such as the apple star crack disease which causes necrotic scabs and spots on apple fruits; and the apple ringspot disease which causes a necrosis of the epidermal and other superficial layers of apple fruits (Figure 89).

Internal necrosis of fruits caused by viruses is usually associated with those viruses that are systemic in the plant. This type of necrosis has been observed on a variety of fruits of herbaceous and woody plants. For example, tobacco mosaic virus causes internal necrosis on immature tomato fruits. Partial or complete necrosis of the pericarp of pepper fruits is caused by cucumber mosaic virus. Some strain of the cherry green ring mottle virus cause necrotic depressions or ringspots on the epidermis and mesocarp of cherry fruits. Necrotic points occur on apricot, peach, and plum fruits infected with the plum pox virus.

In summary, superficial necrosis is especially detrimental to young fruits, which ususally drop. If they remain on the plant, the fruit looses its aesthetic appearance. Internal necrosis alters the composition and often the flavor of affected fruits, making them unmarketable.

F. ROT

Fruit rot can be caused by a large number of pathogens, often in conjunction with weak parasites and saprophytes. Generally there are two types of symptoms associated with fruit rots, change in color and in consistency. Changes in color may range over a wide spectrum, and changes in consistency range from dry to wet rots, which gradations in between.

Wet rot is caused by microorganisms associated with the production of proteolytic enzymes. Such pathogens degrade the intercellular pectic substances resulting in the loss of linkages among cells and their transformation into a semifluid mass. Cowling[2] listed several bacteria belonging to the genera *Erwinia* and *Xanthomonas,* and many fungi, including *Botrytis cinerea, Monilia fructicola, Penicillium digitatum, Pythium* spp., *Rhizopus stolonifer,* and *Sclerotinia sclerotiorum* amoung the causal agents of wet rot of fruits and vegetables. These microorganisms often penetrate plant organs through wounds prior to or following harvest.[2]

Organisms that cause rot and colonize injuries and wounds made by other organisms such as *Colletotrichum lagenarium* on melon and watermelon fruits, *Fusicladium* spp. on apple and pear fruits, and *Oidium tuckeri* and *Plasmopara viticola* on grapes. Similar imjuries are caused by *Pseudomonas syringae* pv. *lachrymans* on cucumber fruits and *Xanthomonas campestris* pv. *vesicatoria* on pepper and tomato fruits.

A classic example of dry rot is the mumification of pome and stone fruits infected with *Monilia* spp. Such degraded, dried, mummified fruits remain firm, sometimes as long as 10 years. Depending upon the color of the dried flowers or fruits, it is often referred to as grey fruit rot.

Rots caused by *Fusarium* spp. on maize and wheat fruits are other examples of grey fruit rot. The mycelium of these pathogens usually begin their development in the ear or spike of the plant. It spreads into the fruits and may penetrate the endosperm and embryo. Infected fruits appear moldy and have a crinkled, but firm consistency. The diagnostic symptoms of dry rot depends upon the organism involved and the structure of the fruit infected. For example, the causal agent of maize fruit rot is *Fusarium monilforme* and is named Fusarium rot, that on wheat grains caused by *F. graminearum* is called white rot.

There is not always a clear destinction between dry and wet rots. Whether the decay takes on the the symptoms of a dry or wet rot depends on external conditions during pathogen development. For example, under conditions of high moisture, *Botrytis cinerea* causes a wet rot and under prolonged dry periods it causes dry rot of grapes. Similarly, *Phytophthora capsici* under warm and dry conditions caused infected pepper fruits to crinkle and desiccate, known as pepper parchment. However, if the disease develops under wet conditions, the fruit is subject to wet rot and the outside becomes covered with mycelial mats producing conidia.

In summary, fruit rots occur in various forms, depending upon damage, injury, moisture and temperature during the course of pathogenesis.

G. COLOR

Fruits among all plant organs, except flowers, can have a rich variety of colors. Usually immature fruits are green because of chlorophyll in the pericarp. As the fruit develops and matures the chlorophyll disappears and carotenoid pigments of orange, red, and yellow; and/or anthocyanin pigments of blue and red appear in various combinations giving color to the fruit. The change in color may aid in the dispersion of fruits and seeds.

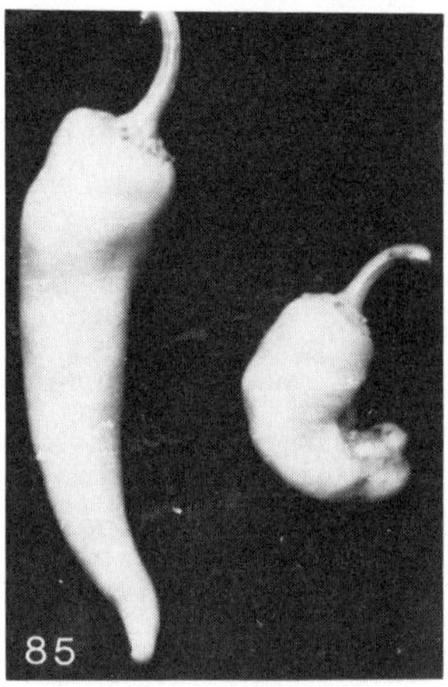

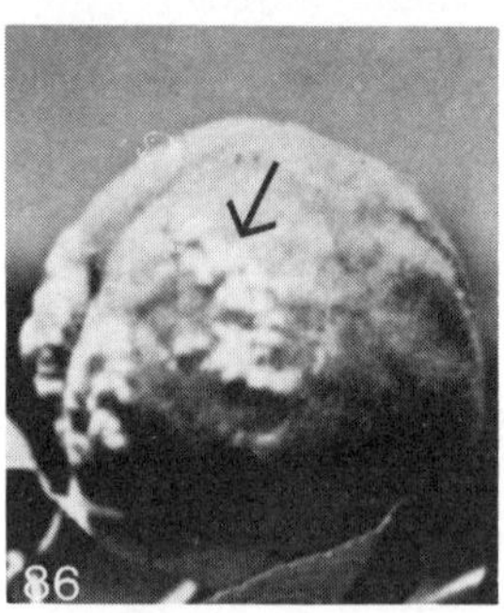

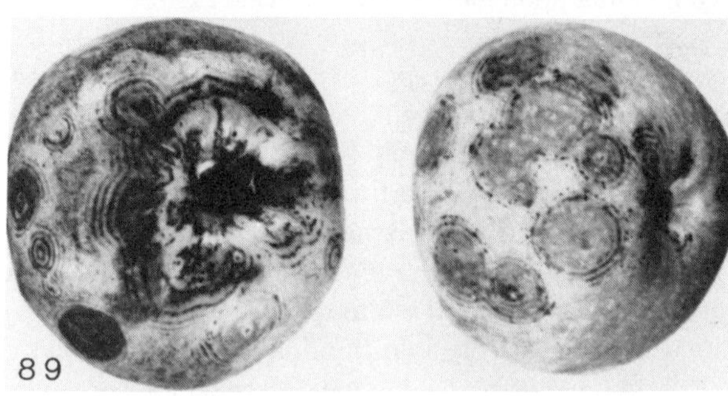

PLATE XVII. FIGURE 84. Sunken lesions on plum (cv. Požegaća) fruits caused by the plum pox virus. FIGURE 85. Stunted and malformed pepper (cv. Turčsijare) fruit (right) caused by the cucumber mosaic virus compared with a uninfected fruit (left). (From Delević, B., Doktorska disertocija, Poljoprivredni fakultet, Beograd (Yugoslavia), 1974.) FIGURE 86. Wartlike outgrowths (arrow) on an unripened peach fruit caused by unknown etiology. (From Boldget, C.E., U. S. Department of Agriculture Handbook No. 437, Washington, DC, 1976.) FIGURE 87. Scabs with fissures on an apple fruit caused by *Fusicladium dendriticum*. FIGURE 88. Cluster of grapes (cv. Belo Zimsko) showing reduced size and number of fruits (right) caused by the grape fanleaf virus compared with a uninfected cluster (left). (From Pejcinovski, F., Doktorska disertacija, Poljoprivredni fakultat, Beograd (Yugoslavia), 1974.) FIGURE 89. Concentric ringspots on apple (cv. Abondanca) fruits caused by an unidentified etiology. (From Posnette, A.F., Technical Communication Bureaux Horticulture, East Maling, UK 30, 95, 1963.)

Almost all pathogens developing on or in fruits causes color changes. Mottling can occur similar to that caused during the natural color change. Premature coloration of infected fruit might be considered as a result of an aging process.

Color changes in infected fruits may occur locally or systemically. Localized changes usually are caused by bacteria and fungi in and around points of infection. Often diseases receive their name from the characteristic color associated with the disease, as for example, sooty fruit spot and scab on apple and pear caused by *Fusicladium dendriticum* and *F. pirinum,* respectively.

Stigmatea mespili causes reddish to black spots on quince fruit; and *Gnomonia juglandis* and *Xanthomonas campestris* pv. *juglandis* both cause tarry spots differing in shape on young walnut fruits. *Phoma lycopersici* causes dark brown spots on tomato fruits; while *Pseudomonas syringae* pv. *tomato* causes black fruit scab and *Xanthomonas campestris* pv. *gardneri* causes a whitish halo spot on tomato fruits.[52]

In plants infected with mycoplasmas or viruses, color changes are more diffused over the surface of either all or parts of the affected fruits. For example, the tomato spotted wilt virus causes pale red and sometimes white patches rimmed with concentric circles on tomato fruits. Plants of *Cucurbita pepo* infected with the cucumber mosaic virus produce fruits have green ringspots in contrast to the normal yellow of the fruit. The apple ringspot disease causes light green to yellow and sometimes red coloration around brown spots or between concentric rings on apple fruits. These reddish zones probably are due to the presence of anthocyanin in the superficial pericarp layers just under the fruit epidemis.[51] Tomato fruits with stolbur disease become pink and only partially red as they ripen.

In summary, color changes accompanying various diseases can be used for diagnosis, but they also alter their aesthetic and market value.

REFERENCES

1. **Mulder, E.A.,** Effect of mineral nutrition on lodging of cereals, *Plant Soil* 5, 246, 1954.
2. **Cowling, B.E.,** The engineering mechanisms of pathogenesis, in *Plant Disease An Advanced Treatise III,* Horsfall, J.G. and Cowling, B.E. Eds., Academic Press, Inc., New York, 1978, 299.
3. **Tanasijević, N. and Šutić, D.,** Bolesti i štetočine lucerke i crvene deteline, *Zadružna knjiga,* Belgrade, 1958.
4. **Hepting, G.H.,** *Diseases of Forest and Shade Trees of the United States,* U.S. Department of Agriculture Handbook 386, Washington, DC, 1971, 1.
5. **Sondheimer, E. and Simpson, W.G.,** Lignin abnormalities or "rubbery apple wood". *Canadian Journal of Biochemistry,* 40, 841, 1962.
6. **Kozlowski, T.T.,** How Healthy Plant Grow, in *Plant Pathology An Advanced Treatise III,* J.G. Horsfall and B.E. Cowling, Eds., Academic Press, New York, 1978, 19.
7. **Campbell, W.P.,** Infection of barley by *Claviceps purpurea, Canadian Journal of Botany,* 36, 615, 1958.
8. **Husain, A. and Kelman, A.,** Tissue is disintegrated. In *Plant Pathology An Advanced Treatise I,* Horsfall, J.G. and Cowling, B.E. Eds., Academic Press, Inc., New York, 1978,143.
9. **Gäumann, E.,** *Pflanziche Infektionsiehre,* Verlag Birkhäuser, Basel, 1946.
10. **Josifović, M.,** *Poljoprivredina Fitopatologija II,* Naučna Knjiga, Belgrade, 1956.
11. **Caldwell, J.,** Some effects of a plant virus on nuclear division, *Annals of Applied Biology,* 39, 98, 1952.
12. **Inouye, T.,** Studies on barley stripe mosaic in Japan. *Sonderdruck aus Berichte des Ohara Institut für Landwirtschaftliche Biologie,* Okayama Universität, Boncl XI, 4, Kuraschiki, Japan, 1962.
13. **Gibbs, A. and Harrison, B.,** *Plant Virology - The Principles,* Edward Arnold Press, Inc. New York, NY, 1976.
14. **Wood, R.K.S.,** *Physiological Plant Pathology,* Blackwell Scientific Publications, Oxford & Edinburgh, U.K., 1967.

15. **Helgeson, P.J.**, Alteration of growth by disease, in *Plant Disease An Advanced Treatise III*, Horsfall, J.G. and Cowling, B.E. Eds., Academic Press, Inc., New York, NY, 1978, 183.
16. **Šutić, D.**, *Viroze biljaka*, Nolit, Belgrade, 1983.
17. **Morvàn, G.**, La chute des bourgeons du poirier une nouvelle maladie a virus, *Annals Epiphyt.*, 16, 153, 1965.
18. **Küster, E.**, *Pathologische Pflanzenanatomie 3*, Verlag von Gustav Fishher, Jena, 1925.
19. **Glišić, Lj.**, *Opšta citologija.*, Unija Bioloških naučnih društava Jugoslavije, Zemun, Belgrade, 1980.
20. **Wilson, L.C.**, Plant teratomas - Who's in control of them? in *Plant Disease An Advanced Treatise III*, Horsfall, J.G. and Cowling, B.E. Eds., Academic Press, Inc. New York, NY, 1978, 215.
21. **Merlo, J.D.**, Crown gall - a unique disease, in *Plant Disease An Advanced Treatise III*, Horsfall, J.G. and Cowling, B.E. Eds., Academic Press, Inc., New York, NY, 1978, 201.
22. **Mathre, E.D.**, Disrupted reproduction, in *Plant Disease An Advanced Treatise III*, Horsfall, J.G. and Cowling, B.E. Eds., Academic Press, Inc., New York, NY, 1978, 257.
23. **Klein, M.**, Safflower phyllody - a mycoplasma disease of *Carthamus tinctorius* in Israel, *Plant Disease Reporter*, 54, 735, 1970.
24. **Sahambi, H.S.**, Studies on *Sesamum* Phyllody virus - virus-vector relationship and host range, in *Proceedings of the First International Symposium on Plant Pathology*, Indian Phytopathological Society, New Delhi, 340, 1980.
25. **Bos, L.**, Plant teratology and plant pathology, *Tijdschr, Plantenziekten*, 63, 225, 1957.
26. **Josifović, M. and Šutić, D.**, Prilog poznavanju porekla i načina nastajanja primarnih zaraza od *Phytophthora infestans* de By., Godišnjak Polj. Fakulteta, 4, 3, 1952.
27. **Osbourne, D.J.**, Defoliation and defoliants. *Nature (London)* 219, 564, 1968.
28. **Zimmerman, H.M. and McDonough, J.**, Dysfunction in the flow of food, in *Plant Disease An Advanced Treatise III*, Horsfall, J.G. and Cowling, B.E. Eds., Academic Press, Inc., New York, NY, 117, 1978.
29. **Kiraly, A.**, Defenses triggered by the invader: Hypersensitivity, in *Plant Disease An Advanced Treatise V*, Horsfall, J.G. and Cowling, B.E. Eds., Academic Press, Inc., New York, NY, 201, 1980.
30. **Cambell, L.C., Huang, J.S., and Payne, A.G.**, Defense at the perimeter; The outer walls and the gates, in *Plant Disease An Advanced Treatise V*, Horsfall, J.G. and Cowling, B.E. Eds., Academic Press, Inc., New York, NY, 103, 1980.
31. **Matthews, F. E. R.**, *Plant Virology*, 2nd ed., Academic Press, Inc., 1981.
32. **Tu, J.C. and Ford, R.E.**, Free amino acids in soybeans infected with soybean mosaic virus, bean pod mottle virus, or both, *Phytopathology*, 60, 660, 1970.
33. **Viglierchio, D. R.**, Nematiodes and other pathogens in auxin-related plant growth disorders, *Botanical Review*, 37, 1, 1971.
34. **Schmelzer, K., Schmidt, H. E., and Wolf, P.**, Zierpflanzen, in *Pflanzliche Virologie III*, M. Klinkowski und Mitarbeiter, Eds., Akademie-Verlag, Berlin, 1977, 1.
35. **Aleksić, Ž., Aleksić, D., and Šutić, D.**, *Bolesti povrca i njihjovo suzbijanje*, Nolit, Belgrade, 1980.
36. **Schmelzer, K.**, Symptomatologie, in *Pflanzliche Virologie I*, Klinkowski, M., Eds., Akademie-Verlag, Berlin, 1980, 21.
37. **Tarr, J.A.S.**, *The Principles of Plant Pathology*, MacMillan & Co., Ltd., London, 1972.
38. **Holmes, F.O.**, Symptomatology of viral diseases in plants, in *Plant Virology*, K.M. Corbett and H.D. Sisler, Eds., University of Florida Press, Gainesville, Fl., 1964, 17.
39. **Nyland, G., Gilmer, M.R., and Moore, J.D.** *Prunus* ringspot group, in *Virus Diseases and Noninfectious Disorders of Stone Fruits in North America*, Agriculture Handbook 437, U.S. Department of Agriculture, Washington, DC, 1976, 104.
40. **Butler, J. E., and Jones, G. S.**, *Plant Pathology*, MacMillian & Co., Ltd, London, 1949.
41. **Josifović, M. and Šutić, D.**, Maladie criblée du Prunier d'oridgine bacteriénne en Yougoslavie, *Phytopathologische Zeitschrift*, 50, 329, 1964.
42. **Arsenijević, M.**, *Bakterioze Biljaka*, Poljoprivredni Fakultet, Novi Sad, Yugoslavia, 1975.
43. **KenKnight, G.**, Peach rosette. in *Virus Diseases and Noninfectious Disorders of Stone Fruits in North America*, Agricultural Handbook No. 437, U.S. Department of Agriculture, Agricultural Research Service, Washington, DC, 1976, 73.
44. **Sands, D. C. and McIntyre, J. L.**, Possible methods to control pear blast (*Pseudomonas syringae*), *Plant Disease Reporter*, 61, 311, 1977.
45. **Murdock, D. J., Nelson, P. E., and Smith, S. H.**, Histopathological examination of Pelargonium infected with tomato ringspot virus, *Phytopathology*, 68, 844, 1976.
46. **Pejkić, B. and Šutić, D.**, Morphologie et germination du pollen du prunier (*Prunus domestica* var. *Požegača*) atteint par le virus de la Sharka. *Annals of Phytopathology H.S.*, 235, 1971.
47. **Cincović, T., Čanak, M., and Kojić, M.**, *Botanika*. Naučna Knjiga, Belgrade, 1980.
48. **Pine, S.T.**, Peach mosaic, in: *Virus Diseases and Noninfectious Disorders in Stone Fruits in North America*, Agriculture Handbook No. 437, U.S. Department of Agriculture, Washington, DC, 1976, 61.

49. **Parish, L. C., Cheney, W. P., and Anthon, W. E.,** Blossom anomaly of cherry, in: *Virus Diseases and Non-infectious Disorders of Stone Fruits in North America,* Agriculture Handbook No. 437, U.S. Department of Agriculture, Washington, DC, 1976, 211.
50. **Gairdner, A. E.,** The inheritance of factors in *Cheiranthus cheiri, Journal of Genetics,* 32, 479, 1936.
51. **Canova, A.,** Apple ring spot, In: *Virus Diseases in Apples and Pears,* Technical Communication Bureaux of Horticulture, East Malling 30, 39, 1963.
52. **Šutić, D.,** Bakterioze crvenog patlidžana, Doktorski Disertacija, Posebno izdanje Instituta za Zastitu Bilja 1, 1957, Beograd.

Chapter 4

PHYSIOLOGY OF DISEASED PLANTS

The study of the physiology of diseased plants (phytopathophysiology) deals with pathological changes in the life activities of diseased plant cells, tissues, and organs. Disorders in the physiology of diseased plants differ in expression and intensity, since they occur under a dynamic system which involves the host, a causal agent, and an environment which can affect both organisms. These disorders result in the partial or complete disruption of the functions of the diseased plant parts and, in some situations, the entire plant.

Changes in these physiological activities are linked with cytopathological and histopathological changes, as well as changes in the structure and organization of the organism involved. All these disorders together make up the primary disfunctions taking place in a diseased plant. Within this framework, functional changes also occur, which represent a reaction to the detrimental influences of the external factors disrupting the biological balance among the infected cells, tissues, and organs.

Various types of disorders occur in all the biochemical and physiological activities of diseased plant cells. Some of the major affects of these disorders are discussed.

I. DISEASED PLANT CELL PERMEABILITY

Permeability is a characteristic of cell membranes which controls the absorption and excretion of substances with the surrounding environment. The permeability of membranes of cell organelles enables the movement and transport of cell constituents taking part in the biological and physioloical activities within the cell. Intact normal cells direct the selection of substances to be absorbed and the order of their intake or excretion. Permeability regulated in such a manner is defined as either semipermeability or selective permeability. Selective permeability is an active process which requires energy. Irreversible loss of selective permeability results from severe injury or death of the cell.

The absorption and excretion of given substances is a function of all cell membranes. The cytoplasmic membranes, such as the plasmalemma or ectoplast, are primarily involved in the selective exchange of nutrients with the external environment. Nutrient exchange among organelles within a cell is performed by the inner cell membranes, such as the tonoplasts or vacuole membranes, the endoplasmic reticulum membranes, plastides, mitochondria and dictyosomes.

The mechanisms of selective permeability is not fully understood; three theories have been proposed according to Wheeler:[1] the absorption, cytosis, and membrane theories. Some explanation and understanding of these theories and how they explain membrane activities in healthy plant cells is necessary before considering the consequences of pathological changes in cell membrane functions.

According to the membrane theory, selective permeability is controlled almost exclusively by the cell membranes. For example, the plasmalemma directs the movement of nutrients between cells and the external environment, while the inner membrane performs an identical function in the cell, simultaneously providing for its own internal spacial identity.

According to the absorption theory, selective permeability is directed by the solubility of the substances in the cytoplasm and their absorptive and chemical binding with cell colloids rather than with the cell membrane.

The cytosis theory explains the movement of nutrients by the encapsulation and transport with the aid of specific cytoplasmic formations known as vesicules or vesiculated excrescences in the form of sacks. Both solid (phagocytosis) and fluid (pinocytosis) parts may be transported in this manner. Although other theories have a certain value, the membrane theory is the most accepted.[1]

On the basis of these theories, it is possible to interpret the disorders in selective permeability of the membranes of infected and diseased plant cells. Based on the membrane theory, disorders in cell permeability would be due mostly, if not exclusively, to changes in the structure and function in the diseased cell membranes during pathogenesis; while based on the absorption theory, water and solute losses would result from alterations of the internal colloids and water characteristics rather than in changes in the membranes. Based on the cytosis theory, permeability changes could be reduced due to the ability of the cytoplasm and organelles to develop initial (endocytosis) and external (exocytosis) vesiculated formations affecting the movement and transport of substances within the cell. From a phytopathological point of view, the membrane theory offers the most plausible explanation of the disorders in general membrane selectivity in infected cells.

Membranes are fundamental structures and carry on fundamental functions in cells. As structural formations they encircle the protoplast, vacuoles and cell organelles, dividing the interior cell components into a number of specialized compartments in which specific metabolic processes take place. Their large surface area in the cell represents large reactive areas in which membranes either directly or indirectly participate in all or almost all cell processes. In the encapsulated intracellular portions of the cell selective absorption and excretion of various components of nutrient exchange take place with the surrounding environment. These components also participate in the cell metabolic processes through their integral or bound enzymes. General membrane function is based on its selective permeability and therefore, any disruption in this permeability is a reliable indication of cell vitality.

Selective permeability changes take place in the course of normal development as well as under the influence of several abiotic and biotic factors. For example, young developing cells show considerable membrane selectivity which gradually increases until the cell is completely differentiated. Cell aging is accompanied by reduced selectivity, which is completely loss with cell death. Abiotic factors that cause injuries, such as extreme temperature, large pH changes, ionization, radiation, etc. increase membrane permeability. Biotic factors, such as a variety of plant pathogens, can increase membrane permeability and alter its selectivity.

As discussed in the section on cytopathological changes, plant pathogens through their enzymatic, toxin and/or mechanical action, cause chemical and structural changes in the membranes resulting in partial or complete dysfunction. It must be stressed that only undamaged membranes remain functional. Any alterations in the chemical or physical properties of the membrane system is manifested in intense exosmosis of inorganic salts and organic compounds from the cell. Altered cell permeability as a characteristic of plants infected with any one of several pathogens was shown by Thatcher[2-4] and Dorohova.[5] According to these authors, altered membrane permeability was observed in cells of plants infected with any one of the following fungal pathogens: *Botrytis cinerea, Fusarium lvcopersici, Phytoththora infestans, Puccinia graminis tritici, Sclerotinia sclerotiorum, and Uromvces fabae.* Later investigations showed that altered cell permeability also occurred in cells of plants infected with various pathogenic bacteria and viruses.

Some authors further confirmed increased cell membrane permeability using an excreated organic matter leaching method[6] as well as an electric conductivity method in infected cells and tissues.[7,8] Kuprevich[6] reported a significant increase in the leaching of organic matter in infected leaves of the following plants: beans infected with *Mycosphaerella pinodes, Cirsium arvense* infected with *Puccinia suaveolens,* oat plants infected with *Puccinia coronifera,* potatoes infected with *Xanthomonas campestris vesicatoria,* and *Ribes nigrum* infected with *Cronartium ribicola.* Leaves infected with these pathogens had 1.5X to twice the quantity of leachates compared to uninfected tissues. In a study on physiologic changes caused by leaf rust in wheat,

Kokin[9] reported that leachates were five times greater in infected than uninfected plant tissue. The susceptibility of individual wheat cultivars to leaf rust fungus depended upon membrane permeability. Similar data was obtained in a study of cotton plants and their susceptibility to the Fusarium wilt fungus.[10]

Wheeler and Black[8] studied permeability changes in oats resistant and susceptible to *Helminthosporium victoriae* based on differences in the electrical conductivity of water in infected tissues. They found that susceptible oat plants and tissues treated with victorin released electrolytes at a faster rate than inoculated or resistant plants treated in the same manner. They concluded that the high rate of electrolyte release was a consequence of increased cytoplasmic membrane permeability.

Increased cell and membrane permeability has been established in many diseased plants regardless of the disease or causal agent. This led Wheeler and Hanchey[11] to conclude that "all researchers, investigating this problem recorded changes in permeability as a disease consequence." According to Wheeler[12] permeability changes represent an early, and in some cases, the first recognizable plant reaction to infection. The altered permeability is accompanied by changes in the charge on the membrane and in the viscosity of the cell protoplast.

Cytoplasmic and cell organelle membranes have an electrochemical potential resulting from different ions unevenly distributed on both sides of the membrane. There is also a difference in the electrochemical potential between membranes and the surrounding environment. This difference also occurs between the cell interior and external environment, with the cell interior being electronegative and the exterior being electropositive.[13] These differences between the interior and exterior of the cell can be measured by inserting the tip of an electrode into the cell and one into a known solution. Those electrical potential differences in plant cells can range from -50 to -150 mV.[14] In their normal state, cell membranes allow ion passage against the density gradient with the maintenance of the differences in the electrochemical potential on both sides of the membranes as well as that between the cell interior and exterior.

Due to increase permeability, infected cells and tissues release electrolytes into the external environment, reducing these membrane potentials. Released electrolytes increase the electrical conductivity of their environment, which may be measured directly in infected and diseased tissues,[7] in leaf extracts,[15] and in electrolytes released from diseased tissues immersed in distilled water.[8] Reduced electrical potential and increased conductivity are a result of altered cell membrane permeability.

Changes in the protoplast viscosity also are connected with changes in the structure and function of the membrane system. Viscosity is a measure of the internal friction in the protoplast resulting from the binding of its macromolecules. Various bonds among the protoplasm macromolecules are continuously disrupted and re-established. Protoplasm viscosity, depending on cell age, is influenced by several external environmental factors.

The viscosity of young cells which are intensely active metabolically, is smaller than in older cells. Extremes in temperature, pH, and other such factors influence the macromolecule bonds, which have a direct effect on protoplasm viscosity.

Rubin and Artsikhovskaya[16] showed that a toxin produced by *Botrytis cinerea* as well as compounds isolated from this toxin, influenced the viscosity of the protoplasm and cell sap of cabbage plants. The extent of the alterations was determined by the susceptibility of the plant to the pathogen. The toxin of *B. cinerea* reduced protoplasm viscosity by four times within 2 hours in susceptible plants. Within 24 hours, cell structure became disorganized to such an extent as to prevent viscosity measurements. However, in resistant plants, viscosity decreased after 24 hours. Reductions in viscosity were also recorded in tissues treated with organic acids and polysaccharides. In contrast, the viscosity of the cell sap increased in infected tissues. In susceptible plants cells, the cell sap viscosity increased by 3.5 times within 44 hours after treatment. Slight viscosity increases were recorded in resistant plants 30 minutes after treatment, but remained practically unaltered after that time. These data suggested that resistant plant tissues are not as susceptible to the processes altering cell viscosity as are susceptible ones.

Altered viscosity affects cytoplasmic movement, which leads to interruptions in the transport of nutrients necessary for the biochemical processes taking place in the cell.

The changes in the physiochemical properties of the protoplasts emphasize the complexity of the various processes in infected cells. Changes in membrane permeability, particularly if irreversible, results in cell death. Also, the alteration in membrane permeability should not be considered as a simple cause and effect relationship. The ability of pathogens to directly or indirectly alter membrane permeability is one of the most important mechanisms used by pathogens to obtain nutrients.

II. CHANGES IN THE MOVEMENT OF WATER IN DISEASED PLANTS

A. AMOUNTS AND MOVEMENT OF WATER IN DISEASED PLANTS

Life began in water and is inseparably linked with it. The entire cell, its organization, structure and activity is based on the presence of water. Terrestrial plants have developed in the course of their evolution, special tissues and mechanisms to ensure an adequate water supply. All cell structures and functions are based on the presence of certain amounts of water and its movement within the cells, tissues, and the entire plant.

Water is the most abundant constituent of living cells. It is present in all cell parts, including walls, cytoplasm and nucleus. Water is bound to the constituents of the cell wall, such as pectin substances, cellulose and hemicellulose, and accounts for 50% of its weight. With some exceptions, plant cell protoplasm contains 75 to 95% water, and water in the vacuoles accounts for 30% of the total water in cells.

Water is found in its bound and free state in the cell. Bound water is primarily found in the protoplasm and accounts for 4 to 7% of the total cell water. In the protoplasm, water binds to protein molecules thus contributing to the maintenance of the normal state of the protoplasm. The largest amounts of cell water is found as free water, which has many functions. Free water serves as a dispersion medium in which the protoplasm colloidal system is maintained. Water is an excellent solvent of many inorganic and organic compounds, and with it movement, provides transport of these compounds within the cell. Water takes part in various biochemical cell processes.

Higher plants, particularly terrestrial ones, obtain their water supply primarily by uptake from the soil. Water movement begins in the soil towards the plant root system and continues through the root tissues into the vascular system and then to all parts of the plant. Some of it eventually moves into the intercellular spaces of leaves and other plant tissues, and passes in the form of vapor through natural openings into the atmosphere. The system represents a continual cycling of water, which is a condition for the survival of all plant organisms, and life in general.

Water movement within this system is based on the energy in the movement of the water molecule. This energy is greater in pure water, which is influenced by no other force, except gravity. The movement of the water molecule is reduced if inhibited by other forces, such as dissolved compounds, colloidal constituent, etc. Free energy of the water molecule is defined as water potential. The status of water in plants is best expressed, according to Kozlowski,[17] as water potential which is the difference between free water energy in the plant and that energy of pure free water. Water movement from the soil through the plant and into the atmosphere is in the direction to the lower water potential which is continuously created by water losses due to transpiration.

Water molecules move by diffusion and thus exert pressure defined as diffusion pressure. The magnitude of diffusion pressure corresponds to a given amount of free energy. Water moves in the direction of a lower diffusion pressure.[18] The overall status of water in a plant is influenced by any disorders in water absorption, movement, and transpiration.

III. CHANGES IN THE ABSORPTION OF WATER BY ROOTS

Plant roots absorb water from the soil. The absorptive ability of the cells participating in water absorption is influenced by abiotic and biotic factors. Examples of abiotic factors are the quantity of water in the soil and its availability, aeration, temperature, soil composition, and the concentration of the soil solutions. Examples of biotic factors include soilborne microorganisms, insects and other pests which can damage root cells and system, thus either partially or completely inhibiting water uptake.

Pathogenic microorganisms which cause damage to root tissues and thus reduce water absorption include the necrotrophic fungi such as *Armillaria mellea* and *Rosellinia necatrix,* which cause white root rot on a wide range of plant species. Other fungi included in this group are *Fusarium* spp., *Verticillium albo-atrum* and *V. dahliae*, which develop in the roots and vascular system of infected plants. Each of these pathogens penetrate through the root hairs and colonize root tissues resulting in root rot. Certain viruses cause root rot; root hair necrosis on grapes is caused by the grapevine fanleaf virus and by the beet necrotic yellow vein virus on sugarbeets. Root cell and tissue necrosis caused by these pathogens disrupts and inhibit the movement of free water from the soil.

IV. CHANGES IN THE MOVEMENT OF WATER IN THE PLANT

Free water, after absorption from the environment, moves along an uninterrupted pathway in the root tissues through the xylem system and finally into the leaf tissues. The quantity and rate of movement within these systems are influenced by any factors in cells or tissues which impede this movement.

For example, water movement through the cell meets with resistance both in its living and nonliving parts. Such resistance is generally greater in the living than nonliving parts, such as the cell wall, intercellular spaces, etc. Pathogens, by causing disorders in the structure and activity of cells and tissues, alter the natural resistance in water transport, which directly affects water uptake and transport in the plant.

V. CHANGES IN THE MOVEMENT OF WATER IN ROOTS

The uptake of water, including nutrients, from the soil is performed in the root growth zone, which is continuously formed in the root meristem tissues. The zone of active water absorption consists of rhizoderm cells of the root epidermal tissue. The water absorbed by these cells moves through primary cortex cells into the central cylinder, where it moves into the xylem system.

Microorganisms associated with the rhizosphere can affect the cells in the root growth zone either mechanically or through the action of enzymes and toxins. Some of the most detrimental soilborne fungal pathogens include species of *Fusarium* and *Verticillium*, as well as many other root rot organisms. These pathogens penetrate directly or indirectly and spread inter-and intracellularly through damaged or dead cells. With the aid of enzyme hydrolysis, these pathogens degrade cell walls causing death of cell protoplasts followed by tissue disintegration.

Root functions also may be disrupted by other soilborne pathogens, such as *Thielaviopsis basicola,* which is a weak parasite causing severe damage when conditions are unfavorable for root development. Water uptake by roots may be inhibited by a variety of mycorrhizal and saprophytic microorganisms developing on roots under stress conditions. Many investigators have reported that the extent of disorders in water transport in infected plants is influenced by environmental conditions following infection. Generally, transpiration decreases with increased severity of wilt symptoms.[19]

VI. CHANGES IN THE MOVEMENT OF WATER IN THE XYLEM

After leaving the root cells and entering the central root cylinder and xylem, water continues its movement through the tracheae and tracheids to the leaf parenchyma tissue cells. This is the longest pathway in the system.

Disruptions in water movement in the xylem most often attributed to pathogens are those that directly colonize the xylem vessels. Examples of these vascular pathogens that invade the xylem from roots are species of *Fusarium* and *Verticillium*. Others are introduced into the xylem by insects, such as the Dutch elm disease fungus, *Ceratocystis ulmi* and the oak wilt fungus, *C. fagacearum*. Several wood rotting fungi reach the xylem through wounds, such as *Stereum purpureum*. The bacterial wilt organism, *Pseudomonas solanacearum* also invades xylem tissues through wounds. A number of investigators have studied the influence of vascular pathogens on water movement in infected plants. Among the various methods used were the study of resistance to water movement in excised infected stem pieces. Using this technique, it was established that resistance to water movement in tomato stems infected with *V. albo-atrum* was 200 times greater than in uninfected stem pieces.[20] It was also observed that vascular pathogens increase this resistance in excised stems by four to 60 times.[21-23] Rubin and Artsikhovskaya[24] cited data obtained by Dimond and Waggoner[25] which showed that the rate of nutrient movement in the xylem of tomato plants infected with the Fusarium wilt fungus was 25 times less than that in uninfected plants, but that transpiration was reduced only by three times in infected plants. Talboys[26] stated that in the case of several vascular diseases studied it has been shown that the resistance to water flow in colonized xylem was substantially higher than in uninfected parts.

Increased resistance to water movement is a consequence of reduced diameter of the conductive elements. This reduction may be influenced by several factors, including the physical presence of fungal hyphae or bacterial cells and slime. In addition, these pathogens may reduce volume with the presence of their metabolites, such as polysaccharides and other compounds resulting from enzymatic action. Polysaccharides frequently are described as wilt toxins, which are formed by both bacteria and fungi. However, these compounds of high molecular weight may be considered as physically clogging the vessels rather than being toxic.

Several pathogenic and nonpathogenic bacteria and fungi recrete extracellular enzymes which degrade pectin and cellulotic components of cell walls, and it has been assumed that the products of the degradation could clog conductive elements. There is insufficient proof to support this supposition.

An important factor in vascular plugging is the formation of gummy substances and tyloses in the conductive tissues. Both result from the reaction of living cells surrounding the xylem vessels. Gummosis develops during the process of cell deterioration and death. These gummy substances enter the xylem elements through pores and other natural openings. These substances, differing in appearance, clog and reduce the diameter of the xylem vessels.

Tyloses are formed by the elongation of contiguous parenchyma cell membranes through pores in the conductive elements, which results in the formation of vesiculated excrescences. Tyloses may develop that are so large and numerous as to completely block a conductive element.

By clogging vessels, gums and tyloses increase the resistance to water movement, which reduces the water potential in the xylem. The extent to which water flow is xylem conductive elements will be reduced is dependent on the number and distribution of the gums and tyloses with each element.

VII. CHANGES IN THE MOVEMENT OF WATER THROUGH LEAF TISSUES

Water reaches leaves through the xylem system where it is further distributed into a network of small veins which eventually terminate in terminal cells. The water then diffuses into other leaf cells and tissues. It has been shown by several investigators that water movement in leaves encounter great, if not the greatest, resistance.[27,28]

According to Duniway,[29] the study of resistance to water movement by pathogens in leaves has not been fully investigated. However, these influences are indisputable. The status of water and water movement in leaves can be greatly influenced directly by foliar pathogens. Also, water movement to leaves can be indirectly influenced by pathological conditions located in the pathway of water movement from roots to the intercellular spaces in leaves.

The most numerous and varying types of pathological changes occur in leaf cells and tissues of infected plants. Such changes may be partial, where damaged cells continue to perform some of their normal functions, or may be complete resulting in cell death. Normal water movement occurs only in undamaged cells. In the situation where the cell suffers any amount of damage, water movement may occur along other pathways.

All leaf diseases result in an alteration of the movement of water and nutrients with in the infected organ, the extent depending on the type of pathogen involved and host reaction.

Cells of susceptible plants infected with viruses, for example, usually react by the production of mosaic symptoms and a prolonged decrease in the loss of vitality. Water transport will remain under the control of the cells and tissues of a virus-infected leaf as long as the cells remain alive, but ceases as the progression of disease severely damages or kills infected cells and tissues. Local necrotic spots and streaks can be caused by certain viruses on hypersensitive plants, such as tobacco mosaic virus in *Nicotiana lutinosa* and grapevine fanleaf virus in *Chenopodium amaranticolor*. Obviously, there is no control of water movement in necrotic tissue, so that large amounts of water can be lost through the damaged tissue by evaporation. The greater the damaged area, the greater the loss of moisture.

Several bacterial and fungal pathogens cause necrosis of leaf cells and tissues, often in the form of necrotic spots and streaks. Such changes are caused by biotrophic fungi, such as *Peronospora parasitica, Phytophthora infestans,* and *Plasmopara viticola,* and the necrotrophic fungi, *Cercospora beticola, Septoria tritici,* and *Polystima rubrum*. Necrotic spots and streaks are also caused by many pathogenic bacteria, including *Pseudomonas syrinae* pv. *phaseolicola,* P. *syrinae* pv. *tabaci,* and *Xanthomonas campestris* pv. *vesicatoria*.

Wilting of leaves is also a symptom associated with reduced water supply. It occurs on leaves as a direct consequence of necrosis of the midrib caused, for example, by *Phomopsis helianthi* on sunflower; and veins caused, for example, by *Xanthomonas campestris* pv. *campestris* on cabbage leaves and by the potato Y virus on potato leaves.

Vein necrosis disrupts the normal water supply within leaves and, without sufficient water, water cells loose turgor and if the water shortage persists, the entire leaf may die. If necrosis occurs in the midribs or vessels of the petiole, the entire leaf blade will wilt and eventually die. Leaves affected by these necrotic processes lose control of water movement coming into the leaf and that lost through transpiration.

VIII. CHANGES IN PLANT TRANSPIRATION

The rate of water movement within a plant depends on transpiration, which is the evaporation of moisture from the mesophyll cells into the intercellular spaces of the leaf, then diffusion through natural openings, such as the hydathodes, lenticels and stomata. Pathogens on infected leaves alter these processes, thus influencing both the amount of water in plant tissues and the extent of its utilization. Changes in the rate of transpiration occur as a consequence of changes in one or all of the following structures: mesophyll cells, stomata and other natural openings, cuticle, or in the atmosphere immediately adjacent to the plant organ.

Pathogens influence transpiration by causing injury to mesophyll cells and by colonizing the intercellular spaces. Injuries to cell walls and membranes usually are caused by biotrophic and necrotrophic bacteria and fungi (see section on structural cytopathological changes). Developing in the intercellular spaces, these pathogens interfere with the evaporation of water vapor.

Phytopathogenic bacteria and fungi can interfere with the normal activity of hydathodes, lenticels, and stomata by blocking transpiration directly, or indirectly by interfering with water movement to the areas of transpiration. For example, biotrophic and necrotrophic fungi penetrate hosts indirectly through the natural openings of leaves and stems thus blocking them and interfering with transpiration. Endophytic, extracellular fungi, such as members of the Hyphomycetes and Peronosporales, block stomata with condiophores which grow through them to produce their conidia on the outside of the tissues.

All phytopathogenic bacteria penetrate their hosts directly, which includes natural openings such as hydathodes, lenticels and stomata. After penetration and reproduction, the stomata aperture and substomatal chambers become blocked by the bacterial growth. Such colonization and pathogenesis interferes with water transpiration. Farrella et al.[30] studying the biotrophic fungus, *Phytophthora infestans,* reported that this pathogen stimulated the opening of stomata prior to the protrusion of its conidiophores through them; and that such stomata remained open longer than those of uninfected leaves when exposed to drying conditions. Similarly, the necrotrophic fungus, *Cladosporium fulvum*, affected stomata n infected tomato leaves increasing stomatal openings by 30%.[24] The same investigators showed a decrease in transpiration occurred on grapevines infected with the biotrophe *Plasmopora viticola*, due to the infection of the stomatal guard cells as well as the xylem vessels.

In summary, any alteration in the water potential caused by changes in transpiration in the plant also influences stomatal activity. Such alterations can be caused by a variety of pathogens which disturb the uptake and transpiration of water in the plant.

Pathogenetic fungi can also damage the cuticle in the course of their pathogenesis, thus significantly increasing water loss. Some examples of such fungi include the subcuticular pathogens, *Venturia inaequalis* on apple leaves and *Diplocarpon rosae* on rose leaves, as well as several rust (Uredinales) and white rust (Albuginaceae) fungi. Due to mechanical damage to the cuticle and walls of epidermal cells, cuticular transpiration is increased several fold. Cuticular transpiration accounts for 5 to 10% of the total in uninfected plants.

Transpiration can be decreased through the interference of microorganisms developing on the surface of plant organs, thus disturbing the normal gaseous exchange between the plant and the atmosphere. Such microorganisms reduce the exchange of various gases, such as carbon dioxide, and water transpiration. Among the fungi of this group are the powdery mildew fungi (Erysiphaceae) and the superficial, saprophytic sooty molds caused by species of *Antenaria* and *Apiosporium*

Studying powdery mildew of barley, Priehradny[31] reported that early infections by the pathogen significantly decreased transpiration in barley leaves grown under light, while in later stages of infection, an increase in transpiration was recorded. The decreased transpiration under light was due to abnormal functioning of the stomatal apparatus. On the other hand, leaves with a heavy covering of sooty mold cuts down on both light and gaseous exchange, thus reducing the transpiration rate.

In summary, disorders and dysfunction of the movement of water in diseased plants can be considered as a general pathological condition. Such disruptions can take place at any point in the water movement pathway beginning with absorption from the rhizosphere to transpiration through natural openings. The mechanisms involved in these disruptions may differ and depend on the pathogen involved and the host reaction to it. Plant cells that are structurally or physiologically altered lose control of the uptake and transport of water usually reducing its availability and utilization. As a result of a reduced or an inconsistent water supply, both nutrients and moisture are deficient and cells become abnormal and eventually die

X. CHANGES IN MINERAL NUTRITION OF DISEASED PLANTS

Mineral nutrition in plants depends on a series of processes for the uptake of those elements essential for normal growth and development. The uptake of minerals ranks among the fundamental functions ensuring plant preservation.

Mineral elements are essential in cell structure and metabolism. They are indispensable as part of the various cell constituents, various inorganic and organic compounds, enzymes, etc. Among these constituents are buffers and coenzymes, which take part in cell functions as well as maintaining the internal osmotic pressure of the cell and membrane permeability. They also act as catalysts, inhibitors and in general, regulators of cell metabolism.

Popvić[18] classified various elements as to their role in plant development; those that are indispensable such as B, Ca, Co, Cu, Fe, K, Mg, Mn, Mo, P, S and Zn; and those that are useful are Cl, Na, and Si. Three elements, B, Cl, and K, although performing certain functions in the cell, do not represent "construction" cell constituents.[32]

Most mineral uptake of plants is through ion absorption from soil solutions through the root system. Ion absorption can take place through the nonsuberized root parts just behind the root tip. A significant amount may also be absorbed through the corky portions of roots by diffusion through lenticels and openings resulting from lateral root death and deterioration.

Kozlowski[17] described the uptake of mineral elements in a number of consecutive states; movement of the ions to the root surface, ion accumulation in root cells, radial ion transport into the xylem, and ion transport from the roots up to the above-ground parts of the plant. Ions reach the leaf veins and enter into the intercellular spaces through the mesophyll cells and are absorbed by the cytoplasmic membranes of cells and transferred into the cytoplasm. Ions may be transferred from cell to cell through the plasmodesmata by symplastic ion movement.[17,32] If the cells absorb more mineral nutrients than needed, the surplus is distributed through the phloem to other plant tissues and organs.

Mineral uptake and distribution of minerals is a continuous plant function and represents an endless process of mineral cycling in nature. It begins with the uptake of minerals from the soil by the plant, becomes part of the constituents of the plants, and then is returned to the soil following plant death and deterioration. However, various abiotic and biotic factors affect the dynamics of this recycling before absorption by the plant and while the minerals are associated with the plant.

Various plant pathogens play a significant role in the cycling during the plant processes. The detrimental influences of pathogens and other microorganisms may take place at various stages of this cycling pathway. Disorders may occur in mineral utilization in the rhizosphere, in reduced root absorption, in a decrease in the ability of the plant to transport nutrients, and in reducing the ability of cells to utilize these substances.

XI. CHANGES IN MINERAL MATTER UTILIZATION IN THE RHIZOSPHERE

The utilization of mineral matter in the soil by a plant is dependent on the formation and development of an adequate and healthy root system, on the presence of certain microorganisms in the soil, and on the chemical and structural qualities of the soil. Soil pathogens and numerous saprophytic microorganisms are capable of influencing soil composition in the rhizosphere.

Some nonpathogenic soil microorganisms transform the mineral constituents of the soil into unusable forms resulting in deficiencies in plant nutrition. Huber[33] stressed that various soil microorganisms may cause nutrient immobilization by iron oxidation in which the ferrous form is transformed into the ferric form by sulphate anion reduction to sulphur, or by utilizing degraded plant residues. Huber[33] also stated that certain bacteria oxidize magnesium in the plant rhizosphere of certain oat cultivars, which deprives these plants of magnesium, resulting in a disease called gray rot. Symptoms do not appear on cultivars resistant to this deficiency, whose rhizosphere is unsuitable for the development of these bacteria. Steinberg[34] pointed out that the bacterium *Bacillus cereus* may have a similar influence in the tobacco rhizosphere, resulting in the disease called leaf frenching. Calcium, magnesium, phosphorus, and potassium deficiencies result from the presence of large amounts of free amino acids in leaf tissues and cause the symptoms of leaf frenching in tobacco leaves.[33] By inoculating clover seedlings with a suspension of rhizosphere microorganisms, Hale[35] showed a substantial reduction in the uptake and movement of calcium and rubidium in these plants.

Microorganisms associated with soil compete for nutrients in the rhizosphere of plants. When mineral nutrients are at a low level, particularly in low fertility soils, they are less accessible to plant. Butler and Jones[36] stated that certain inorganic substances, such as nitrogen, are indispensable to both microorganisms and higher plants. In addition to nitrogen, phosphorus and potassium play an indispensable role in the nutrition of both fungi and higher plants, whereas lesser amounts of calcium, iron, magnesium and sulphur are required. Aiyar[37] showed, for example, that soil microorganisms immobilized and utilized calcium to such an extent to induce calcium deficiency on rice plants, regardless of a soil analysis indicating presence of sufficient calcium. He also maintained that large quantities of phosphorus brought into solution by bacteria in the rhizosphere may immobilize iron, in the form of ferriphosphate. This results in iron deficiency symptoms developing on plants even though a soil analysis may indicate an adequate level of iron present.

The relationships between soil microorganisms and higher plants in mineral nutrition are very complicated, depending on numerous soil factors and, therefore, deserve understanding by plant pathologists and physiologists.

In contrast, other rhizosphere microorganisms enhance the mineral nutrition of plants. For example, certain mycorrhizal fungi play an important role in the mineral nutrition of plants in subfertile soils. Mycorrhizal fungi may enhance the availability of phosphorus and other elements which would otherwise be unaccessible to plants.[33] He also stated that mycorrhizal fungi may increase the availability of the following elements: aluminum, boron, copper, iron, magnesium, nitrogen, phosphorus, sodium, strontium and zinc; and decrease that of manganese and potassium. Mycorrhizal fungi occur on many plant species, particularly forest trees.

Plant nutrition is also enhanced by the symbionic relationship between legumes and rhizobacteria, *Bradyrhizobium* and *Rhizobium* spp. In these associations, the rhizobacteria bind atmospheric nitrogen for the use of the plant, while the plant binds atmospheric carbon dioxide to meet the needs of the bacteria.

XII. CHANGES IN ABSORPTION OF MINERALS

One of the major functions of plant roots is to transport dissolved minerals and other nutrients from the soil rhizosphere into the plant, where they are used in the growth and development of the plant. This activity is performed through absorption, which is dependent on an extensive spread of the root system in the soil constantly seeking fresh sources of nutrients and water to meet the needs of the growing plant. Various factors affect the rate and energy of absorption. Root pathogens as well as animals are two of many factors that may be particularly detrimental to this function.

In general, root pathogens can partially reduce or completely inhibit root absorption of many nutrients, including calcium, magnesium, phosphorus, potassium and sodium. Certain pathological changes in roots caused by pathogens can have a greater affect on minerals that are less abundant or less accessible to the roots, such as phosphorus, than those in ample supply. Among the pathological changes caused by pathogens, root stunting, degradation, and death are particularly important.

Root hairs, which are the most active component in nutrient absorption, are delicate and thus, highly susceptible to damage. Partial or complete root hair necrosis can be caused by numerous pathogens, as for example, beet necrotic yellow vein virus on roots of sugar beets, which caused severe root stunting when plants are infected early. The stolbur mycoplasma causes, at first, root hair necrosis and later necrosis of the entire root system of infected pepper plants. Such damage to the roots results in nutritional disorders followed by plant dwarfing and severe yield reductions.

Root degradation such as that which occurs with various root rots and white root rots, seriously reduces the absorptive activity. It has been estimated that soilborne pathogens, including nematodes, can reduce absorption of nutrients and water by affected root systems from 5 to 10% or more depending on soil conditions, microorganisms involved, and available nutrients.[33]

The most detrimental pathogens belong to a group that cause postemergence damping-off and root rot of seedlings. Often in such root rots, decay progresses rapidly and nutrient uptake is quickly interrupted and young plants die quickly. Examples of pathogens causing this type of disease are *Pythium debaryanum* and *Rhizoctonia solani*. Another group, made up of *Fusarium* spp. which contain vascular pathogens, generally cause a more progressive root rot of herbaceous plants resulting in a slower death or sometimes recovery. Similarly, root rots of woody plants caused by *Armillaria mellea* and *Rosellinia nectrix,* cause a progressive root rot (see section on pathological changes in roots).

The underdevelopment of roots is another disorder resulting in the altered or decreased uptake of nutrients. This disorder occurs as a result of abnormal cell formation and differentiation during the rapid growth period. All root pathogens are capable of preventing normal root formation, thus reducing the absorptive surfaces. The influence of various pathogens on root size differs depending on the type of pathogen involved, host susceptibility, and in part on the stage of root development when infection takes place. For example, many pathogens, such as the obligate parasitic fungi which infect primarily the above-ground parts of a plant, do not infect roots and therefore do not affect nutrient uptake.[33] However, Martin and Hendrix[247] reported a significant reduction in the number and size of wheat roots infected with *Puccinia striiformis*. Similar observations were recorded on barley plants infected with the powdery mildew fungus and on tobacco plants infected with the yellow dwarf mosaic virus.[39,40] Roots of tobacco plants infected with the yellow dwarf virus were reduced to 50% of the controls with a concomitant decrease in nitrogen in leaves.[33]

Finally, plants with large and deeply developed root systems are often more successful in combating disorders of nutrient uptake. Such plants can form new roots and sometimes new root systems to counteract those that are infected or have died.

XIII. CHANGES IN TRANSPORT OF MINERALS

All mineral elements are transported from roots by sap through the xylem system to all parts of the plant. In order to perform such a function, the xylem has developed into a complex tissue. Various pathogens, primarily fungal and bacterial ones have evolved as vascular and tracheiphylic pathogens colonizing xylem tissues. These pathogens through degradation of the xylem cell constituents and toxic activity cause necrosis in the system. Such disorders result in disturbances or even prevention of normal nutrient transport through the affect xylem components. This group of specialized pathogens include several fungal genera, such as *Fusarium* and *Verticillium* spp., as well as bacterial genera, such as *Corynebacterium, Erwinia, Pseudomonas,* and *Xanthomonas* spp.

Tyloses and resinous substances are formed in xylem vessels in response to colonization by these pathogens and are examples of specific barriers which inhibit mineral transport in addition to other functions of the vascular elements (see section on pathological changes in xylem).

Because of this disruption in the transport of nutrients, plants develop symptoms of chlorosis, stunting, leaf rolling and general dwarfing of various organs. Typical symptoms of such changes occur on pepper plants infected with *Verticillium albo-atrum,* potato plants infected with *Pseudomonas solanacearum*, and other crops with other vascular pathogens. Tyloses have been observed, for example in barley plants infected with the barley yellow dwarf virus and in stone fruit trees infected with *Clasterosporium carpophilum.*

Structurally altered xylem cells as well as partially or completely blocked conductive elements restrict or prevent the upward movement of minerals to various parts of the plant. Therefore, various mineral deficiency symptoms develop in infected plants, regardless of the availability in the soil. For example, leaves of tomato plants infected with the tobacco mosaic virus have a significant decrease of up to 50% on a dry matter basis of copper, manganese and zinc; and a slight influence of aluminum, boron, calcium, iron, magnesium, and potassium.[41] It has not been established whether changes in the levels of these elements are a consequence of reduced root absorption or of transport in the xylem.[42]

Finally, a general symptom of nutrient transport disorders as well as abnormal nutrition are reduced yields and in extreme cases, death of plants (see section on plant cell pathogens).

XIV. CHANGES IN NUTRIENT UTILIZATION IN THE CELL

Mineral nutrients leave the xylem and move into the intercellular spaces of the various plant tissues. The nutrients then move from the intercellular through the cytoplasmic membrane into the cytoplasm where they may be used in the life sustaining functions of the cell. In order for the mineral ions to move against the concentration gradients through the cytoplasmic membrane and into the cytoplasm, energy is required. The extent of nutrient utilization in the cell is influenced by several factors including selective cell membrane permeability and ion immobilization, both of which can be affected in infected or damaged tissues.

Nutrient ion transport from the intercellular spaces and into the cell is performed with the aid of plasmalemma permeability. Therefore, any disorder in the structure of plasmalemma will influence cell permeability. Only undamaged cell membranes remain functional. Any lasting or irreversible disruption in the selective permeability of cell membranes eventually leads to death of the affected cell or cells.

Selective permeability may be altered under the influence of various pathological factors. Plant pathogens with their potential enzymatic, toxin and/or mechanical activity, may cause chemical and structural changes in the cell membranes resulting in their dysfunction. Any alteration or damage to membrane permeability allows for leakage from the damaged cells providing a source of nutrients for the pathogen involved. Such cell leakage is a general occurrence in the infection process. Changes in membrane permeability are also accompanied by changes in the electrical potential of the membrane, viscosity in the protoplast and other chang s, which are the most detrimental of cytopathological changes (see section on cell membrane structural changes).

Most minerals may be cycled and recycled many times during the life of a plant; and sometimes certain minerals will be accumulated in excess in certain areas, where they may be immobilized and stored excluding them from further use. Such an accumulation of minerals have been observed associated with infected and necrotic areas. Such mineral accumulations are associated with changes in cell membrane permeability. The accumulation of mineral matter associated with infections may be toxic to the pathogen involved or may stimulate the activity of secondary meristematic tissue, which often is associated with infections and wounds and forms secondary peridermal tissues. Thus, even though the accumulation of mineral nutrients may be detrimental to plant nutrition, such an accumulation may also rep esent a protective mechanism against pathogen attack.

XV. PHOTOSYNTHESIS IN DISEASED PLANTS

In a process that is not totally understood, chlorophyll synthesizes carbohydrates using light energy, atmospheric carbon dioxide and water in green plants. The photosynthetic bound light energy is basic source energy for the formation of overall organic matter synthesized by autotrophic and heterotrophic organisms. The photosynthetic process has always attracted and continues to attract the attention of biologists, biochemists, biophysicists, cytologists, physiologists and other researchers.

Infection by plant pathogens can significantly influence the photosynthetic process. Changes in the photosynthesis due to pathogens represent one of the most important areas of research by plant pathologists and biologists. Pathological changes may influence photosynthesis at any one of its various steps in the chemical process. A brief survey of some of the features of the photosynthetic process is necessary.

The fundamental process of photosynthesis consists of the conversion of light energy into chemical energy and the utilization of the chemical energy for the conversion of atmospheric carbon dioxide into carbohydrate.

The absorption of light (sunlight) is performed by a large number of molecules of a variety of compounds. The ability of such absorption is characteristic for molecules of various pigments present in photosynthetic cells. Green pigments, the chlorophylls, play a leading role in sunlight absorption with supplementary or auxillary pigments, such as the carotenoids playing secondary roles.

Electrons, part of the structure of atoms, absorb photons the smallest quantum of light. A molecule is made up of electronegative electrons circling around the electropositive atomic nucleus. In stable atoms, the number of electronegative electrons is equal to their energy charge which in turn is dependent on the distance of their pathway from the nucleus; those electrons close to the nucleus have less energy than those more remote.

The deviation of an electron from an electronegative particle from the nucleus requires external energy. In the case of photosynthesis, this external energy is the absorbed photons of light. Electrons so deviated to the external pathway tend to return to the original lower pathway with the result that the atom is in a state of instability or excitement. When an electron returns from a higher to a lower energy state, the energy which has been absorbed from the photons is released in the form of light energy (fluorescence) with the simultaneous return of molecules to their original state.

In chloroplasts, the absorbed light energy is converted to chemical energy through a process of photosynthetic phosphorylation, or photophosphorylation. During this process, energy-rich adenosine triphosphate (ATP) and reduced nicotinamide adenine dinucleotide phosphate (NADPH) are formed as follows: energy-rich electrons in the excited chlorophyll molecule leave the pathway of these molecules and are transferred within the chloroplast to a series of electron carrier compounds through the mediation of corresponding oxidoreduction enzymes. The released energy-rich electrons in this series are accepted by ferredoxin transforming it into its reduced form; the reduced ferredoxin transfers electrons to the next acceptor, most likely to be a flavoprotein, which in turn transfers them to cytochromes b and f.[43] This electron transport is performed on the basis of differences in the redox potential of the compounds taking part in the transport system. Each pair of equal or equivalent electrons leaving the chlorophyll molecule releases along its transport pathway at least 20,930 joules of energy before returning once again to the chlorophyll molecule. The energy thus released from the electrons is built into the ATP and NADPH compounds as chemical energy.

Depending on the circular electron flux in the electron transport system, photosynthetic phosphorylation may be cyclic or noncyclic. In cyclic phosphorylation the electrons which originally left the chlorophyll molecule and has passed through the entire system, returns once again to these molecules. Two ATP molecules are formed from adenosine diphosphate (ADP) and phosphate (P) per each pair of electrons passing along this pathway. Only ATP as an energy-rich compound is formed in cyclic phosphorylation:

$$\textbf{XVI. LIGHT + ADP + PI —— ATP + } \mathbf{H_2O}$$

Cyclic phosphorylation occurs as a system of light energy accumulation for transformation into ATP.

During noncyclic phosphorylation, NADPH and ATP are formed as energy compounds in the chloroplasts of photosynthetic cells as follows:

$$\text{light} + \text{NADP} + \text{ADP} + \text{PI} + 2H_2O \rightarrow$$

$$\rightarrow \text{NADPH} + H^+ + 1/2\ O_2 + \text{ATP} + H_2O$$

Electrons released from excited chlorophyll molecules are transferred to ferredoxin and reduced as in the case of cyclic phosphorylation. Electrons then are transferred from reduced ferredoxin to pyridine nucletide ($NADP^+$, nicotinamide adenine dinucleotide phosphate), which transforms them into its reduced form (NADPH). Electrons are used for the $NADP^+$ reduction and therefore, distinction from cyclic phosphorylation, they do not return into chlorophyll molecules.

In noncyclic phosphorylation, chlorophyll compensates for the lost electrons by accepting electrons from chlorophyll b molecules previously excited by photons. Electrons are transported between the two pigments through the mediation of a series of carrier compounds simultaneously forming ATP molecules, rich in chemical energy. Compensation for electrons lost from chlorophyll b is accomplished, according to some opinions, by electrons whose donors are hydroxyls (OH^-) formed by light degradation, water photolysis (Hill reaction).

Thus, electrons are transferred from water to chlorophyll molecules and then, through the mediation of the absorbed sunlight to $NADPH^+$ which is reduced to NADPH, take part in the noncyclic electron transport. In its reduced form, pyridine nucleotide is accumulated in the chloroplasts where it serves in the reduction of CO_2 during photosynthesis. In a completed phosphorylation cycle one NADPH molecule and one or two ATP molecules are formed.[43]

Compounds synthetized during photosynthetic phosphorylation, i.e. ATP and NADPH, are used in the second stage of photosynthesis for the reduction of CO_2 and its conversion into the carbohydrate form:

$$6CO_2 + 18\ \text{ATP} + 12H_2O + 12\text{NADPH} + 12\ H^+ \rightarrow$$

$$\rightarrow C_6H_{12}O6 + 18\text{Pi} + 18\text{ADP} + 12\text{NADP}^+$$

Atmospheric CO_2 diffuses into the substomatal cavity and into the cytoplasm through the cell membrane and finally into the chloroplast stromata through the chloroplast membrane. There CO_2 is bound to certain substances in the chloroplast stroma and, based on the mode of its initial binding, three metabolic types of plants are distinguished.[17,44] Most CO_2 binding is performed by the reductive pentose phosphate cycle (type C_3). In this cycle CO_2 is bound first to ribulose-l, 5-diphosphate with the aid of the enzyme carboxydismutase (ribulose-l, 5-diphosphate carboxylase). According to Goodman et al.[45] this cycle consists of the carboxylation, reduction, and a regeneration stage.

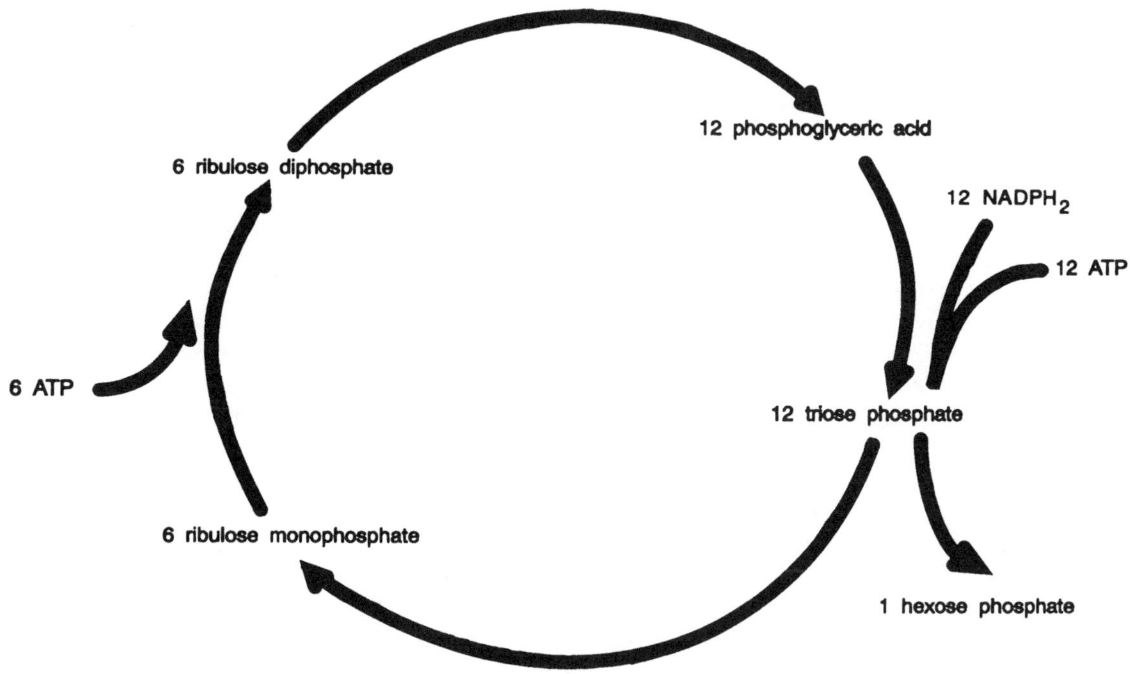

FIGURE 90. Cycle mechanism for the synthesis of hexose phosphate from carbon dioxide by photoautotrops.

In the carboxylation stage ribulose-5-phosphate converts into ribulose-l, 5-diphosphate which accepts CO_2 and then provides two molecules of phosphoglyceric acid:

ribulose-5-phosphate + ATP→
ribulose-l, 5-diphosphate + ADP;

ribulose-l, 5-diphosphate + CO_2→
2, 3-phosphoglyceric acid

In the reduction stage the binding energy results in the synthesis of triose phosphate molecules which practically completes the conversion of CO_2 into sugar.

The major part of triose phosphates formed (glyceraldehyde-3-phosphate and dihydroxyacetone phosphate) is used for the resynthesis or regeneration stage of ribulose-l, 5-diphosphate taking part in continuous CO_2 binding. The other part of synthetized triose phosphate passes through the chloroplast membrane into the cytoplasm,[46] where it is transformed into sugar and other carbohydrates (starch). The main features of the cycle mechanism for the synthesis of hexose phosphate from CO_2 by photoautotrophy is shown (Figure 90).

Carbon compounds synthetized in the process of photosynthesis, represent fundamental compounds for direct and indirect organic matter synthesis in living organisms.

XVII. PATHOLOGICAL CHANGES IN PHOTOSYNTHESIS

Pathological changes effecting photosynthesis are a regular occurrence in infected green plants and occur due to direct infection of photosynthetic cells, or as a consequence of infection of other cells and tissues which indirectly influence photosynthetic activity. The dynamics of such changes differ and depend on the type of pathogen, host susceptibility and environmental conditions during pathogenesis.

With their enzymatic and toxic activity necrotrophic bacteria and fungi cause early and rapid death of photosynthetic cells and tissues which, in turn, results in a rapid reduction and arrest of photosynthesis. Cell death due to pathogen action results in the direct loss of photosynthetically-active plant parts.[47] Necrotrophic pathogens are capable of saprophytic development in necrotic cells and therefore, become independent of the energy system in the photosynthetic cell. By using nutrient constituents of necrotic cells, these pathogens form their own energy systems. The facultative parasitism by necrotrophic pathogens is based on such activity. According to Rubin and Artsikhovskaya[16] facultative parasites reduce plant photosynthetic activity at the beginning of infection, but becomes progressively less with disease development.

In contrast, a living cell represents an irreplaceable source of energy for growth and development of biotrophic pathogens. Biotrophes depend on host cell metabolism for survival and use the products of photosynthesis as well as the energy system of the infected cell for their growth. Photosynthetic disorders in such cases are slower ensuring the completion of the life cycle of the pathogen involved.

In view of their dependence on host cells, biotrophes usually do not alter the photosynthetic activity of the infected plants at the beginning of infection and may even increase it in some cases. However, as infection continues to develop photosynthesis becomes progressively reduced compared to noninfected plants; the rate dependent on the pathogen involved.

The dynamics of photosynthetic changes caused by biotrophic pathogens depends on the parasitic relationship established between the host cells and the pathogen involved. For example, during the early epiphytic development of powdery mildew fungi there is little or no reduction in photosynthetic activity, but a more rapid one. Allen[48] showed that wheat plants infected with *Erysiphe graminis* had a high rate of photosynthetic activity at infection, which was later reduced to 10% of that in healthy plants.

An initial photosynthetic rate increase followed by a pronounced reduction in later stages of pathogen development is characteristic of other biotrophic intercellular pathogens. For example, wheat plants infected with *Puccinia triticina* showed a significant increase in photosynthesis at the beginning of infection, followed by a reduction as infection progressed.[49] Wood[47] obtained similar results on beans infected with *Uromyces phaseoli*. According to Kuprevich[6] rust fungi which form or fail to form diffusive mycelia in its initial stages of development do not cause a significant reduction in the photosynthetic processes up to the time of sporulation. Montalbini and Buchanan[50] also observed that photosynthesis does not undergo significant changes in plants during the early stages of infection by biotrophic fungi. Dorohova[5] recorded a stimulation of photosynthesis in potato leaves the first day following infection by *Phytophthora infestans*.

Viruses, in contrast to other biotrophes, are intracellular obligate parasites which multiply exclusively using the chemical constituents and energy of the host cells. Therefore, the products of photosynthesis are indispensable for the synthesis of virus particles. The reduction of photosynthetic activity during the course of viral disease development represents one of the most detrimental in diseased plants. Certain dynamic changes take place in photosynthesis in virus-infected plants which result from this host-parasite relationship. Generally, this relationship is characterized by the maintenance of photosynthetic activity during the period of rapid virus replication, which is followed by a progressive decrease until the cell dies.

According to some authors, tobacco mosaic virus caused an increase in photosynthetic activity in infected tobacco plants.[51,52] Other studies on the protoplasts isolated from Chinese cabbage (*Brassica chinensis*) infected with the turnip yellow mosaic virus showed that photosynthesis increased during active virus replication and decreased with the occurrence of symptoms.[53]

Many examples concerned with the decrease of photosynthesis associated with disease have been published.[53] For example, Goodman et al.[45] stated that the rate of photosynthesis may be reduced as much as 20 to 25%, and Lovisolo et al.[54] stated it may be as much as 20 to 50%. Šutić[53] stated that the plant viruses causing yellows and leaf roll diseases are more detrimental to the photosynthetic process than those that cause mosaic-type diseases.

A. SPECIAL CHANGES IN PHOTOSYNTHESIS

The photosynthetic process involves a number of biochemical stages resulting with the formation of carbohydrates. Any disruption in any of the biochemical processes influence those succeeding, thus influencing the entire process. Those effects on specific stages in the photosynthetic process have been the subject of numerous investigations, some of which will be considered.

Changes in the penetration of light. Any obstruction in the pathway of light to the photosynthetic centers can seriously affect the process. Many examples of such interruptions by parasitic and saprophytic fungi which develop on the surface of photosynthetic organs can be given. Mycelia of the saprophytic fungi, *Antenaria* spp., and *Capnodium* spp., colonize the honey dew from insects and form dense colonies on the surface of herbaceous organs of a wide variety of susceptible plants, such as lemon, olive, orange, and plum. The colonies appear at first as sparse then become more dense, black areas sometimes covering the entire plant organ. It is referred to as sooty mold. These superficial colonies differ in consistency and thickness depending on environmental conditions. The larger and more dense the colony the less light can penetrate and the presumed greater effect on photosynthesis. There is little reliable data on the direct influence sooty mold may have on light absorption, but it is probably similar to that cause by other light-inhibiting factors. According to some authors, sooty mold in combination with reducing CO_2 absorption may significantly reduce photosynthetic activity.[6] This condition, if allowed to persist can eventually damage whole orchards of fruit trees and even cause the death of trees within a few years.

Pathogenic organisms may also affect the penetration of light, such as the powdery mildew fungi (Erysiphales). These pathogens develop almost exclusively on the surface of herbaceous organs, where the mycelium forms a white to ashy colored growth. This mycelial layer may both reflect light and act as a barrier to light penetration. Rubin and Artsikhovskaya[16] emphasized that when the mycelial growth of these fungi cover the surface of a leaf which is photosynthetically active that it inevitably reduces the rate of photosynthesis. Kuprevich[6] stated that the degree of reduction in the photosynthetic rate depended on the thickness of the mycelial deposition on the leaf surface. Any changes in the photosynthetic rate associated with the powdery mildew fungi may also be influenced by other factors, such as the effect of the infection process by these fungi, as well as restricting the gaseous exchange of CO_2 and O_2

In order to utilize solar energy most efficiently, the leaves of plants are shaped and arranged to provide the greatest exposure to sunlight. However, this architecture can be threatened by various pathogens that significantly alter leaf shape and arrangement. For example, any leaf malformations, such as leaf rolling or epinasty, reduces the exposure of that organ to light. Leaf rolling, which begins at the leaf margins and continues to the midrib will prevent light from reaching almost the entire leaf surface, can be caused by certain viruses; cherry leaf roll, grapevine leaf roll, and potato leaf roll viruses. Potato leaf roll virus caused a pronounced reduction in photosynthesis at all stages of disease development.[6]

Similar interference with light penetration can be expected in plants infected with pathogens that cause leaf epinasty. Leaf epinasty is characterized by the turning down or bending of leaf or leaflet blades along the midrib from the apical point downward. Leaf epinasty occurs on tomato plants infected with *Fusarium* or *Verticillium* spp., and on Spy 227 apples by the Spy 227 apple virus. Epinasty ranks among the most detrimental disorders on leaves as related to its effect on photosynthesis.

Changes in leaf arrangement in infected plants can also alter light intensity, resulting in the reduction of light penetration. Changes in leaf arrangement occurs in witches' broom diseases, such as that caused on cherry trees by *Taphrina cerasi* and on plum trees by *T. insititiae;* in rosette diseases, caused on peach by the peach rosette mosaic virus and on lily by the lily rosette virus; in stem proliferation diseases, such as caused on pepper plants by the cucumber mosaic virus and on apples by the proliferation mycoplasma. With all of these diseases the leaves are irregularly arranged, usually are smaller than normal, often chloratic, and fall off prematurely (see section on pathological changes in leaves).

Changes in light absorption. The absorption of light after the penetration into the photosynthetic cells is performed primarily by the green pigments, the chlorophylls (see section on photosynthesis in plants). Absorption efficiency is based, in part, on the quantity of chlorophyll present and its continuous formation in the cells. Disruption of light absorption can be the result of reduced quantities of chlorophyll in the photosynthetic cells and tissues. This reduction may be caused by a variety of factors including; insufficient chlorophyll synthesis, loss due to chloroplast degradation, necrosis of photosynthetic cells and tissues, reduced surface of photosynthetic organs and tissues, and the loss of leaves.

The reduction in the synthesis of photosynthetic pigments often occurs in plants infected by completely or partially systemic pathogens. Infections caused by such pathogens affect individual parts or whole plants and can spread throughout the plant influencing chlorophyll synthesis anywhere in the infected plant. The greatest reduction in pigment formation may be expected in young leaves developing after infection. Examples of biotrophic fungi that reduced chlorophyll reduction are *Peronospora tabacina* on tobacco and *P. schachtii* on sugar beets; and *Uromyces pisi* on plants of *Euphorbia cyperissias*. Among the necrotrophic fungi that reduce chlorophyll formation are the vascular pathogens in the genera *Fusarium* and *Verticillium*, which disturb or prevent the normal functioning of vascular tissues through various means.

Plant mycoplasmas and viruses which cause systemic infections may have particular importance as inhibitors of chlorophyll synthesis, since they continue to spread in plant parts during their development and growth. Matthews[42] stated that the inhibiting action of pigment formation by viruses probably is more common on young leaves developing simultaneously with virus particle replication than on mature leaves. Chlorophyll a and b degradation was reported in sugar beet leaves of plants infected with the beet necrosis yellow vein virus.[55] Also, chlorophyll a content was reduced in Chinese cabbage leaves infected with the turnip yellow mosaic virus.[42] In this disease and others, hormones may have had a stimulative role in the degradation of the chlorophyll molecules. Matthews[42] stated that ethylene is increased in leaves with chlorotic spots caused by viruses in some plants. Increased amounts of ethylene have been reported in the chlorotic spots on plants of *Tetragonia expansa* infected with the bean yellow mosaic virus,[56] as well as in cucumber cotyledons infected with the cucumber mosaic virus[57] (see section on chromatic changes in chlorophyll).

The absorption of light by chlorophyll depends on the structural integrity and normal functioning of the chloroplasts. As a general rule only uninjured chloroplasts remain biologically active. However, chloroplasts rank as some of the most susceptible to pathological changes among cell organelles. Their development may be disturbed or prevented by a variety of abiotic and biotic factors. Various disorders may occur in their structure and chemical composition resulting in detrimental influences on their functions. Various investigations have shown that such disorders may occur in all parts of the chloroplast, including the grana, lamellae and membranes. The consequence of such disturbances is the inhibition of its normal functions (see section structural chloroplast changes).

Numerous investigations have shown that the amount of chlorophyll in an infected plant may be reduced both as a consequence of damage to the chloroplast and the chlorophyll. Chlorophyll reduction depends on a number of factors such as the rate of parasitic development and host plant susceptibility. Reduction in quantities of chlorophyll reduces photosynthetic activity, but numerous studies have shown that there is no linear correlation between reduced chlorophyll quantities and a reduction in photosynthesis. This correlation may approach a linear one in the case of infection of plants by necrotrophic rather than by biotrophic pathogens. In studies on wheat plants infected with the powdery mildew fungus, Allen[48] found high rates of photosynthesis in early stages of infection regardless of reduced quantities of chlorophyll, whereas in later stages of infection photosynthesis decreased at a faster rate than did the quantities of chlorophyll. These and similar examples indicate that photosynthesis does not depend entirely on the amount of measurable chlorophyll present, but on other changes taking place in the infected cell.

Pathogens causing death of plant cells and tissues have a very influential effect on the reduction of chlorophyll. In such cases the reduction in chlorophyll content is approximately proportional to the size of the necrotic area involved. The consequence of such a reaction is a reduction in the photosynthetic rate proportional to the loss in chlorophyll.

Examples of such changes in the amount of chlorophyll and the rate of photosynthesis are found in diseases caused by facultative parasites, such as the phytopathogenic bacteria *Pseudomonas syringae* pv. *lacrymans, P. svringae* pv. *tabaci, Xanthomonas campestris* pv. *malvacearum* and *X. campestris* pv. *vesicatoria*; and among the necrotrophic fungi such as *Alternaria solani, Cercospora beticola,* and *Exerohilum turcicum.* Rapid death of cells and tissues accompanied by destruction of chlorophyll also occurs in hypersensitive reactions by some plants infected with obligate parasites, such as by some American grape cultivars infected with *Plasmopara viticola* and plants of *Nicotiana glutinosa* infected with the tobacco mosaic virus.

Also, the quantity of chlorophyll and number of photosynthetic cells are lower in leaves which are reduced in size compared to normal leaves. For example, a decrease in chlorophyll occurs in filliform or thread-like leaves of tobacco and tomato plants infected with the cucumber mosaic virus. Also, defoliation, caused by many pathogens, results in the total loss of photosynthetic organs with the result of complete loss of photosynthesis.

Changes in photosynthetic phosphorlation. The level of photosynthetic phosphorylation depends on the quantity of chlorophyll present, since the electrons of the chlorophyll molecule circle in the chloroplasts as carriers of light energy, i.e. the rate of phosphorylation is dependent on the amount of chlorophyll. However, this relationship is influenced by pathogens and their interaction with their hosts.

Photosynthetic phosphorylation is altered to a greater extent by necrotrophes than by biotrophes, the former causing a more rapid cell and tissue necrosis with a corresponding rapid decrease and eventual destruction of chlorophyll. However, studies of phosphorylation in plants infected by obligate parasites have shown that there is a gradual reduction in phosphorylation with can be studied over longer periods of time.

It was reported that the biotrophe *Uromyces fabae* reduced noncyclic photophosphorylation in the chloroplasts of *Vicia faba* by 29 to 30% with a decrease in the cyclic phosphorylation of 10%.[50] In contrast, no changes in phosphorylation were recorded in chloroplasts isolated from an oat plant infected with rust.[58] However, Magyarosy et al.[59,60] reported that both powdery mildew and rust fungi preferentially inhibited noncyclic photophosphorylation. This inhibition resulted from the blocking of electron transport from water to NADP. For example, the noncyclic photophosphorylation in chloroplasts of sugar beet leaves infected with *Ervsiphe polgoni* was reduced 50% while that of the cyclic phosphorylation remained unaltered. Also, in chloroplasts isolated from cotton leaves of plants infected with *Verticillium albo-atrum*, the Hill reaction was less efficient than in those of uninfected plants.[61]

The effect of virus infection on photophosphorylation differs among various host and parasites combinations and environmental conditions. It is presumed that the energy obtained through increased phosphorylation in the initial stages of a virus infection is used for virus particle replication. However, this activity is reduced following a period of intensive virus multiplication. For example, Goffeau and Bové[62] established that both the cyclic and noncyclic photophosphorylation and the Hill reaction increased in chloroplasts isolated from Chinese cabbage leaves infected with the turnip yellow mosaic virus during the period of intensive virus multiplication. Also, Zaitlin and Jagendorf[63] showed phosphorylation and the Hill reaction in chloroplasts of tobacco leaves infected with tobacco mosaic virus changed during the course of infection. Infection by the virus did not significantly influence these chemical reactions until approximately seven days after infection. However, during the period of seven to nine days after infection, both reactions were reduced by 40%. These results show that the activity of enzymes associated with the infection decreased at a faster rate than the amount of chlorophyll. Spike and Stout[64] showed a 50% reduction in the efficiency of the Hill reaction per unit weight of chlorophyll in chloroplasts of sugar beet leaves infected with the beet yellows virus.

Thus, the effect of disease on photophosphorylation depends on the amounts of chlorophyll and the relation between the host and the pathogen involved. Reduced activity usually occurs in the later stages of pathogenesis reducing the energy necessary for normal biosynthetic activity and forcing the infected plant to use its reserves as an energy source.

Changes in carbon dioxide intake. Atmospheric CO_2 is the only source of carbon used in the photosynthetic process to produce triose phosphate which leads to the production of sugar which is used for the synthesis of all carbon compounds found in living organisms. Atmospheric CO_2 passes through the stomatal aperatures and through diffusion reaches the chloroplasts of the photosynthetic cells.

In infected or injured plant parts the prevention or inhibition of CO_2 intake may take place at any point. The first interruption may take place at the initial point of entry into the plant, i.e. through the stomatal aperatures; the second major interruption can take place during the passage through the plasmalemma, cytoplasm and chloroplast membranes to the chloroplast stromata. Disruptions at the initial point of entry can be caused by epiphytic and endophytic pathogens which prevent the opening or block the stomatal aperature thus preventing or reducing the intake of CO_2.

Epiphytic microorganisms which reduce or prevent the intake of CO_2 include the agents of sooty mold and powdery mildew which grow over epidermal tissues. These rank as the most detrimental to CO_2 intake.

In a study comparing the rate of photosynthesis in leaf halves with and without sooty mold, Kuprevich[6] found a reduction in affected leaf halves ranging from 8 to 62% depending on the quantity of mycelium present and the host. For example, the percentage reduction in photosynthesis by host plant was; *Citrus aurantium*-25%, *Gardenia thunbergia*-8%, *Nerium oleander*-17%, *Psidium aromaticum*-62%, and *Sciadophyllum ellipticum*-12%.

Powdery mildew fungi not only develop mycelial colonies on the surface of epidermal tissues, but their haustoria establish a parasitic relationship with the epidermal cells causing their death in the course of pathogenesis. Such development negatively influences the functions of the epidermal cells including that of the stomatal cells. Therefore, the powdery mildew fungi represent an important group of pathogens capable of altering the intake of CO_2. Reduction in CO_2 intake increased with the progression of the disease and the amount of leaf infection.[65]

Last[65] studied the influence of *Erysiphe graminis* on CO_2 intake (mg/dm^2/h) by barley leaves with 5, 15, or 30% of the leaf surface covered with mycelial growth at 5% infection, the amount of absorbed CO_2 was significantly decreased with the spread of infection over the leaf surface. In three trials where 15 and 30% of the leaf surface was infected, CO_2 intake was reduced between 28 to 41%, 44 and 57%, respectively.

Most biotrophic and necrotrophic pathogens can enter their hosts through stomata and some of these in the Hyphomycetales and Peronosporaceae, grow outward through the stomata from infected tissues below in order to sporulate. In both cases, the function of the stomata is impared and the intake of CO_2 reduced. The extent of reduced CO_2 intake will depend on the pathogen involved, the extent of infection and the susceptibility of the epidermal tissue to the pathogen (see section on structural changes in cell walls and in transpiration).

Vascular pathogens, although usually considered disruptive to normal vascular activity, may also influence stomatal cell functions. Duniway[29] showed that tomato plants infected with *Fusarium* sp., had an increase in resistance to CO_2 diffusion; and Duniway and Slathyer[66] further showed that tomato plants infected with *F. oxvsporum* f. sp. *lycopersici* had an increased resistance to intracellular diffusion of CO_2 associated with increased resistance to stomatal diffusion of CO_2.

Viruses also may have a detrimental influence on the diffusion of CO_2, although the data concerning this phenomenon is sparse. For example, it was found in beet leaves infected with the beet yellows virus that the number of stomata was 16% less than in noninfected leaves.[67] This virus caused disorders in the stomatal apparatus resulting in the restriction in the uptake of CO_2 and thus, represents one of the most important limiting factors of photosynthetic activity.[68] The tomato aspermy virus was shown to significantly reduce the exchange of CO_2, but without significantly influencing diffusion rates of CO_2 or transpiration.[69]

Internal diffusion of CO_2 compared to external diffusion of the gas takes place in a relatively closed system between the cell wall through the plasmalemma, cytoplasm and chloroplast membranes to the chloroplast stromata, where the reduction of CO_2 takes place. Normal diffusion of this gas takes place only as long as this entire system is undamaged. Plant pathogens through the action of enzymes and/or toxins and by mechanical damage partially or completely degrade and destroy the membrane system of cells and cell organelles (see section on cytolytic changes, structural cytopathological changes and permeability changes in diseased plant cells). With the degradation of chloroplast membranes, organelles loose their integrity resulting in free passage of their various constituents into the cytoplasm. This results in the loss of normal spatial activity between chloroplasts and cytoplasm. In such a structural disorganization, disorders in the internal diffusion of CO_2 takes place preventing the fixation of the gas for photosynthesis.

Fixation of CO_2 takes place in the chloroplast stromata where activity is dependent on the extent of chloroplast damage in infected photosynthetic cells. Any degeneration of chloroplasts is accompanied by a loss of chlorophyll which in turn results in a corresponding reduction of fixation of CO_2. The extent of the reduction of CO_2 fixation in diseased plants depends on the rate of chlorophyll reduction (chlorosis) and cell death (necrosis).

Based on such symptoms, plants infected with necrotrophic fungi reduce CO_2 fixation at a faster rate at the initial stages of infection than those infected with biotrophic ones. In plants infected with biotrophic pathogens, where the processes of pathogenesis are slower than those caused by necrotrophic ones, the rate in the reduction of CO_2 reduction is slower. For example, reduction in CO_2 fixation in barley plants infected with *Erysiphe graminis* f. sp. *hordei* takes place six days after infection when the first symptoms appear; and in bean plants, late infection with rust fungi the rate of CO_2 fixation is reduced by a third to half compared to uninfected ones, which is attributed to the intensity of infection, amount of chlorophyll degradation and destruction of cell organelles.[70] In Pinto bean plants infected with the bacterium *Pseudomonas svringae* pv. *phaseolicola*, CO_2 fixation was rapidly reduced to 15% of that of uninfected plants in a brief period suggesting a rapid if not total destruction of chloroplasts.[71] According to Mathews[42] there are numerous other examples of reduced CO_2 fixation in virus-infected plants displaying symptoms of mosaic and/or yellows. Such reductions usually are recorded within a few days after infection takes place.

Changes in CO_2 photoassimilation. Photoassimilation is the conversion of CO_2 into triose phosphate, a fundamental compound for chloroplast and cell metabolism, by the reductive pentose phosphate cycle (Calvin photosynthesis cycle). Disorders in this cycle occur due to deficiencies of ATP and NADPH. For the conversion of one CO_2 molecule into triose phosphate, two molecules of NADPH + H^+ + 2ATP molecules are required.[43] Insufficient quantities of any of these compounds result in a corresponding reduction of photoassimilation activity (also see section on photosynthetic phosphorylation).

Disorders in photoassimilation may also occur with an insufficient supply of CO_2. As previously stressed, these disturbances pertain to the external and internal diffusion of CO_2 to photosynthetic cells and their chloroplasts. Deficiencies of CO_2 may significantly reduce or block photoassimilation activity.

Disruptions in the Calvin cycle can also be caused by disorders in the activity of individual enzymes involved in it. According to Kosuge,[57] some of these enzymes are activated by the presence of light, such as fructose-l, 6-diphosphate phosphatase, glyceraldehyde-3-phosphate dehydrogenase, ribulose-l, 5-diphosphate carboxylase, and ribulose-5-phosphate kinase. Some plant pathogens reduce the amount of light required for photosynthesis (see section on light penetration). Insufficient quantities of light limit the activity of these enzymes as well as other photosynthetic processes.

One of the most studied enzymes in the reductive pentose phosphate CO_2 assimilation cycle is ribulose-l, 5-diphosphate carboxylase.[72] The quantity of this enzyme is reduced in virus-infected plant cells due to the degradation and disruption in protein synthesis. Matthews[42] stated that the activity of this carboxylase could be reduced by 20% in a virus-infected plant. Hirai and Wildman[73] explained that disruption in the synthesis of this enzyme by viral infections was due to the blocking of protein synthesis in the chloroplasts. In such a circumstances, there is a detrimental influence on the activity of other enzymes in the reductive pentose phosphate cycle.

More specifically, Crosbie and Matthews[74] showed that in Chinese cabbage leaves infected with the turnip mosaic virus that the content of the primary constituents of chloroplasts was reduced: chlorophyll a, carboxylase-l, 5-diphosphate, and ribosome 48s. These authors also showed that these changes occurred primarily after the completion of virus particle replication. Reduced ribulose-l, 5-diphosphate carboxylase (carbosydisumtase) activity was also recorded in plants infected with mycoplasmas.

The activity of the same enzyme was reduced in wheat plants infected with a rust pathogen at 12 days after infection.[75] No data are available on the mechanisms involved in the reduction of ribulose diphosphate carboyxlase activity in diseased plants, although it is possible that, in addition to other factors, any disorders in protein synthesis in the advanced stages of infection play an important role.

Changes in the chloroplasts and photosynthesis. Chloroplasts rank among the most sensitive of all cell organelles to various functional, organizational, or structural changes, which may be caused by both abiotic and biotic factors. The influence of these factors on chloroplasts has been described in many investigations.

Plant pathogens cause numerous and variable changes in chloroplasts. These changes may be reduced to either a decrease in the number of chloroplasts or to alterations in their function, organization, and structure. The various components of chloroplasts, such as lamellae, membranes and stromata, are degraded in the course of pathogenesis. In the disorganized internal structure, chloroplasts and their components are released into other parts of the cell. Degraded and damaged chloroplasts become inactive and, since they cannot be formed again, a general reduction in photosynthesis takes place.

Chlorophylls also are very susceptible to the influence of plant pathogens. As a general rule, the quantity of chlorophyll and other pigments is altered in all infected photosynthetic cells, thus influencing the structural unity of these organelles (see sections on changes in structural chloroplasts and chromatic chlorophyll).

Necrotic processes affect photosynthetic cells and tissues by reducing the total quantity of chloroplasts and chlorophyll in photosynthetic organs. Necrosis of large portions of tissues can result in the death of the affected organ, such as leaves, with the disruption of photosynthesis.

In conclusion, all plant pathogens have a detrimental influence on photosynthesis. Pathological changes may occur in all photosynthetic organelles and processes of infected plants. The most detrimental consequence of such changes is the decrease production of metabolites resulting from photosynthesis. Thus, with a deficiency of these products, infected plants use reserve nutrients for their respiration, resulting in a degradation of total plant growth and reproduction.

XVIII. RESPIRATION IN DISEASED PLANTS

Respiration is a series of oxidation-reduction reactions in which organic matter is degraded to yield energy and carbon compounds which are indispensable for cell formation, function, and preservation. Various disorders in respiration occur in damaged and infected plant cells. These disorders may affect all parts of the respiration cycle, such as enzyme activity, respiration pathways, oxidative phosphorylation, and mitochondria activity. Some examples of the effects that plant diseases have on respiration will be presented.

Plant respiration. Respiration or biological oxidation is accomplished through a series of interlinked oxidation-reduction reactions mediated by various enzymes and enzymes systems. The enzymes involved can be classified into three main groups; those that directly influence the respiratory substrate and the isolation of hydrogen; those involved in the electron transport from hydrogen donors to acceptors; and those that react directly with oxygen and stimulate its binding with hydrogen from the substrate.

The first group of enzymes include the dehydrogenase enzymes, which act directly with the substrate isolating hydrogen. The most well known of these are the anaerobic dehydrogenases, such as nicotinamide adenine dinucleotide (NAD) and nicotinamide adenine dinucleotide phosphate (NADP).

The second group of enzymes include those which do not react directly with the substrate, but serve as mediators in hydrogen transport from reduced NAD oxidized by accepting hydrogen and transporting it to a new acceptor. Activated oxygen, which together with hydrogen, forms hydrogen peroxide and water, may also serve as an acceptor. Flavoprotein enzymes, for example, such as dehydrogenase or oxidase in the form of flavin adenine dinucleotide (FAD) belong to this group.

The third group of enzymes include the oxidases. These enzymes react directly with atmospheric oxygen and catalyze the binding of oxygen with the substrate hydrogen resulting in the formation of hydrogen peroxide or water. The cytochrome enzymes belong to this group.

The oxidase group also includes some specific oxidases, such as the polyphenol oxidases (oxidizing O-diphenols into O-quinones), tyrosinase (oxidizing tyrosine into melonoid compounds), and peroxidase and catalase (degrading hydrogen peroxide into oxygen and water).

Finally, respiration of anaerobic organisms may, depending on the sequence of enzyme activity, be schematically represented as hydrogen, i.e. its electron transport from substrates, which are oxidized, to oxygen, which is reduced as follows:

$$\text{NAD} \rightarrow \text{flavins} \rightarrow \text{cytochromes} \rightarrow O_2$$

Carbohydrates are the major respiratory compounds. They ensure the greatest quantity of energy for immediate use in the metabolic processes.[76] One of the simplest substrates is glucose whose degradation during respiration is expressed in the following general formula:

$$C_6H_{12}O_6 + 6O_2 = 6CO_2 + 6H_2O$$

However, the degradation of glucose is complex requiring a series of reactions, which can be divided into two major stages. The first, the glycolysis or Embden-Meyerhof-Parnas (EMP) glycolic system, in which each glucose molecule is converted into two pyruvic acid molecules; and the second, in which pyruvic acid degradation takes place. Pyruvic acid is degraded in two different reactions. The first takes place under aerobic conditions where pyruvic acid is degraded in the Krebs cycle with CO_2 and water as final products. This type of degradation takes place in the presence of atmospheric O_2 in plant tissues. However, under anaerobic conditions, pyruvic acid is degraded by fermentation, yielding lactic acid, alcohols, and other molecules.

Glycolysis takes place in a number of stages.[43] The first stage consists of the conversion of glucose into its active form, i.e. glucose-6-phosphate, followed by the transformation of the active form of his hexose into two triose phosphate mols. By dehydrogenation, the triose phosphates are transformed into phosphopgyceric acid, which is further converted into pyruvic acid. Pyruvic acid then is subject to further degradation. Thus, pyruvic acid plays a key role in carbohydrate metabolism.

In the Krebs cycle, acetyl-SCoA or the active acetate, which is formed by oxidative decarboyxlation of pyruvic acid, is degraded into CO_2 and water. Changes in pyruvic acid begin with oxidative carboxylation in which this acid gives rise to acetyl-coenzyme A (acetyl CoA). The acetyl-radicale further alters it, in a cyclic series of mutually-linked tricarboxylic and discarboxylic acids, transforming one into the other accompanied by CO_2 and H separations in the form of NADH and FAD-2H.

Briefly and in summary, pyruvic acid oxidation into CO_2 and water via acetyl A-SCoA and the Krebs cycle is coupled with the production of reduced NADH + H^+ (or NADPH $^+$ H^+) and FAD·H_2 oxidized in the process of oxidative phosphorylation into NAD^+ and FAD accompanied by the abundant formation of ATP from ADP and inorganic phosphate. In the course of the oxidation of 1.0 mol of glucose, there is a total yield of 38 ATP mols, of which two ATPs are formed in reactions resulting with pyruvic acid synthesis.[43]

In the course of respiration, ATP formation in oxidation reactions is defined as oxidative phosphorylation, playing an irreplaceable role in the supplying the cell with energy. Inorganic phosphorus and ADP take part in the ATP synthesis, and therefore are indispensable constituents of cell respiration with the primary reaction being:

$$ADP + H_3PO_4 <\!\!\longrightarrow ATP + H_2O$$

Respiration can also take place aerobically with the process known as the pentose phosphate cycle, the hexose monophosphate cycle, or Warburg-Dickens-Horecker (WDH) pathway of glucose oxidation.

Two specific traits of the pentose phosphate cycle need to be emphasized. The first is the glucose oxidation cycle with the separation of CO_2; and the second, the pentose phosphate or ribulose phosphate cycle, after which this cycle has been named. In order for oxidation to take place, glucose must be activated by conversion into glucose-6-phosphate, the initial oxidative substance in this cycle.

The pentose phosphate pathway of glucose-6-phosphate oxidation consists of two stages.[43] In the first stage, the transition of glucose-6-phosphate into ribulose-5-phosphate takes place with the separation of 2NADPH + H^+ and CO_2, while in the second stage, there is a resynthesis of glucose-6-phosphate from xylose-5-phosphate and ribose-5-phosphate. If the newly formed NADPH + H^+ are included in the process of oxidative phosphorylation, then three ATP molecules are formed from one of NADPH + H^+ oxidized into $NADP^+$ and water:

$$NADH^+ + H^+ + 1/2\ O_2 + 3ADP + 3H_3PO_4 \longrightarrow 3ATP + NAD^+ + 4H_2O$$

For the oxidation of 1.0 mol glucose-6-phosphate along the WDH pathway into CO_2 and water, it is necessary that the cycle be repeated five times, or that 6.0 mols of glucose-6-phosphate be included in the process since CO_2 is separated during the course of one cycle:

$$6 \text{ glucose-6-phosphate} + 6CO_2 \longrightarrow 6 \text{ ribulose-5-phosphate} + 6CO_2 + 6H_2O$$

Hexose phosphate resynthesis from pentose phosphate may be expressed as the following:

$$6 \text{ pentose phosphate } (5 \times 6C) = 5\text{-hexose phosphate } (6 \times 5C)$$

During pentose phosphate oxidation of 1 mol glucose-6-phosphate to CO_2 and water, a total of 36 mols of ATP is formed. However, the actual yield of energy is 36 mols of ATP, since 1 mol ATP is used for the phosphorylation of glucose into glucose-6-phosphate.

The importance of the pentose phosphate cycle is in the formation of pentose phosphates and for the synthesis of nucleic acids and nucleotides. The pentose phosphate cycle is also the primary source of phenolic compounds, many of which are associated with defense mechanisms in plants.

From a phytopathological point of view, it should be stressed that respiration is partially performed in the cytoplasm as well as in the mitochondria. Glišić[13] pointed out that the glycolytic processes yielding pyruvic acid as their final product take place in the cytoplasm and when it reaches the mitochondria it is subject to aerobic oxidation in the Krebs cycle. Mitrochondria contain the entire enzyme complex necessary for the production of electron donors from a given substrate, for electron release, and for their binding by oxidative phosphorylation. The primary role of mitochondria is the formation of energy-rich phosphatic compounds (ATP), to ensure the energy required for all types of cellular activity.

XIX. PATHOLOGICAL CHANGES IN RESPIRATION

Respiration provides the energy and materials necessary for the formation of organic compounds during the course of ontogenetic plant development. Based on the use of respiration products, Kosuge[67] classified crop respiration as either "maintenance" or "development". Maintenance respiration ensures the subsistence and replacement of various plant structures undergoing constant change, while products of developmental respiration are necessary for plant development and growth, particularly fructification.

From the phytopathological point of view, respiration due to disease could be added to that of normal respiratory processes. As a metabolic activity, respiration is highly susceptible to various changes in diseased plants. Disease respiration changes could be divided into general and individual changes, where general changes are characteristic or commonly associated with a large number of diseases and individual changes are associated with a particular or individual process in respiration.

General changes in respiration. It has been verified that the intensity of respiration changes of diseased plant varies with the course of pathogenesis. In most cases investigated, respiration increases during the initial stages of infection, and later progressively decreases below the rate of that recorded for healthy plants. The intense respiration found at the beginning of pathogenesis is explained as a defense mechanism in which the increase in the formation of metabolites takes place either due directly to the activity of the parasite, to the preservation or re-establishment of the disturbed cell structure and balance by the host, or a combination of both.

For example, in diseases resulting in local lesions, changes in respiration occur in the infected cells, as well as in those immediately adjacent, but not invaded by the pathogen. These respiratory changes in contiguous tissues are caused by toxin substances produced by the pathogen reaching uninfected tissues from the points of infection. Causal agents of local lesions include most phytopathogenic fungi and most of the bacteria, with the exception of vascular bacteria. Many viruses cause local infections only in hypersensitive plants.

Vascular pathogens may be responsible for changes in respiration at points remote from points of infection. Such changes are usually caused by toxic metabolites of the pathogen involved which is transported through the vascular tissues. Several bacteria and fungi which cause vascular disease with wilt symptoms are included in this group. In susceptible plants, some viruses cause general or systemic infections with the result that respiratory disorders represent a general reaction to infection.

The dynamics of respiration, with its ascending and descending flows, is particularly characteristic of necrotic processes in infected plants. The more intense the necrotic process, the greater the effect on respiration. Pathogens that cause cell and tissue necrosis usually are necrotrophic, although some biotrophic fungi may cause necrosis. Most pathogenic bacteria cause local necrotic spots and most viruses are agents of rapid tissue necrosis at infection points in hypersensitive plants.

The rate of respiration influences and recurrently affect the degree of plant sensitivity to a given pathogen. In resistant plants, some reactive systems are replaced with new ones, which are capable of preventing further detrimental effects of the pathogen begun at the beginning of pathogenesis. This type of reaction, according to Rubin and Artsikhovskaya[24] is characterized by an exchange of energy and reduced hydrolase activity of the pathogen. Respiration in resistant plants, which may be reduced during pathogenesis, can be re-established as a normal biochemical-physiological process. Therefore, plant resistance begins with the intensity and normalization of this process.

Generally, increased respiration has been observed in plants infected by a variety of plant pathogens. Most studies have been done on plants infected with fungi and fewer on those infected with bacteria and viruses. Goodman et al.[45] emphasized that the initial increase in respiration after infection usually occurs with the first appearance of symptoms, reached its peak at the time of sporulation by the pathogen, and then decreased after this time. The increase in respiration as an indicator of disease has been observed for many diseases caused both by biotrophic and necrotrophic fungi.

For example, studies on wheat plants infected with *Erysiphe graminis* showed that respiration increased five to six times at 5 to 7 days after infection compared to uninfected plants.[77] Studying the same host and pathogen, Millerd and Scott[78] reported that oxygen intake at 2, 4, 6, and 8 days after infection increased by 26, 66, 218, and 167%, respectively. In contrast, in resistant plants, respiration increased at 1, 2, 4, and 6 days after infection by 36, 19, 20 and 26%, respectively. In other studies it was reported that in wheat plants infected with the biotrophe, *Puccinia graminis* var. *tritici*, oxygen intake was increased by two to three times that of uninfected plants.[79] Heitefuss and Fuchs[80] reported transpiration increases of 50 to 100% in cabbage leaves infected with *Peronospora parasitica*. Respiration in susceptible oat plants infected by the necrotrophe, *Helminthosporium victoriae*, increased to twice that of uninfected plants at 4 days after infection.[81] Similar results were obtained in tomato plants infected with the vascular fungus, *Fusarium oxysporum* f. sp. *lycopersici* at 9 to 14 days after infection compared to uninfected plants.[82]

It should be remembered that the respiration rate of pathogens colonizing plant tissues may also contribute to respiration measurement and, therefore, the data pertaining to respiration rates may sometimes be misleading. However, although the data may not be perfectly accurate, the fact that increased respiration occurs cannot be denied. This has been confirmed by data showing that respiration rate in noninfected tissues close to fungal colonies of powdery mildew and rust fungi is also substantially increased.[45] Similar data were recorded from tissues surrounding areas of infection caused by necrotrophic fungi.

The influence of phytopathogenic bacteria on host respiration has not been studied as thoroughly as that of phytopathogenic fungi. Phytopathogenic bacteria develop almost exclusively in intercellular spaces, degrading pectic substances between cells, and therefore, their influence on respiration and other biochemical activities differ from those characteristics of phytopathogenic fungi. One of the major differences is the hypersensitive reaction cause by bacteria in the intercellular spaces. Because of the environment in which bacteria survive and even with their enzymatic activity, their influence on the constituents of living cells is less than that of fungi.

Most of the data concerned with the influence of phytobacteria on plant respiration have come from studies on the effects of *Agrobacterium tumefaciens* and *Pseudomonas solanacearum*. A review of the data from studies on these and other bacteria concluded that phytobacteria had a significant effect on respiration rates in infected plants.[45]

Most phytopathogenic bacteria cause a rapid decay and/or necrosis of plant cells and tissues. Therefore, the influence of bacteria on plant respiration may be expected to be approximately the same as that associated with tissue necrosis caused by other abiotic and biotic factors.

Changes in the respiration of plants infected with a virus are linked primarily with the biosynthetic processes of virus replication in infected plant cells. During the replication process some of the infected cells, in addition to forming new virus particles, form polyamine, a compound not normally found in uninfected cells.[83] Results of other studies on the effect of virus infection on plant cell respiration differ, due in part to different host-parasite combinations and experimental conditions. Matthews[42] suggested that these differences may be divided into two primary groups; those viruses that produce necrotic symptoms and those that do not.

For example, Takahashi and Hirai[84,85] studying the effect of tobacco mosaic on the respiration of epidermal tissue strands in tobacco plants, found a short term increase in respiration at 1 or 2 days after infection and prior to any significant virus replication. Following an increase in virus multiplication, respiration rates dropped below those of uninfected plants. In another study, it was found that in tobacco leaves infected with four strains of potato mosaic virus a significant increase in respiration followed the development of symptoms, with the level of increase concomitant with symptom intensity.[86] Similar results were reported from studies on tobacco leaves infected with the potato mosaic and tobacco etch viruses.[87]

In plants infected with viruses that cause a necrotic reaction an increase in respiration occurs simultaneously with the formation of localized spots in hypersensitive host plants. For example, the rate of respiration increased in leaves of *Nicotiana glutinosa* infected with the tobaccc mosaic virus at the time of local lesion formation on inoculated leaves.[88]

There are a large number of other examples that could be given, but a few examples have been chosen for illustration. Owen[89] showed that there was a 30% increase in respiration in tobacco leaves infected with the potato mosaic virus at the time of mosaic symptom development. Also, it was shown that there was a 30% increase in the respiration rate in the leaves of *Nicotiana glutinosa* infected with the tobacco mosaic virus with the percent depending upon the number of necrotic spots on inoculated leaves. Roberts and Corbett[90] found that the respiration rate in *N. glutinosa* increased 67% in necrotic leaves infected with the tobacco ringspot virus; and Jensen[91] reported an increase in respiration rate of 270% in the same host inoculated with the barley yellow dwarf virus. Tu and Ford[92] recorded a 6% and 32% increase in respiration rates in maize leaves inoculated with the maize dwarf mosaic virus at 6 days and 9 days after inoculation, respectively. Based on a series of studies, Lovisolo et al.[54] showed that respiration rate increases of between 20 and 200% occurred as a result of systemic infection and then decreased approaching normal values with the beginning of chronic symptoms of virus-infected plants. In conclusion, similar to the results from studies on diseases caused by bacteria and fungi, viruses stimulate an initial increase in the rate of respiration in infected plants.

Many examples could be given showing an increase in respiration associated with mechanical damage to plants. As a result of plant damage, whether mechanical or infectious, any damage to cell structure, changes in the internal spatial relationships, or in membrane systems which results in an uncontrolled mixing of cell constituents will be accompanied by cell death.

In disrupted cells and plant parts, numerous biosynthetic reactions take place, sometimes resulting in cell death, sometimes resulting in wound healing. Such intensive biosynthetic activity may explain ncreases in respiration rates noted by many researchers and could be considered as a general rather than an atypical phenomenon.

A. INDIVIDUAL CHANGES IN RESPIRATION

Changes in oxygen *uptake*. During the respiration process, oxygen is absorbed from the environment, distributed in the plant and used for the process of organic compound degradation finally resulting in the formation of two final products, CO_2 and water. A normal or an average rate of respiration takes place in the normal atmosphere which contains 21% oxygen under 0.2 atmospheres of pressure. Generally, respiration rates in plants will decrease with a reduction in the quantity of oxygen when the concentration is reduced by half of the normal atmospheric concentration, and will increase only a limited extent when oxygen concentrations are increased.

The intake of atmospheric oxygen is also influenced by the condition of epidermal tissues. A decrease in oxygen uptake can result from colonization of pathogens on the plant surface (see section on pathological changes in epidermal tissues). Limited data are available on the detrimental effects of surface-borne microorganisms on the rate of respiration. Certain epiphytic microorganisms, such as those that cause sooty mold, and others, such as *Cladosporium fulvum* are frequently cited.

Epiphytic fungi, by covering the leaf surface, may reduce the intake of oxygen.[16] Ectotrophic fungi can cause a decrease in the respiration rates by 9% in *Citrus auratus*, 12% in plants of *Gardenia thunberis,* and 23% in plants of *Psidium aromaticum.*[93] *Cladosporium fulvum* reduced the respiration rate in tomato plants.[94] This fungus can caused a significant reduction in the respiration rate in only slightly infected tomato plants.[24]

The effect of virus infection on oxygen intact has not been thoroughly studied. Kuprevich[6] suggested that the amount of CO_2 released is reduced in plants during the first stages of infection by the potato leaf roll virus indicating reduction in oxygen uptake. Exceptions to this phenomenon have been recorded in young plants up until development of characteristic leaf roll symptoms. Respiration in infected plants decreased between 65 and 75% of that of uninfected plants. It was presumed by the author that the same reduction occurred in plants infected with yellow-type viruses.

Respiratory pathways in diseased plants. Many investigators have shown that glucose degradation can take place through several different pathways in microorganisms and higher plants. The glycolysis or EMP pathway of sugar conversion into pyruvic acid, plays an important role in supplying plant cells with chemical energy and intermediary products for the biosynthesis of carbon compounds. The important respiratory pathway in plant metabolism is also the pentose phosphate or WDH pathway. The intermediary products of this pathway are used for the synthesis of some organic compounds important to the life of the cell (see section on plant respiration).

The intensity of respiration may be expressed quantitatively showing the relationship between pentose phosphate (WDH) and glycolysis (EMP) pathways in the symbol of C6/Cl or the rate of $^{14}CO_2$ released from glucose-6-^{14}C/rate of $^{14}CO_2$ released from glucose-1-^{14}C, which should be equal to 1. However, if the respiration is performed in the pentose phosphate cycle, the C6/Cl ratio remains below 1, since CO_2 formation from glucose in this case is performed at a slower rate.

The rate of respiration in an infected plant is characterized by a greater participated of the pentose phosphate pathway. This increase has been shown to occur in plants infected with pathogens belonging to various groups, and therefore, it is an alternate pathway which occurs generally in infected plants.

The activation of the pentose phosphate cycle generally occurs late in the course of pathogenesis. For example, in plants infected with one of the biotrophic fungi, such as the causal agents of rust on beans, sunflowers and wheat plants, and of powdery mildew on barley plants, the increase in the incidence of this cycle occurs during sporulation of the pathogen.[45] Daly et al.[95] showed that a sudden increase in the occurrence of this cycle in rust-infected plants was evident only in the later stages of disease development, particularly when symptoms became clearly visible with uredosorus formation.

The pentose phosphate cycle has an important function in the course of pathogenesis, and plays a fundamental role in the formation of intermediary products that are important in the defense and recovery of infected cells and tissues. The activation of this cycle represents one of the most characteristic features of the metabolism in a diseased plant.

Wood[47] related the increase in the pentose phosphate cycle to respiration of rust fungi. During the early stages of infection by rust fungi, it seems that their activity alters host cell respiration, which becomes more and more similar to that of the pathogens involved. This has been proved by data showing that with continuous pathogen development in host tissue, the metabolism of the affected cells become closely correlated with the carbohydrate metabolism of the pathogen. Thus, changes in host cell metabolism to approach that of the pathogen is necessary for the establishment of parasitism.

Increased pentose phosphate activity in wheat leaves infected with *Puccinia graminis* var. *tritici* resulted in the reduction of the C6/Cl ratio from 1.0 to 0.3 compared to that in healthy leaves, which ranged from 0.5 to 0.6.[96] This increase may be attributed in part to the metabolism of the pathogen involved. It has been reported, for example, that respiration in diseased tissues of bean leaves infected with *Colletotrichum* sp. increased three times that of the control and the C6/Cl ratio was reduced; and in tissue surrounding areas, the rate of respiration was twice as high with C6/Cl ratios similar to noninfected tissues.[97] However, Beckman[98] stated that the C6/Cl ratio may be affected by pathogen metabolism as well as host metabolism, but the latter probably had a greater influence.

Other evidence on the action of the pentose phosphate cycle in plant disease was shown by Scott[99] in his studies of tumor tissues on sugar beet roots caused by *Arobacterium tumefaciens*. In special studies carried out later, a C6/Cl ratio below 1.0 was recorded, indicating that the metabolism of some glucose molecules was accomplished in the pentose phosphate cycle.[45] Other evidence came from the studies of Farkas et al.[100] who showed that increased enzymatic activity (glucose-6-phosphate dehydrogenase) took place in tobacco leaves inoculated with *Pseudomonas svrinae* pv. *tabaci*.

Glucose metabolic pathways have also been studied in virus-infected plants with varying results. Baur et al.[101] showed that glucose degradation in healthy tobacco leaves was performed 80% by glycolysis and 20% along the pentose phosphate pathway. However, tobacco plants infected with the tobacco mosaic virus did not alter the ratio of the systems even at 5 days after inoculation prior to systemic symptom production, nor after 21 days after systemic symptoms had fully developed. In contrast, Bell[102] reported an increase in the pentose phosphate cycle in necrotic spots on two cultivars of beans infected with the alfalfa mosaic and southern bean mosaic viruses. Also increased respiration was recorded in necrotic spots on tobacco leaves infected with the tobacco mosaic *Xanthi* virus in the glycolytic and pentose phosphate oxidation pathways.[103] Based on these and similar data, Mathews[42] concluded that in plants systemically infected with a virus, regardless of the details studied, that there are no recognizable changes in respiration pathways. When necrotic leaf spots occur on hypersensitive plants, any increase in respiration rates takes place with at least partial participation of the pentose phosphate pathway.

The increase in pentose phosphate activity should be considered as a nonspecific plant reaction, since it occurs both in infected and damaged plants. Similar increases have been recorded in normal plant development, particularly during the interruption of plant dormancy.[104] The activation of the pentose phosphate cycle takes place shortly after plant organ wounding, reaching a peak about 24 hours after the injury.[98] Increased pentose phosphate cycle activity was recorded in slices of potato tubers as well as in wounds on carrot, sugar beets and sweet potatoes.[105,106] Thus, it is evident that plant injuries can stimulate identical or similar changes in respiratory metabolism. These changes represent an immediate reaction of injured cells in which intermediary metabolic products of importance for both defense of the cell and its preservation and maintenance are formed.

In conclusion, the pentose phosphate cycle is important in the life of both the healthy and diseased or damaged plant since it takes part in the development of nucleic acids.[24] Also, this cycle from a phytopathological point of view, is of particular importance since it plays a defensive role in protecting plant cells and tissues against injuries regardless of the cause.

Respiration Enzyme Activity in Diseased Plants. Enzymes catalyze the biochemical activity in cells from their formation until extinction. Enzymes can alter the course of cell development depending on the needs of cytogenesis. However, if foreign factors disturb cell structure and function, this can affect enzyme activity. Such changes often are oriented toward the establishment and preservation of the disturbed cell's integrity.

Changes in the respiration enzymes that occur in damaged or diseased plant cells has had a great deal of study because of the importance of the role of enzymes in respiration. It has been shown that any alterations in the quantity or activity of enzymes or enzymes systems indicate the presence of diseased or damaged cells.

Most investigations have been concerned with respiration enzymes in diseased plants that affect oxidase and dehydrogenase activity in the pentose phosphate cycle and will be discussed in some detail.

Oxidases. Oxidases take part in the terminal respiration processes and are defined as terminal oxidases. This group of enzymes includes ascorbic acid, catalases, oxidases, peroxidases, and polyphenoloxidases.

Ascorbic Acid Oxidase. Ascorbic acid oxidase is a copper proteid, which catalyzes ascobic acid oxidation. In leaf parts, as well as in plant tissue extracts, ascorbic acid oxidation also may be catalyzed by other oxidases, such as cytochrome oxidase and peroxidase. This should be considered when determining the specific role of ascorbic acid oxidase in the oxidation of this acid.

The physiological function of ascorbic acid oxidase in plants during pathogenesis is not clearly understood. Some authors believe that this oxidase has no significance as a terminal oxidase in healthy plants, except in cereals where it maybe plays a primary role.[47] Also, the role of ascorbic acid oxidase in electron transport is not certain.[45] Daly[75] suggested that the most probable function pertained to the maintenance of the redox-potential by transporting surplus electrons into oxygen and water formation:

$$\text{NADPH} \longrightarrow \text{glutathione} \longrightarrow \text{ascorbic acid} \longrightarrow O_2$$

Results from a variety of studies have shown varying activities of ascorbic acid oxidase in infected plants; some have found an increase and others a decrease, or little or no change. Kiraly and Farkaš[108,109] observed an increase in ascorbic acid oxidase activity in wheat plants infected with *Puccinia graminis* var. *tritici*. Similarly, Frić and Majernik[110] recorded increased activity of this enzyme in barley plants infected with *Erysiphe graminis* var. *hordei*. An increase in ascorbic acid oxidase was found in cabbage plants infected with *Fusarium oxysporum* f. sp. *conglutinans*, with the extent of the activity related to plant susceptibility and the mode of measuring the activity of the enzyme.[111] Increased activity of the enzyme was found in susceptible oat plants treated with victorin, a toxin produced by *Helminthosporium victoriae*. The activity of this enzyme in the sap of treated plants was two to four times greater than in the sap of untreated susceptible or treated resistant oat plants.[47]

In contrast, however, Sanwal[112] concluded that the activity of ascorbic acid oxidase was less pronounced in tomato leaves of plants infected with *Fusarim oxysporum* f. *lycopersici* than in uninfected plants. What reduction occurred took place in the period between the seventh and tenth day after inoculation, which was followed by an increase in activity. Similarly, identical or reduced activity of this oxidase enzyme was recorded in bean plants infected with *Colletotrichum lindemuthianum*,[113] as well as in cabbage plants infected with *Peronospora parasitica*.[114]

Maine[115] reported that a variety of enzymes participated in ascorbic acid oxidation in tobacco tissues infected with *Pseudomonas solanacearum* and that ascorbic acid oxidase and some polyphenoloxidases took part in the increased oxidation of ascorbic acid in infected tissues.

In conclusion, the influence of plant infection on ascorbic acid oxidase activity differs depending on the relationship established between various hosts and pathogens.

Polyphenoloxidase. Like ascorbic acid oxidase, polyphenoloxidases are copper proteids. They are important in the terminal stages of oxidation since they catalyze the oxidation of ascorbic acid and phenolic compounds with oxygen molecules. Polyphenoloxidase activity has been recorded in many cell components, such as the chloroplasts, cytoplasm and other cell inclusions.[116-119]

Oxidized phenolic compounds, in the absence of reducing agents, are bound to the protein constituents of the plant cell forming dark-colored melanoid pigments; their occurrence being influenced by points of injury or infection. A general example of this process can be found in hypersensitive reactions, which arrest development of biotrophic and some necrotrophic pathogens. On the basis of many investigations, it has been shown that increased polyphenoloxidase activity is associated with plant resistance to pathogens, similar to that of peroxidase and phenolic compounds.

A distinctive reaction by plant tissues to damage or infection is the increase in polyphenoxidase activity. Wood[47] suggested that the increase in polyphenoloxidase activity is most frequently due to cell and tissue damage caused by pathogens in host cells and is the direct result of the formation of oxidized phenols, which in turn is accompanied by an increase in the uptake of atmospheric oxygen.

Intense polyphenoloxidase activity has been established in many plants infected with a variety of plant pathogens. For example, the activation of this oxidase occurred in plants infected with species of one of the following fungi: *Ceratocystis, Cochliobolus, Fusarium, Helicobasidium, Phytophthora*, and *Thielavia*.[120] Similarly, intense activity by this enzyme was recorded in plant tissues infected with various phytopathogenic bacteria. For example, in tumors on tomato plants infected with *Arobacterium tumefaciens*, the activity was increased 130% over that of uninfected tissues.[121] Also, polyphenoloxidase activity doubled in tomato stems infected with *Pseudomonas solanacearum* accompanied by a significant increase in oxygen uptake following infection.[115]

Virus infections also are accompanied by increases in polyphenoloxidase activity. In some cases this influence is variable or not pronounced because of differences in reactions of host plants as well as variations in experimental conditions.[45] Martin[38] showed that an increase in polyphenoloxidase activity in tobacco leaves infected with the tobacco mosaic virus. Tumaru et al.[122] found similar changes occurred in tobacco leaves infected with the cucumber mosaic virus. Likewise, the cucumber mosaic virus stimulated a triple increase in the activity of this enzyme in plants of *Datura stramonium* infected with the virus at two to three days following infection.[123]

Increased polyphenoloxidase activity reduces the quantity of polyphenols in tissues surrounding necrotic lesions caused by various plant viruses. This reduction is accompanied by an increase in quinone-type compounds, which are oxidized polyphenolic derivatives. Quinones are phytotoxic to plant cells and the influence of these compounds may, *inter alia*, serve to explain the development of local necrosis or hypersensitive reaction caused by many plant viruses.

Peroxidase. Peroxidase is a hemoproteid containing iron as a prosthetic group in the porphyric ring. It represents a complex of more than 20 isozymes enabling it to catalyze a variety of oxidation reactions, since isozymes have a variety of biochemical properties. These compounds form and catalyze the oxidation of various substrates in the presence of hydrogen peroxide, such as aromatic substances, ascorbic acid, and phenolic compounds forming quinone and water. Peroxidase activity may be expressed by the following formula:

$$AH_2 + H_2O \text{ >peroxidase-> } 2H_2O + A$$

where AH_2 = the phenolic compound, such as chlorogenic or caffeic acid.

The presence of peroxidase has been established in various parts of the plant cell; in nuclei[124], in mitochondria,[125-128] in ribosomes,[129] in cell walls,[130] and in cell membranes.[131] Many researchers have established that there is an increase in peroxide activity in infected plant parts and that this activity is frequently connected with the oxidation of phenolic compounds, which are toxic to many plant pathogens. Such increases have been associated with infection by bacteria, fungi, and viruses. For example, intensified activity of this enzyme was recorded in susceptible cabbage plants infected with *Fusarium oxvsporum* f. *conglutinans,* with increases of 18 to 33 at nine days and 94 to 130% higher at 17 days after inoculation when compared to uninoculated control.[47] Other examples of increased activity are: in potato tubers infected with *Synchytrium endobioticum*, in cotton plants infected with *Verticillium albo-atrum*, in citrus fruits infected with *Penicillium italicum,* in tobacco leaves infected with *Cladosporium fulvum,* in rice leaves infected with *Piricularia oryzae*, and in barley leaves infected with *Erysiphe graminis.*[16,47]

Peroxidase activity was 120% higher in tumor tissue on tomato plants infected with *Arobacterium tumefaciens* than in contiguous healthy stem tissues.[121] Rudolph and Stahmann[132] reported greater activity of this oxidase in leaves of a resistant bean cultivar infected with a virulent strain of *Pseudomonas svringae* pv. *phaseolicola* than in susceptible ones, leading to the conclusion that peroxidase activity increased the resistance of the host to this pathogen.

Martin[38] recorded increased activity of this oxidase in tobacco leaves infected with the tobacco mosaic virus; and Menke and Walker[133] reported identical increases in cucumber plants infected with the cucumber mosaic virus. Increased peroxidase activity was also recorded in barley plants infected with the barley yellow dwarf virus.[134] Tošić[135] found intense peroxidase activity in wheat leaves infected with the wheat streak and brome mosaic viruses. A number of other authors have found an increase in the activity of this enzyme in maize leaves infected with the maize dwarf mosaic virus and in plants doubly-infected with the maize dwarf mosaic and brome mosaic viruses.[136,137]

Catalase. Catalase, like peroxidase, is a hemoproteid containing iron as a prosthetic group in the porphyric ring. Its functions, likewise, is important for the metabolisms of living cells, since it catalyzes toxic hydrogenperoxide into water and oxygen:

$$H_2O_2 \text{> catalase>} H_2O + 1/2\ O_2$$

Catalase if found in plant cells in glycosisomes and peroxisomes. The exact role of catalase as a respiratory enzyme in the metabolism of diseased plants is not clear and therefore, difficult to determine precisely its role in respiration. Nagy[121] showed a 160% increase in catalase activity in tumors on tomatoes infected with *Arobacterium tumefaciens* compared to uninfected tissues. Higher catalase activity was also recorded in bean plants infected with a virulent strain of *Pseudomonas syrinae* pv. *phaseolicola.*[132] Thus, it would seem that virulent pathogens stimulated increased catalase activity in bean plants and reduced the efficiency of the natural defense mechanisms in the plant.

Other investigators have shown that catalase activity decreased in virus-infected plants. For example, Tošić[135] proved that there was reduced activity of this enzyme in wheat leaves infected with the wheat streak mosaic and brome mosaic viruses. Serova et al.[138] obtained similar results while studying a mixture of the tobacco and potato mosaic viruses on inoculated tomato plants, as well as in cucumber plants infected with the cucumber green mottle and aucuba mosaic viruses. The reduced catalase activity was recorded in maize leaves infected with the maize dwarf mosaic virus and in plants doubly infected with the maize dwarf mosaic and brome mosaic viruses.[136,137]

Glycolic Acid Oxidase. This enzyme affects a comparatively small number of substrates and is located in peroxisomes. Its primary function is to oxidize glycolic acid into glyoxylic acid during respiration. Generally, it is believed that any weakening of the activity of this enzyme results in a reduction in respiration and an increase in photosynthetic CO_2 assimilation. Such activity has been recorded in plant infected with *Puccinia graminis* var. *tritici*,[140] *Fusarium lycopersici*,[141] and the cucumber mosaic virus.[139] The precise mechanism responsible for the oxidase reduction in photosynthetic tissues is not known.

Pentosephosphate respiration pathway enzymes. Studies on the pentose phosphate respiration pathway have been confined to two enzymes, glucose-6-phosphate dehydrogenase and 6-phosphogluconate dehydrogenase. Glucose-6-phosphate dehydrogenase catalyses, in the presence of the coenzyme $NADP^+$ or NAD as hydrogen acceptors, changes glucose-6-phosphate into 6-phosphogluconlactone. During the process, 6-phosphogluconic acid is converted into ribulose-5-phosphate, which in a strict sense, ends the pentose phosphate cycle.[43] Most studies have shown that glucose-6-phosphate dehydrogenase and 6-phosphogluconate dehydrogenase activity increased in diseased plants, which serves as an intermediate proof for the importance of the pentose phosphate pathway in the respiration of diseased plants.

Dehdrogenase activity has been shown to occur in plants infected with fungi, bacteria and viruses. Substantial increases in activity of these enzymes has been described in wheat leaves infected with *Puccinia graminis* var. *tritici* when compared to uninfected plants. The higher dehydrogenase activity is associated with increased respiration and it occurs in susceptible but not in resistant leaves.[142-143] Also, increased dehydrogenase activity was reported in maize leaves infected with *Bipolaris zeicola*.[144] A lesser increase in the activity of these enzymes was reported in barley leaves infected with *Erysiphe graminis*.[145] Rubin and Artsikhovskaya[24] found the activity of both dehydrogenases was two to four times greater in potato tubes infected with *Phytophthora infestans* compared with uninfected ones; and that certain facultative and obligate parasites may inhibit the activity of these enzymes, resulting in the death of infected tissues. Obligate parasites may cause cell death in tissues of resistant plants and facultative parasites may cause cell death in susceptible ones.

For example, Scott et al.[99] discovered that the activity of pentose phosphate cycle enzymes was two to three times higher in sugar beet tissues infected with *Arobacterium tumefaciens*. The same authors, stated reduced activity of some enzymes mediating hexose degradation along the glycolysis pathway in similar tumor tissues. Farkaš et al.[100] recorded intense glucose-6-phosphate dehydrogenase activity in tobacco leaves infected with *Pseudomonas syringae* pv. *tabaci*. The activity of this enzyme in the halo part of leaf spots (in the absence of the bacterium) was 250% greater than in affected tissues. In tobacco tissues inoculated with the same bacterium, dehydrogenase activity was almost 200% higher than in those spray-inoculated with water.

Data concerned with dehydrogenase activity in virus-infected plants is sparse and what has been reported is contradictory. For example, data from some investigations have shown that an increase in the activity of the enzymes involved with the pentose phosphate cycle was associated with the yellow zones around leaf spots on tobacco leaves caused by inoculation with the tobacco mosaic virus.[146,147] Dwurazna and Weintraub[86] studied the activity of this particular cycle in tobacco leaves infected with various strains of the potato mosaic virus. Due to some methodological procedures, their results were challenged by Matthews.[42]

Similar to infected tissues, glycose-6-phosphate dehydrogenase and 6-phosphogluconate dehydrogenase increase in mechanically damaged or injured tissues. This particular activity is accompanied by an increase in the pentose phosphate respiration cycle. It was shown that glucose-6-phosphate dehydrogenase activity increased some 100% in potato tube slices at 48 hours after cutting.[148] Also, increased dehydrogenase activity was reported in damaged carrot, sugar beet and sweet potato tissues.[106] Thus, it is evident that any increase in the pentose phosphate cycle with accompanying increased enzyme activity represents an nonspecific phenomenon associated with both mechanically damaged and infected plant tissues.

Changes in oxidative phosphorylation. Oxidative phosphorylation is a process in which chemical energy becomes isolated by the oxidation of a substrate during respiration and in which this energy becomes bound and preserved in energy-rich phosphatic compounds, such as adenosine triphosphate (ATP). It is a process in which oxidative phosphorylation involves glucose oxidation coupled with the addition of phosphate to ADP to produce ATP. Almost 50% of the total energy released in oxidation-reduction reactions is converted into the chemical energy stored in the ATP molecule during oxidative phosphorylation. The energy contained in ATP is the only form which may be used directly in the exchange of energy required for metabolic processes in the living cell. The phosphorylation reaction is closely associated with respiration.

From a phytopathological point of view, it is important to emphasize that oxidative phosphorylation takes place in the Krebs cycle and the rate is dependent on the concentration of the oxidative substrate and the presence of inorganic phosphorus (Pi) in the form of ADP as well as oxygen in the plant cells and tissues.

Inorganic phosphorus and ADP are indispensable in the formation of the ATP molecules. Pi is usually present in sufficient quantities but the supply of ADP in plant cells is usually limited, and if the total amount of Pi converted to ATP, oxidative phosphorylation ceases.

In addition to other factors, oxygen is indispensable for the synthesis of ATP. The absence of oxygen will also result in the disruption of oxidative phosphorylation. Oxygen intake greatly depends on the amounts of Pi and ADP present in the mitochondria. In the absence of these substances, oxygen intake becomes insufficient to meet needs, and vice versa, when the intake of oxygen is increased, an increase in respiration of 40 or more times can result.[13] The ratio between the amount of oxygen absorbed in the absence of Pi and ADP and those absorbed in the presence of these substances differ and are defined as the "respiratory control index".

In studying the effect of pathogens on plant respiration, it is important to understand the effect of toxins and other compounds produced by microorganisms and viruses on oxidative phosphorylation in mitochondria. In the situation where electron transport and energy release continues, however, the energy is not bound to ATP molecules, but released in the form of heat, thus, the coupled reaction between substrate oxidation and phosphorylation is missing.

The uncoupling of oxidative phosphorylation may be the result of various compounds produced by a variety of causal agents, such as 2,4-dinitrophenol (DNP). The detrimental influence of DNP has been studied by applying it to healthy plants.

The uncoupling of substrate oxidation may also be caused by various toxins produced by fungi and bacteria. In addition to various pathogen toxins, ethylene may also cause the same breakdown.[149] With other detrimental effects on the activity and structure of mitochondria and other organelles in which phosphorylation occurs, these metabolites also cause changes in the mechanisms of oxidative phosphorylation.

Data pertaining to the uncoupling of oxidative phosphorylation in plants infected with biotrophic and necrotrophic fungi has been reported. Among biotrophic fungi, the list includes *Erysiphe graminis* and *Puccinia graminis* var. *tritici* on wheat plants.[47] Among necrotrophic fungi, the following are included: *Helminthosporium* on oats,[150] *Gibberella saubinetii* on potato tubers,[151] and *Cercospora personata* on peanuts.[152] It was reported by several authors that ipomeamarone, isolated from sweet potatoes infected with *Ceratostomella fimbriata*, caused an increase in the respiration rate by 30 to 50% in uninfected sweet potato tissues accompanied by an uncoupled oxidation.[24]

The intensification of respiration accompanied by uncoupled oxidative phosphorylation is due to the absence of the synthesis of ATP and the mechanisms involved in the transport and utilization of energy. Unbound energy is released as heat in the uncoupled oxidation process, which results in the increase in the temperature of diseased plant tissues. According to Yarwood,[153] temperature increases in wheat tissues infected with powdery mildew and rust fungi ranged from 0.3 to 0.7°C.

Another characteristic response of plant tissues invaded by fungi and bacteria is the elimination of the Pasteur effect. The Pasteur effect is a phenomenon in which the presence of oxygen suppresses anaerobic respiration. The influence of the Pasteur effect on respiration is calculated by the ratio between anaerobic and aerobic respiration. In anaerobic respiration, one molecule of glucose plus two of pyruvic acid yields two molecules of CO_2, while in aerobic respiration it yields $6CO_2$, which is numerically expressed as a ratio of 2/6, which equals 0.33. If this ratio increases, it is an indication of an increase in anaerobic respiration and the abolishment of the Pasteur effect. Detrimental consequences of such changes are the increase the enzymatic degradation of substrate molecules, thus yielding less energy than that produced by oxidation under aerobic conditions. As a consequence, a greater use of glucose by glycolysis would be required, since there would be a reduction in the amount of oxygen present. One role of the Pasteur effect is to indicate the rate of economical carbohydrate consumption in a cell during respiration.

There are several examples of the reduction of the Pasteur effect being accompanied by an increase in respiration in infected plants. Allen[149] showed that an increase in respiration in wheat leaves infected with *Erysiphe graminis* was accompanied by a simultaneous decrease in the Pasteur effect. An inability to measure the Pasteur effect in diseased plants, particularly those infected by the rust fungi, is considered as indirect proof of uncoupled oxidation.[154]

Changes in mitochondria and respiration. Mitochondria are organelles which must be present in cells for aerobic respiration to occur. Their primary function is to perform aerobic substrate oxidation and to synthesize energy rich phosphatic compounds, such as ATP, which is indispensable for the life of the cell. Mitochondria are highly sensitive to various changes caused by abiotic and biotic factors. They are easily changed functionally, morphologically, and structurally and thus, are one of the most sensitive indicators of any change in the condition of the cell.

Any plant pathogen that disturbs a plant cell will cause changes in the mitochondria of the affected cells. The various toxins and other metabolites produced by plant pathogens probably have the most rapid and detrimental influence on mitochondria. Changes in mitochondria can be caused by mechanical damage or injury as well as toxic substances. Symptoms of these changes usually are similar regardless of the causal agent involved and are considered as being nonspecific.

The various pathological changes that occur in mitochondria may be classified as follows, increases in their number, and changes in their shape, structure, constituents, and spatial organization (see section on structural changes in mitochondria).

The formation of a greater number than normal mitochondria occurs as a general reaction in infected and diseased cells and is representative of the increase in cell respiration during the initial stages of disease development. This increase is part of the biochemical and physiological reaction of ensuring sufficient energy for the increase in requirements due to infection. For example, two to three times as many mitochondria in cells of local lesions occurred on leaves of *Nicotiana glutinosa* infected with the tobacco mosaic virus than in normal cells.[155]

In addition, the size and shape of mitochondria may change due to the influence of some plant pathogens as well as abiotic factors. Structural changes in mitochondria are accompanied by a disorganization of their parts and either a reduction or cessation of their function. Only those mitochondria whose integrity remains intact can preserve their functionality. Irreversible pathological changes inevitably result in the interruption of normal function and finally cell death.

XX. PROTEIN SYNTHESIS IN DISEASED PLANTS

Proteins are organic nitrogen compounds composed of amino acids or of amino acids and other chemical compounds. Approximately 20 different amino acids mutually linked with peptide bonds into long polypeptide chains take part in the structure of proteins. Simple proteins are built of only amino acids, while more complex proteins contain other substances in addition to amino acids. For example, nucleoproteins are built of a complex of nucleic acids, and most enzymes are complex proteins composed of protein and nonprotein or coenzymes. Other proteins in combination with lipids form the complex lipoproteins.

Of all constituents of the living cell, proteins are of particular importance in cell function, organization and structure. Proteins assist in the life functions of the cell, and therefore, life is linked with protein molecule formation and activity.

From a phytopathological point of view, general protein synthesis is very important in determining the effect of disease on plants, including the location and mechanism of protein synthesis as well as the role of the cell genotype.

Ribonucleic acids (RNA) directly participate in protein synthesis in the plant cell. Messenger RNA (mRNA) is formed in the nucleus under the control of corresponding deoxyribonucleic acids (DNA), which carry the genetic information in the cell. RNA is formed through the mediation of RNA polymerase. RNA polymerase after becoming attached to a given location on a DNA strand, stimulates the formation of a chain of RNA. When the RNA chain has completed its development, the RNA polymerase is released from the DNA strand.[156] The newly formed RNA chain is complementary to the original DNA strand, but contains uracil instead of thymine.

The RNA base contains the genetic information from the DNA in its sequence and therefore is known as mRNA. The transfer of genetic information from DNA to mRNA is defined as a genetic information transcription. The mRNA molecules pass into the cytoplasm and together with the ribosomes, form the protein biosynthesis system. Protein synthesis begins with the transfer of amino acids according to a given sequence to the mRNA, here they are bound to polypeptide chains. Amino acid transfer to the mRNA is performed by transport molecules of RNA or tRNA.

The sequence of amino acids in the polypeptide chain is determined by the genetic information contained in the mRNA; genetic information for the selection of an amino acid is composed of three nucleotides or bases, known as triplets or codons contained in the mRNA molecule.

The amino acids for polypeptide chain formation are bound to their tRNA, which transfers them to the ribosomes. The tRNA molecules also contain triplet defined as an anticodon or anticod which is recognized by mRNA. Thus, the interaction between the codon and anticodon serves to determine the location of any given amino acid in the polypeptide chain being synthesized. The transfer of amino acids, according to a sequence determined by codons and anticodons, is defined as the translation in protein synthesis.[13] When protein molecules are completely formed, they leave their mRNA and ribosomes, thus marking the end of protein synthesis.

Transcription   Translation
DNA - - - -> mRNA - - - -> protein (enzyme)→|
↓
Replication
↓
DNA

In conclusion, it should be stressed that the genetic information or message from the DNA cell, through the mediation of transcription and translation, directs and determines the biosynthesis of proteins in the living cell and constantly provides for the maintenance of the cell function and structure.

Proteins and plant cell parasitation. Protein compounds in host cells play an essential role both in the infection process by the pathogen and in the processes involved in cell parasitism. Of particular interest has been the role of proteins in pathogen recognition during the host penetration process. Recognition processes take place following penetration of pathogens and prior to plant cell infection. During this stage of pathogenesis, plant cells react to the presence of pathogens as foreign bodies in their environment.

Pathogen recognition in the initial stages of pathogenesis may be specific or nonspecific.[157] Specific recognition of pathogen constituents is performed by specific recipient cells.[158-60] Nonspecific recognition includes plant cell reactions to chemical and physical injuries in the course of pathogen penetration. In the case of recognition by specific recipient proteins in the plant cell membrane, these proteins may react with macromolecules found in the cell membranes or those which are excreted into plant cells.[157] It is not certain whether identical injuries can induce either a specific or a nonspecific recognition.[157]

Proteins perform their role in pathogen recognition only in living tissues involving biochemical relationships between the pathogen and plant cell. Samborski et al.[156] pointed out that most investigations of protein synthesis whether in a host-pathogen system including soluble proteins or in more specific relationship, such as a gene for gene system, mutual interactions occurred from the formation of insoluble peptide and glycoprotein on the surface of plant cells. These recognition processes, including recipients on the cell surface, were shown to occur using bacteriophages, viruses, and cells of *Arobacterium tumefaciens*.[156]

Sequeira[161] in his studies on the recognition process between *Pseudomonas solanacearum* and the mesophyll cells in potato and tobacco leaves, showed that factors in three membrane systems take part; lipopolysaccharides in the external bacterial membrane, extracellular polysaccharides on the external bacterial cell coats, and lectins as a protein constituent of host cell walls. Lectins in particular bind with saccharides of the bacterial lipopolysaccharides and polysaccharides. He pointed out that recognition might represent a general phenomenon, occurring in other mutual relations between bacteria and plant cells.[161]

Investigations of other host-bacterial systems, such as with *Bradyrhizobium* spp. found that the bacterial cells were capable of binding to lectin on the root surface of susceptible plants.[162] Thus, *B. japonicum* may bind to the soybean lectin with their capsules or polysaccharide constituents. Many investigators have studied the role of lectin in pathogenic phytobacteria recognition and concluded lectin provides a number of polyvalent carbohydrate binding sites.

There is limited data on the recognition processes between plant cells and phytopathogenic fungi. It could be expected that general recognition would occur between host plant cells and obligate parasites. This was shown to be the situation between wheat plants and rust fungi, where it was observed from histological studies that there was a specific gene recognition between the fungal cell walls and the host cell plasmamembrane.[163]

In the course of recognition, when the host cells and the pathogen are in direct contact, biochemical processes take place leading to the establishment of compatible or incompatible relations between the pathogen and its host. If these relations indicate a compatible one, then infection takes place; if incompatible, infection does not take place and the plant continues its normal functions.

Compatible and incompatible relations are manifested and easily studied in leaf inoculations with homologous-pathogenic and heterologous-nonpathogenic bacteria in one plant species.[161,164] Homologous bacteria replicate freely in the fluid in the intercellular spaces of inoculated leaves and cause partial cell wall degradation. In infected plant tissue, pathogenesis develops normally advanced by pectolytic and cellulolytic bacterial enzymes.

In contrast, in leaves inoculated with heterologous bacteria, they cause a hypersensitive reaction characterized by rapid host cell death. Thus, bacterial cell populations decrease and bacterial development in the plant tissues is prevented. This type of reaction has been described, for example, on tobacco leaves inoculated with *Pseudomonas svringae* pv. *pisi* and avirulent and heterologous strains of *Pseudomonas solanacearum*.[165,166] It was stressed by Sequeira[161] that on the basis of this reaction, that in the course of their evolution, plants developed mechanisms enabling them to prevent multiplication of many microorganisms reaching the plant interior.

When a compatible relationship exists between a pathogen and host cells, amino acids and proteins, in addition to lipids, starch and sugars, play an important role in the defensive host reactions following infection. The role of proteins in this reaction can have a significant influence on the expression of host susceptibility. Defensive mechanisms involving protein molecules in host cell protection are influenced by the ability of these molecules to continue to perform various life functions in the plant cell. Some of these proteins are capable of autonomously altering and redirecting cell processes depending on the influence of external or internal detrimental factors.

An example of a defensive mechanism by proteins is their presence in the structure of several enzymes whose increase in number and activity is associated with the initial stages of infection (see section on photosynthesis in diseased plants and diseased-plant respiration). Increased enzyme activity immediately following infection provides energy and materials necessary for the reestablishment of the disturbed balance in the function and structure of affected cells. Catalytically-active proteins, as found in enzymes, stimulate chemical reactions resulting in the formation of protective substances, such as phytoalexins and phytoncides, and others of a chemical or mechanical nature.[167]

Serological studies of disease resistant and susceptible plants showed that there were antigenic differences between resistant and susceptible plants.[167] These data confirm the role of proteins in disease resistance in plants, since most proteins are highly active antigens. It was shown that protein lectin materials of high molecular weight are capable of causing reactions to serum antibodies with certain antigens in plant tissues.[167]

Proteins of the composition found in the external structure of plant and animal pathogens perform as antigens, inducing the formation of specific antibodies in certain animals. Upon contact with corresponding antigens, antibodies inactivate them preventing infection . Thus, certain protein antibodies can play an important role in natural and artificial immunization in man and animals.

Certain causal agents of plant disease have antigentic traits which are unable to form antibodies in plant tissues, but do form them in animals. Hundreds of antisera for different plant viruses have been developed and studied.[53] Plant virus antisera are used for the detection and identification of many plant viruses, in the diagnosis of virus diseases, and in studies on plant virus classification.

Another example of qualitative alteration in protein synthesis is the possibility of the production of protective proteins of the interferon-type found in plants in reaction to virus infections. Several authors have noted the possibility of the presence of interferonlike materials in local and systemic virus infections.[53,168,169] Ladigina et al.[170] isolated an antiviral protein from mitochondrial membranes of tobacco plants resistant to the tobacco mosaic virus. However, other authors believe that this protein is formed in the mitochondria only during the development of cell necrosis, thus decreasing the synthesis of new virus by 50%.[171]

The increase in the number of ribosomes in infected plant cells and tissues results in an increase in protein synthesis and their role in the postinfection period in plants.[172-174] An increase in the number of ribosomes is a prerequisite to increased protein synthesis.

Some types of proteins are capable of preventing or restricting pathogen enzyme activity and thus, arrest the infection process. This was shown by Albersheim and Anderson[175] who isolated proteins that prevented the activity of exopolygalacturonase in several plant pathogenic fungi from the cell walls of some plants. The prevention of the activity of the enzyme, which degrades pectin substances in the middle lamella of cell walls, may arrest further development of pathogenesis. Akseenova et al.[167] stated that based on their physical properties, these inhibitory proteins resemble plant hemoglutins and glycoproteins which can cause agglutination in cells.

In conclusion, it should be stressed that plant proteins play a key role in pathogenesis. In a reactive plant system, the pathogen and protein constituents are carriers of biochemical and physiological processes influencing plant resistance to pathogens.

Special changes in protein synthesis. Pathological changes in protein synthesis can occur at all stages in the process of protein synthesis. All such changes basically are related to disturbances in the synthesis of amino acids as well as that of nucleic acids and ribosome formation and function in the cell.

Changes in amino acid synthesis. A generalized reaction by infected or damaged plant cells is a change in quantities and composition of amino acids in the cells. The initial changes in amino acid synthesis may occur as a result of deficiencies in nitrogen and carbohydrate metabolites, which are the major building blocks of these compounds. Plants use ammonia nitrogen from degraded nitrogen compounds or bound atmospheric nitrogen, or nitrate nitrogen from the soil for amino acid synthesis.

Ammonia and nitrates, absorbed by the root system, are transported by the flow of water in which they are dissolved into the xylem and then to various organs of the plant. Only ammonia ions take part in amino acid synthesis and therefore, nitrate utilization is possible only following reduction to ammonia.

Ammonia reacts with carbohydrate metabolic products in plant cells to form corresponding amino acids in the reduction processes. These metabolic products include, for example, keto-acids, which can be direct acceptors of ammonia in protein synthesis.[43] Pyruvic acid, for example, reacts with ammonia yielding alanine as well as other compounds. Free ammonia in certain concentrations is toxic to plant cells, and therefore, its detrimental influence is eliminated by inclusion into organic forms.

Plant pathogens are biological factors which alter or prevent the uptake of nitrogen compounds from the soil as well as their transport through the plant. Root and xylem pathogens are of particular importance with root pathogens causing various deformations, necrosis, and rots of root parts or entire root systems. When such diseases occur, nitrogen absorption from the soil and transport in the root system are partially or totally prevented.

Disorders in nitrogen nutrition are caused by vascular bacteria and fungi. These specialized pathogens colonize the xylem and may cause pathological, structural and physiological changes in its cells. During early stages of infection these pathogens may prevent normal development of these tissues and cause cell necrosis. Some of these pathogens cause pathological formations in tracheae and tracheids, such as resins and tyloses, which can completely block vessels and thus exclude the transport of nitrogen and other nutrients (see sections on xylem pathological changes and root pathological changes and disorders in the transport of minerals).

Disorders in amino acid synthesis may also be caused by a deficiency of carbohydrate metabolites, which are the second most important constituents of amino acid synthesis. When a deficiency of carbohydrate metabolites occur, free ammonia ions can accumulate and in this form have a detrimental influence on plant cells and tissues. Therefore, normal amino acid synthesis depends on adequate photosynthesis for carbohydrate production and respiration for metabolic carbohydrate production.

The change in amino acid content in infected or damaged plant cells is influenced simultaneously by protein formation and degradation processes. Pathogenic fungi degrade cell proteins to amino acids for energy or growth by the action of proteolytic enzymes. These amino acids, then are released and activated in infected cells. In infected cells containing reduced carbohydrates due to reduced photosynthetic activity, amino acids become a source of energy for respiration, which is reflected directly on the quantities in these cells.

Alterations of amino acids may also be reflected in the quantity and type of amino acids present in infected plants. The most important factors influencing the quantity of amino acids in an infected host cells are: mode of parasitism, whether by obligate or facultative parasites; the stage of pathogenesis; and the level of plant resistance to a given pathogen. In susceptible plants, pathogens may alter both the quantity and composition of soluble nitrogen constituents.[176] For example, Shaw and Colotelo[177] showed that in susceptible wheat plants infected with *Puccinia graminis* var. *tritici* that there was an increase in total protein and soluble nitrogen, as well as in the ratio between soluble and insoluble nitrogen with progression of disease compared to uninfected plants. Also, that the amount of protein nitrogen in infected plants doubled, while the amount of amino acids and amino-constituents increased by four times. Aspartic and glutamic acids, serine, and glutamine were the more important among these compounds. In contrast to susceptible plants, the amount of protein and soluble nitrogen in infected resistant plants was smaller. Fuchs and Rohringer[178] found that wheat plants infected with the same fungus, compared to uninfected plants, contained no histidine, leucine or asparagine. Increased levels of glycine betaine and choline have been recorded in rust-infected wheat plants, otherwise not present in resistant plants.[179] The intensification of amino acid and amide synthesis or their absence are dependent on the extent of the altered enzyme activity of these compounds.

In other studies, alanine, asparagine and serine increased and amino butyric acid, phenylalanine and bound amino acids decreased in wheat roots infected with *Helminthosporium sativum*.[24] Also, changes in the quantity of various amino acids were recorded in potato tubers infected with *Phytophthora infestans*.[180] Tumors on potato tubers infected with *Synchytrium endobioticum* contained 15 to 21 amino acids, while tissues of the tubers itself contained only nine to ten amino acids, and in lower concentrations than in diseased tubers.[181]

Pathogenic bacteria, similar to pathogenic fungi, alter the quantity and composition of amino acids in infected cells. *Pseudomonas syringae* pv. *phaseolicola* significantly altered the amino acid composition and quantity in leaves of susceptible bean plants, and that these changes occurred both in infected and in the contiguous noninfected, chlorotic tissues.[182] In addition, there was significant accumulation of ornitin and histidine in infected and chlorotic tissues, which were not present in healthy tissues. The level of ornitin was five times greater in inoculated than in healthy tissues. Infections on the first trifoliolate leaf in resistant bean plants inoculated with this bacterium did not cause any alterations in the content of the amino acids or amides. As with most other phytopathogenic bacteria, *P. syringae* pv. *phaseolicola* is an intercellular pathogen, and therefore, any changes in amino acid metabolism of host cells may be attributed to bacteria toxins either inactivating or inhibiting biosynthetic enzymes for amino acid synthesis.

Arobacterium tumefaciens causes various alterations in protein and nonprotein nitrogen metabolism in normal and tumor tissues of sugar beets. Neish and Hibbert[183] found that normal and tumor tissues of sugar beets contained the same levels of nonprotein nitrogen, but that tumors contain three times more protein and six times more water soluble protein than normal tissues. According to Klein[184] the greatest increase of protein and soluble nitrogen occurred in crown gall tissues of tomatoes during later stages of development, i.e. 27 to 35 days after inoculation, probably associated with rapid cell division.

Infection by viruses have shown varying influence on the levels of free amino acids and amides in host plants. The increased presence of these compounds was recorded in plant infected with many viruses, including the potato leaf roll virus, which stimulated an increase in proline, and the potato streak virus (PVY), which stimulated an accumulation of proline as well as the amides, asparagine and glutamine.[165]

The effect of various viruses on amino acids and amides is variable. In wheat plants infected with the wheat streak mosaic and brome mosaic viruses, Tošić[135] identified 18 free amino acids and two amides of identical quantities both in inoculated and uninoculated plants. Commoner and Nehari[185] established that reduced levels of free amino acids and amides in plants soon after inoculation with the tobacco mosaic virus and attributed this reduction to rapid virus particle synthesis. Following this first burst of virus synthesis, the content of free amino acids and amides returned to levels recorded in uninoculated plants.

In summary, the possible role of amino acids in disease resistance in plants should be emphasized. Several authors have shown that some amino acids, such as DL-serine, DL-histidine, methionine and isoleucine, may induce resistance in cucumber to *Cladosporium cucumerinum* and in apple plants to *Venturia inequalis*.[165]

The toxic action of amino acids and amides has been described in virus-infected plants also (see section on toxic compounds in host cells). The toxic action of these compounds may be associated with the toxic action of increased levels of ammonia when amino acid metabolism is disrupted. Symptoms associated with virus infections also may be due to altered concentrations of these compounds in leaves of plants infected with plant viruses.

Changes in nucleic acid transcription. Transcription is a process in which genetic information from DNA is transcribed into mRNA. Plant pathologists are interested in the effect pathogens have on the transcription process. Several studies have shown that plant viruses in particular, and biotrophic fungi, such as the rust and powdery mildew fungi, may affect transcription. Although the exact mechanism of this action is not completely understood, it is known, that the effect occurs when pathogens affect the composition and function of chromatin, the main substance in host cell chromosome and gene structure.

Plant viruses, because they are intracellular and have a measurable effect on transcription, merit particular attention. Since plant viruses have no metabolism of their own, replication must be accomplished at the expense of host cell constituents and energy. Following DNA or RNA replication, the process of transformation takes place when virus mRNA is formed. Virus nucleic acid transcription is a pathological process which takes place outside of the control of host cell nucleic acids. In plant cells infected with a virus, transcription of normal cell and virus nucleic acids may take place simultaneously.

The formation of mRNA in host cells differs depending on the virus involved.[79] For example, the tobacco mosaic and brome mosaic viruses genomes may function as mRNA; while for another virus, mRNA is formed in infected cells either by transcription either from RNA or DNA, depending on which of these nucleic acids is present in the composition of the virus particle. Examples of this type of replication are in single- and double-stranded viruses, such as the lettuce necrotic yellows and clover wound tumor viruses, respectively (see section on cytogenetic changes).

Phytopathogenic bacteria and fungi have their own metabolism and therefore, may have little or no effect on plant cell transcription. However, these microorganisms establish a form of recognition with host cells through chemical affinities. For example, *Agrobacterium tumefaciens* alters the host cell genotype by transporting bacterial DNA into the host cell DNA, where both are replicated, probably through the mediation of host DNA polymerase.[156] Such genetically altered cells divide and proliferate forming tumors in the absence of the bacterium.

In plant cells infected by fungi, the parasitic relationship is established by the transfer of information from the fungus to the cell. According to Samborski et al.[156] this message transfer could be performed either by a mutual recognition process taking place on the plasma membrane or by the transport of molecules through the plasma membrane, perhaps into specific host or pathogen organelles.

Transferred messages influence the synthesis of certain proteins and can play an important role in the establishment of parasitic relationships in the host-parasite relationship. This mechanism of mutual exchange of information is undoubtedly highly developed in biotrophic fungi, such as members of the Erysiphales and Uredinales, all of which are obligate parasites.

Changes in amino acid translation. Any altered message for protein synthesis in infected or damaged cells is accompanied by a corresponding change in the selection and transfer of individual amino acids during the translation process. All such changes in protein quantity and composition implies a change in the biosynthetic activity involved in the process of translation. Therefore, deviations from the normal translation may be expected when new proteins are formed in infected and damaged cells, which are not found in healthy cells.

Several proteins not ordinarily found in healthy cells have been reported in plant cells infected with several pathogenic organisms. For example, three protein antigens were described in cabbage plants infected with *Fusarium oxysporum* f. sp. *conlutinans*, which were not found in healthy cells.[186] New proteins were also described in sweet potato plants infected with *Ceratocystis fimbriata* that were not described for healthy plants.[172]

Proteins which are part of the capsid of virus particles also are formed in virus-infected cells. Thus, high concentrations of virus protein have been found in cells of plants infected with such the potato and tobacco mosaic viruses and contribute to the total protein of the leaf. However, infection by many other viruses results in lesser concentrations of virus protein and generally, have a lesser influence on total protein of infected plant tissues and a lesser influence on the protein metabolism of infected cells. Other proteins, such as X-proteins, have been found in plants infected with the potato and tobacco mosaic viruses.

The level of protein usually increases during the initial stages of infection, which is accompanied by an increase in the energy exchange, and the synthesis of phenolic and other compounds. Some of these compounds are part of a defense mechanism against the pathogen. Increased protein synthesis occurs primarily in plants resistant to viruses, which underscores the active role of protein in disease resistance in plants. The translation process for the synthesis of both normal cell proteins and virus proteins takes place in the plant cell (see section on proteins and plant cell parasitation and changes in amino acid synthesis).

Ribosome changes and protein synthesis. Ribosomes are macromolecular organelles found in the cytoplasm of most plant cells. They are scattered within the cytoplasm and often attached to the endoplasma reticulum. Many types of proteins may be synthesized on them, depending on the mRNA, which also carries genetic information. The number of ribosomes may vary within a cell, not because of division of older ribosomes, but because of the direct formation of new ones. Any change in the number of ribosomes is an important indicator of changes in protein synthesis. Kahl[106] stated that any increase in the ribosome number is one of several characteristic changes associated with diseased or damaged tissues.

It was reported that the number of ribosomes increased in cells of resistant cabbage (cv. Amager) tissues by 40% when infected by *Botrytis cinerea,* while those in susceptible plants infected by the same pathogen increased 15% when compared with uninfected plants.[167] The same authors showed tht the number of ribosomes in resistant cucumber plants infected by *Erysiphe cichoracearum* increased by 30%, while their number decreased by 24% in susceptible plants infected with the same pathogen.[167]

In virus-infected plants, Matthews[42] stated that the influence of virus infections on ribosomal RNA synthesis and ribosome concentrations depended on the type of virus and virus strain, conditions following infection, susceptibility of the host plant and the type of tissue infected. Changes in the number of ribosomes have been recorded in both locally- and systemically-infected plants. Early infections of tobacco plants with the tobacco mosaic virus may prevent ribosomal RNA synthesis in the cytoplasm, while the influence on the ribosomal RNA in the cytoplasm may be less pronounced.[187] For example, in the yellow-green zones in Chinese cabbage plants infected with the turnip mosaic virus, ribosome concentration (68S) was significantly greater than in the dark mosaic zones.[188] Those viruses that result in complete chlorosis or whiting of infected tissues have an even greater effect. Thus, viral infections have a significant effect, in general, on protein synthesis.

Ribosomes accumulation was found in tissues surrounding local necrotic spots or lesions in hypersensitive plants; with injury and the presence of contiguous uninfected cells being indispensable for such reactions.[189]

Several authors have observed an increase in mitochondria, ribosomes, and endoplasmic reticula in incised tissues both in respect to their number and their activity at the time of incubation.[157] Ribosome accumulation and mitochondria formation usually occur in the healthy cells surrounding the points of infection or injury.

Finally, it should be stressed that various changes in protein synthesis in damaged or injured cells is a complex process, and that parasitation and plant protection are based on these changes in protein synthesis.

XXI. PHENOLIC COMPOUND METABOLISM IN DISEASED PLANTS

Phenolic compounds are complex organic compounds important in cell structure and function. In plant pathology, they have been studied because of their toxic action against disease agents and as substances to be used in plant protection, in addition to understanding the biology of plant cells.

Phenols are organic compounds composed of one or more benzene rings. Hoffmann et al.[190] classified them into several groups — simple phenols, phenylpropanes, and flavan derivatives. Simple phenols, such as catechol and hydroquinones, consist of a benzene ring composed of one or more hydroxyl groups and depending on certain conditions, yield substitute groups, such as phenylcarbonic acid derivatives like benzoic acids. Phenylpropanes, such as cinnamic acid derivatives like coumarin and lignin, are composed of an aromatic system with a single side chain consisting of three carbon atoms. Flavan derivatives or flavonoids have a flavon carrier from aromatic rings A and B which are linked with oxygen in heterocyclic compounds. Several phenolic compounds are found in the cell in the form of glucosides or sugar esters.

According to Goodman et al.[45] phenolic compounds comprise anthocyanins, anthoxanthins, coumarin products, heteroglucosides, hydroxybensoic acids, hydrocinnamic acid esters, leuco-anthocyanins, quinic and shicimic acid sugar esters. An important group of phenolic compounds includes the plant pigments, anthocyanins and flavonoids or anthoxatins. These pigments are glucosides composed of aglucones and corresponding sugars.

Anthocyanins are glucosides having anthocyanidins as their aglucone and some hexoses, such as galactose, glucose or pentose, as their sugar component. The anthocyanins, such as cyanidin, delphinidin and pelargonidin, are the most widespread of plant pigments. Anthocyanins vary in their water solubility with the glucoside acid form being more soluble than the aglucon anthocyanidin form. Anthocyanins consists primarily of blue and red pigments with some black ones.

Flavonoids and flavans are glucosides composed of certain phenolic compounds with saccharides as the carbonate component. They are known in a variety of chemical forms, such as flavons, flavonols, flavanonils and isoflavons.[43] Flavonoids rank among the most widespread of pigments; are carriers of orange and yellow colors, but in some cases are colorless. They are often present in flowers and fruits as well as lignified plant cells.

Aromatic amino acids are formed from carbohydrates and are precursors of phenolic synthesis.[191] Phenolic synthesis in higher plants is performed from carbohydrates along the acetate and shicimic pathways.[45]

Shicimic acid is formed by linkage of the three-membered carbon compound phosphophenol and four-membered carbon compound erythrose phosphate, which hold a key position on the shicimic acid pathway. Prephenic acid which is formed from shicimic acid following a series of products, is further converted into phenylalanine and tyrosine, precursors of various aromatic compounds. These precursors give rise to a variety of aromatic compounds through to cinnamic acid finally yielding coumarin, flavonoids and lignin.

Phenolic compound synthesis is performed from polyacetic acids along the acetate pathway by condensation of acetate units. Flavonoids in higher plants are probably formed in this manner.[45]

Lignin makes up between 15 and 35% of the structural materials of plants.[192] It is the second most important substance in the structure of plants following cellulose. It is a polymer compound composed of phenylpropane, whose synthesis begins with schicimic acid yielding conypheryl alcohol, which is the primary metabolic product used in lignin formation. Lignin fills the spaces between cellulose fibers in plant cell walls and thus contributes to the linkage between these fibers, strengthening cell structure and providing protection against outside forces. The group of phenolic compounds that includes the coumarins are formed from metabolic products in plant cells and spread throughout undamaged plants without being linked to any particular tissue. They are often found in cell vacuoles.[193] In clover plants, the precursors of coumarin are glucosides, such as cis-trans-O-hydroxycinnamic acid.[194] When sweet clover plants are damaged by beetles, *Epicauta* spp., beta- glucoside is activated in damaged leaves which catalyzes glucoside hydrolysis yielding coumaric acid, which give rise to coumarin. Glucoside and coumarin interferes with the nutrition of insects.[195] Thus, their formation contributes to the resistance of attacked plants. The coumarin compound, scopoletin, frequently is present in higher plants with a tendency to accumulate around spots or areas of localized infection on plants. It was presumed that its synthesis begins with chlorogenic acid, which plays an active role in the formation of localized spots.[120] This presumption may be justified because of the active part chlorogenic acid plays in respiration and its occurrence in tissues around localized spots in initial stages of their development.

The phytoalexins, another group of phenolic compounds, also play a protective role in injured plant parts. Phytoalexins are cell products of higher plants and occur in association with plant injuries caused by abiotic and biotic agents. Generally, phytoalexins are not found in healthy plant cells and tissues, but traces have been reported in such tissues.[148] Phytoalexin accumulation occurs as a consequence of being induced in injured plants (see section on phytoalexins).

Certain functions of phenolic compounds need to be emphasized, among them their participation in amino acid metabolism. Aromatic amino acids play a key role in the synthesis of phenolic compounds. For example, along the pathway from carbohydrates to complex molecules, hydroxylation of monophenols through the mediation of creolase (phenol-hydrolase) results in the formation of complex phenols, such as flavonoids and lignin. The darkening of plant parts is a result of the oxidation of polyphenols by catecholase enzymes (O-diphenol dehydrogenase). The metabolic phenol products such as the pigments anthocyanins, flavonoids and melanoids are of significant importance in plant life.

In summary, phenols play an important role in the exchange of energy since they participate in oxidative processes of plants. Several other phenolic compounds, such as chlorogenic acid, pyrocatechin, and pyrogalol, serve as electron and hydrogen donors and acceptors in the terminal oxidation reaction catalyzed by polyphenoloxidase. Phenols and their oxidized products also influence auxin production metabolism, thus having an effect on plant growth and development.

XXII. PATHOLOGICAL CHANGES IN PHENOLIC COMPOUND METABOLISM

Normal phenolic compound metabolism, i.e. their synthesis and degradation, is based on the organization and function of internal cellular space. Any disruptions in this organization results in the breakdown of normal metabolic processes, thus altering the role and reaction of phenolic compounds. Any pathological changes in phenolic metabolism is associated with injuries to plant cells or tissues. The changes that take place are similar regardless of the causal agent. Similar to disruptions in the metabolism of other organic compounds, interruption may occur at any stage in phenol metabolism. The role of phenols in plant resistance has been studied by biologists and plant pathologists leading to understanding of resistance to diseases in plants.

Quantitative changes in phenolic compounds. The major changes in phenolic metabolism of damaged or infected plants occur in the formation of greater quantities of phenolic compounds than found in unaffected plants (Figure 92). Generally, the accumulation of phenol takes place in any living tissue infected by a pathogenic agent. Goodman et al.[45] stated that phenol accumulation differed in susceptible and hypersensitive plants. In the latter, the accumulation of these compounds accumulated at a faster rate than in susceptible ones. For example, greater quantities of phenol was found in tissues of hypersensitive wheat plants infected with *Puccinia graminis* var. *tritici*,[196] in apple tissues infected with either *Podosphaera leucotricha* or *Venturia inaequalis*,[197] and in resistant cotton plants infected with *Verticillium albo-atrum*.[120]

Changes in the levels of chlorogenic acid in diseased tissues also have been studied since it is one of the primary oxidation substrates taking part in the development of necrosis associated with damaged or diseased cells and tissues (Figure 92). There are a number of studies showing

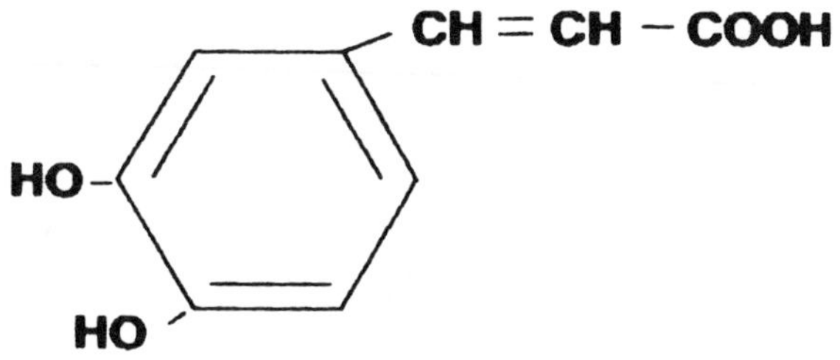

Caffeic acid

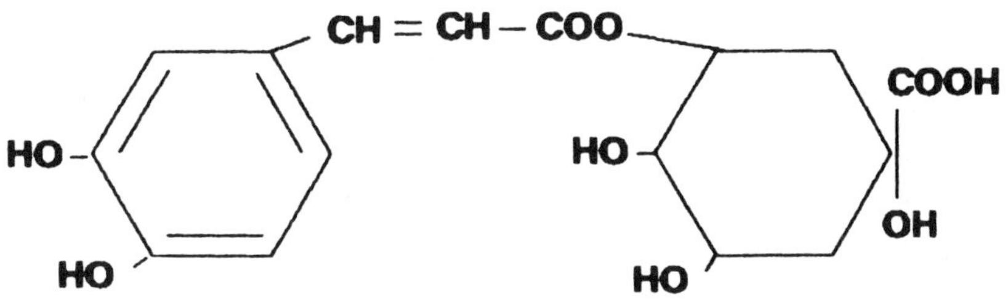

Chlorogenic acid

FIGURE 91. Chemical structure of caffeic and chlorogenic acids.

the accumulation of chlorogenic acid and its constituents, caffeic and quinic acids in infected tissues (Figure 91).[24] For example, the level of chlorogenic acid more than doubled in carrot root tissue infected with *Thielaviopsis basicola* and three phenolic compounds not found in uninfected tissues also were found. Ladigina and Edreeva[198] reported increased amounts of chlorogenic acid in tobacco leaves infected with *Peronospora tabacina,* as well as increased quantities of this acid in necrotic spots caused by physiological factors.

In early stages of plant infection, chlorogenic acids increase in infected tissues, then in the final stages of infection, the level drops to less than that found in uninfected tissues. The quantity of this acid also increase in uninfected contiguous tissues around points of infection until the final stages of the disease. Ladigina and Edreeva[198] presumed that the quantity of chlorogenic acid did not increase due to tissue injury, but to synthesis activity in the affected cells.

Reduced levels of chlorogenic acid in tissues in the late stages of pathogenesis may be explained by the degradation of the acid to its constituent parts, caffeic and quinic acids. Reductions also may occur due to the oxidation of chlorogenic acid accompanied by the formation of quinones.

Quninones are highly reactive, but relatively unstable and therefore do not accumulate in its free state. They bind with various components of protoplasm, such as proteins, forming dark-colored pigments and take part in the eventual death of plant cells and tissues. Since the

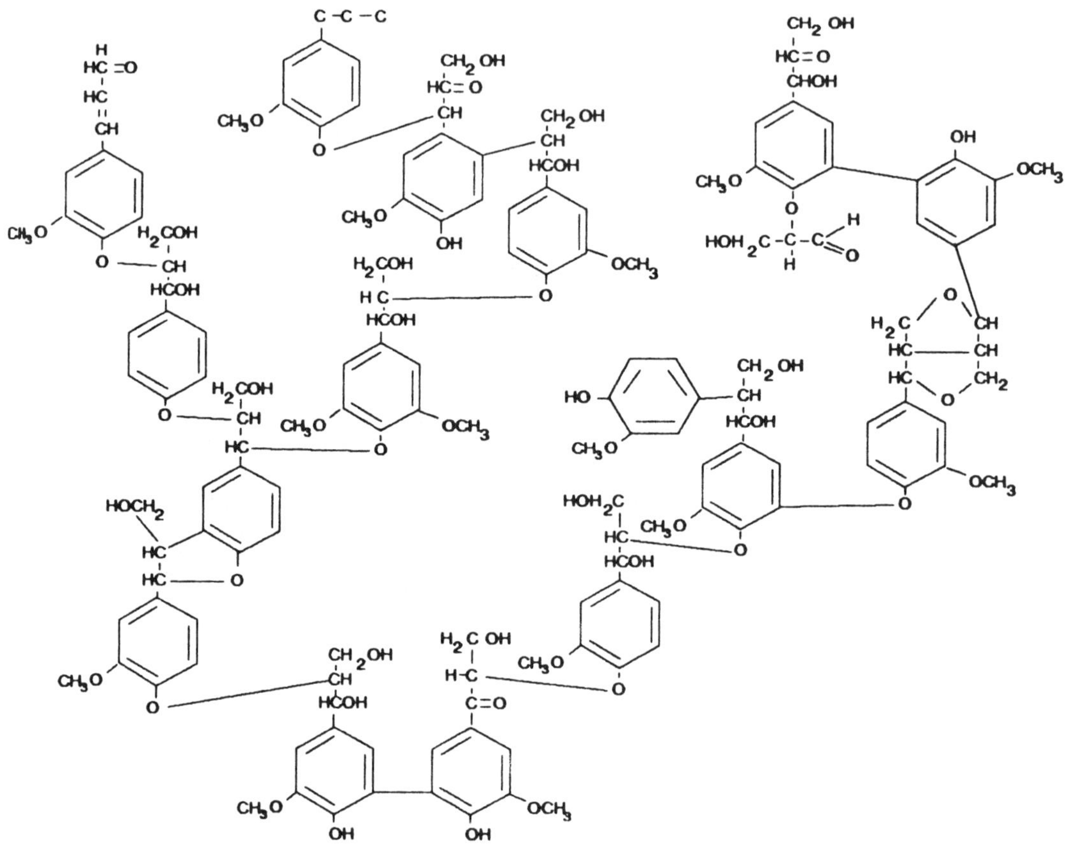

FIGURE 92. Proposed model of a typical softwood lignin (After Adler, E., *Wood Science Technology*, 11, 169, 1977.)

development of obligate parasites and some facultative parasites ceases with the development of necrosis, these compounds contribute to the development of resistance to various pathogens.

The presence of chlorogenic acid in carrot plants uninfected and infected with *Phoma rostrupii* was established by Handobina et al.[199] However, the levels of this acid in diseased plants was considerably higher than in uninfected plants; it was 40 times higher in necrotic tissues associated with the fungus, and 20 times higher in uninfected tissues adjacent to the lesions, and 2.5 times higher in tissues 7 to 8 cm from the necrotic spots.

Chlorogenic acid, as well as scopolin glucosides and its aglucone scopoletin accumulation has been observed in tobacco plants infected with *Pseudomonas solanacearum*.[200] The amount of scopoletin was by 10 times higher than in healthy tissues. Increased arbutin quantities in pear tissues following infection by *Erwinia amylovora* are converted, due to the activity of beta-glucosidase, into hydroquinone displaying an antibacterial influence following its oxidation.[190]

In virus-infected plants, an increase in the formation of phenolic compounds such as scopoletin and several other unidentified fluorescent materials, was recorded in tissues surrounding necrotic spots.[45] The presence of quinones is common in the necrotic tissues of hypersensitive host plants and various viruses. For example, aglucones of flavonoid compounds prevent the transmission of some viruses to indicator plants, thus preventing diagnostic determinations and research.[190] In cherry leaves infected with the cherry rusty mottle virus, Geissman[201] reported at least three times the amount of chempherol and quercetin, and in some cases, twice the levels of caffeic acid when compared to uninfected leaves. The level of chlorogenic acid also was higher in infected leaves.

It has been assumed that aromatic compounds, such as anthocyanins, aromatic amino acids, coumarin products, flavonoids, glucosides, phenols, and polyphenols accumulate in and around infected and injured plant tissues.[202] These compounds may be fungistatic and contribute to the defense reaction of plants.

Changes in the phenolic compounds, anthocyanin and flavonoids, are associated with color changes in diseased or damaged plant parts in which these pigments are present. The changes in coloration are associated particularly with diseases caused by mycoplasmas and viruses, which are characteristic and used for diagnoses (see section on chromatic anthocyanin and flavonoid changes).

In some diseases, there is a reduction in the synthesis of some phenols. For example, the apple rubbery wood mycoplasma reduces the lignin content of cells walls in the apple cv. Lord Lambourne.[203] As a result, the branches of infected trees become pliable, a characteristic symptom of the disease.

Qualitative phenolic compound changes. Phenolic compounds present in infected or damaged plant often are not present in uninfected plant tissue These qualitative changes alter the metabolism of such plants. The most pronounced changes of this type include the production of phytoalexins formed in plant parts only following disease or damage. Phytoalexin synthesis of these compounds in higher plants act somewhat like antibiotics protecting plants from the spread and further damage by pathogens.

Phytoncides, distinct from phytoalexins, are formed in healthy as well as diseased plants and have a wider range of action against unfavorable external factors.

Beuchat[204] stated that the possibility exists for the formation of phytoalexins in all healthy plants and is realized either as a result of a response to enzymes produced by pathogenic bacteria and fungi, or due to changes in host cells caused by pathogenic fungi. Phytoalexins, as induced plant products, have been recorded in over 100 plant species belonging to 21 families.[205] Several phytoalexins have been isolated and chemically identified, such as capsidiol, $C_{15}H_{24}O_2$, found in pepper plants infected with *Botrytis cinerea;* phaseollin, $C_{20}H_{18}O_4$, found in bean plants infected with *Monilia fructicola*; pisatin, $C_{17}H_{14}O_2$, found in pea plants infected with *M. fructicola;* and rishitin, $C_{14}H_{22}O_2$, found in potato tubers infected with *Phytophthora infestans* (see section on phytoalexins).

Qualitative differences in the composition of other phenolic compounds may occur in diseased plants. For example, scopolin glucoside aglucone (scopoletin) occurs in necrotic tissue of tobacco plants infected with *Peronospora tabacina*, in contrast to the presence of chlorogenic acid and rutin in uninfected tissues.[198] The accumulation of scopoletin was also recorded in tobacco tissues infected with other pathogens and in potato plants infected with the downy mildew fungus.[198] Caffeic acid, usually not present in healthy plant tissues, was found in extracts of carrot roots infected with *Phoma rostruptii*.[199] The phytoalexin, isocoumarin (3-methyl-6-metoxyhydroxy3,4-dioxyiocoumarin) also was found in the infected carrot root tissues remote from the necrotic points of infection.

These and other examples substantiate that complex alterations take in the metabolism of diseased and damaged plants either in the form of changes in the quantity of phenolic compounds or in the formation of new ones. Since phenolic compounds regularly occur and accumulate in necrotic plant tissues, it proves that their synthesis occurs in the area and is probably a consequence of cell necrobiosis.

Changes in phenoloxidase activity. Increased phenoloxidase activity was found generally in plants infected with a range of pathogens. The occurrence and activity of this enzyme in infected or damaged plant parts is another primary characteristic of pathological changes in phenolic compound metabolism (see section on oxidases).

No fully reliable explanations have been made for the increased activity of phenoloxidases in diseased plants Handobina et al.[199] presumed that one could be the synthesis of phenoloxidase by the substrate. Other authors believe that the increase could be due to increased amounts of phenolic compounds which stimulate the activity of these enzymes. The normal separation of phenolic compounds from phenoloxidases may be disrupted during the disturbance of the internal organization of a diseased plant.[45,198] Based on this interpretation, phenolic compounds in a healthy plant are found almost solely in cell vacuoles, and are separated from phenoloxidases which occur naturally in the cytoplasm, cell organelles and cell walls. In healthy cells and tissues, the influence of these various constituents takes place through the mediation of the intracellular membranes. Any damage to this inner structure due to disease or injury would release the phenoloxidase which would react with the phenolic compounds resulting in the oxidation of the latter groups. The oxidized phenols, which may be bound to proteins, would cause protoplasm coagulation and eventual cell death and necrosis of tissues.

In summary, the metabolism of phenolic compounds would be described by a number of interlinked changes.[190] These phenolic compounds accumulate in damaged on infected tissues, followed by increased glucosidase activity, resulting in the release of phenols, bound in glucosides, into toxic aglucones. This process is accompanied by increased oxidation of phenolic compounds by the activation of polyphenoloxidase, and with quinone binding with protein cell constituents, together cause cell and tissue necrosis. Phytoalexin formation also takes place as a defense mechanism particularly in hypersensitive reactions by plants to injury or infection.

XXIII. ANTIPARASITE ACTION OF PHENOLIC COMPONDS

Phenolic compounds and their byproducts differ from other organic compounds in a cell by playing a role in the direct defence of plants against external factors. These compounds participate directly in the defense of plants through a series of reactions in the regeneration and healing of damaged or infected plant parts.

The study and understanding of phenolic compounds is very important to plant pathologists because of their action against plant pathogens, which is manifested in pathogen inhibition or extinction. Depending on the mode of defensive action phenolic compounds may be classified as either active or passive substances against plant parasites.

Passive action of phenolic compounds aginst plant pathoens. Passive phenolic compounds are those found in the structure of cell walls which act as a barrier to pathogen penetration and spread. Characteristic compounds of this nature include lignin(Figure 93), suberin, terminal metabolic cell products, which are part of the permanent cell wall structure. Lignin permeates all cell wall layers forming complexes with cellulose through glucoside bonds, providing mechanical resistance to cell walls. In contrast, suberin is excreted in the form of fine lamellae and deposited on the inner layers of the primary wall, thus, providing a resistant layer between the protoplast and the external environment.

Lignin and suberin, although chemically different, perform similar protective functions in the preservation and survival of cells and, in general, of plants. Suberin is a lipid and lignin a complex phenolic compound. Both are classified chemically and mechanically as resistant cell components (see section on phenolic compounds in plants and structural cytopathological changes).

Lignin also is included as part of the structure of highly specialized cells and tissues. It is the main constituent of stone cells, tracheae and tracheids, sclerenchyma tissue, and wood fibers. The primary root cortex exodermis and the endodermis surrounding the central cylinder, cell layers are permeated with lignin or cork. Lignin is also a primary constituent of the cell wall structure of the woody portions of perennial plants. Lignin accumulation in cell walls often occurs as a consequence of cell damage or infection, and is also a characteristic of cell aging.

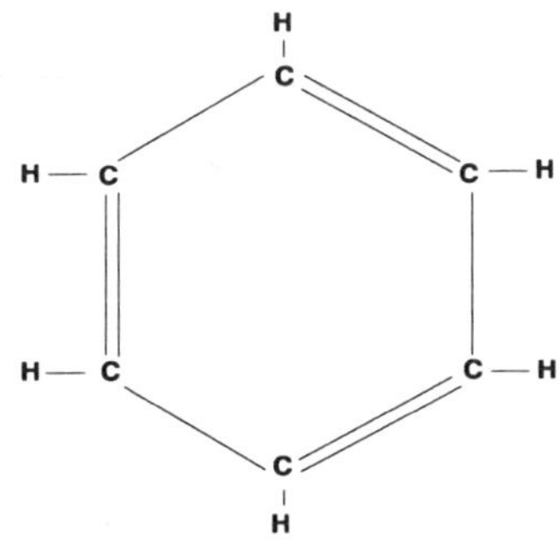

FIGURE 93. Benzene ring with hydrogens substituted by atoms or other elements or by groups of atoms.

Lignin is one of the most important, if not the most important, barrier preventing or delaying pathogen development in plant cells and tissues. It is highly resistant to many pathogen enzymes, and by binding with cellulose and hemicellulose into complexes, lignins prevent or inhibit the activation of enzymes that degrade the polysaccharide components of cell walls.

Following lignification of the middle lamella, phytopathogenic bacteria cannot grow and develop in parenchymal tissue cells. Even the most pathogenic fungi having colonized herbaceous plant parts may neither mechanically nor enzymatically affect lignified cell walls. Thus, the presence of lignin, as well as suberin in plant structures act as passive factors in plant resistance to some pathogens. For example, McClure and Robbins[206] showed that resistance in mature cucumber seedlings to infection by *Pythium* spp. was accompanied by the formation of lignin in all the cell walls surrounding points of infection.

Lignified and suberized endodermis cell walls of the primary root cortex may arrest the development of some vascular fungi and prevent penetration into the vascular system. For example, hyphae of *Fusarium nivale* and *Gibberella zeae* frequently spread through the primary root cortex of certain wheat cultivars and then cease their development when they reach the endodermis layer which contains cells permeated with lignin, thus, preventing the pathogens from reaching the vascular system.[15] The lignified ring around the pericycle of the central cylinder and the lignified xylem elements in vascular system present a barrier to the spreading of *Fusarium roseum* in carnation stems.[207] Gäumann[15] described the linear arrangement of the uredosori of several wheat rust fungi on wheat leaves and stems being determined by lignified sclerenchyma tissues in these plant parts, in contrast to the free development of the pathogens in tissues containing only collenchyma and parenchyma tissues. Also, the accumulation of callose and lignin, stimulated by plant damage, tended to localize infection by viruses giving rise to the hypersensitive reaction in leaves of several plants.[208]

A specialized group of fungal pathogens which cause white wood rot, is characterized by their ability to degrade lignin. This group includes many fungi of the Hymenomycetes. According to Mount[209] members of this group can degrade lignin, sometimes more rapidly than polysaccharide constituents. The extent of this degradation depends on a complex system of enzymes and coenzymes.[210] Although lignin-degrading enzymes have not been isolated from this group of fungi, a number of stages in lignin degradation have been described. The stages include the elimination of methoxyl groups, side chain polymer oxidation, oxidative degradation of aromatic rings in the lignin polymer, and the release of oliphatic molecules following ring separation.[209,210] Because of this ability to chemically degrade lignin, the white wood rot fungi are considered as an ecologically separate group of fungi specialized for the development in woody tissues of perennial plants.

Active action of phenolic compounds against plant pathogens. The antiparasite activity of phenolic compounds against plant pathogens is considered a direct one between the pathogen and the compounds. Their mode of action is based in part on the composition and structure of free phenolic compounds, particularly the oxidized phenolic compounds. An example of a group of phenolic compounds that have antiparasite activity is the phytoalexins. Induced lignification of cell walls of tissues surrounding points of infection is another form of an active defense by plants against pathogens.

Phenolic compounds may have a static or toxic effect on plant pathogens depending on the composition, concentration, and mode of action. Examples of a static influence and plant diseases are phytoalexin production and lignification of cell walls. Examples of toxic action are various oxidized phenolic compounds which cause death of the cells of both the host and the pathogen.

Free phenols. This group of compounds are not bound to other compounds in the plant and as free molecules can act directly against plant pathogens. The efficiency of these compounds as antiparasite activity depends on the arrangement of their chemical composition and structure. Greathouse and Bigler[211] studying the influence of 45 phenolic and other related compounds on the growth of *Phymatotrichum omnivorum,* showed that the antifungal activity of these materials depended on the number and kind of chemical constituents in the benzene ring (Figure 93). They found that simple substitutions in the benzene ring could decrease the biological activity according to the following order:

$$OH > H > COOH > COH > OCH_3$$

When two substitutions were made, the activity depended on the nature of the added groups and their position in the molecule. The ortho- and para-dihydroxybenzoic acids were the more active than the meta-positions of these acids; while the para- and meta-forms of methylphenolic and methylbenzoic acids were more active than the ortho-forms. It was suggested that their biological activity depended in part on the dissociation ability and solubility of these compounds. In their soluble form, the free phenols spread throughout plant tissues and can play an active role in the defense of plants against pathogens and inhibit or prevent infection.

Direct action, for example, of phenolic compounds against various pathogens has been established in culture and in plants. Murkerjee and Kundu[212] studied 18 phenolic compounds of differing chemical structure on their effect on spore germination and growth of *Alternaria solani, Curvularia lanata*, and *Helminthosporium oryzae.* They reported that 10 of these compounds showed antimicrobial activity against these fungi, with pentachlorophenol, picric acid, pyrogalol and tannic acid, preventing spore germination. Some dihydroxyphenols and trihydroxyphenols showed only inhibitory action, with the positions of the hydroxyl groups in the benzene ring playing a more important role in their activity than the level of hydroxylation.

A frequently cited example of the efficiency of free phenolic compounds against a plant pathogen under natural conditions is the work of Walker and Stahmann.[213] They showed that the free phenolic compounds, catechol and pyrocatechuic acid in the outer leaves of colored onion bulbs, prevented the penetration of *Colletotrichum circinans.* These compounds produced by the bulb and dissolved in water droplets spread over the surface of the outer leaves preventing spore germination and thus, infection. The nonliving outer protective leaves with their phenolic compounds prevented the infection of the inner living leaves. This research became an example of active protection of living parts by nonliving plant parts, similar to the bark in woody plants. Another example is the discovery that tannins, which are composed of a mixture of pyrogalol and pyrocatechin, provided resistance to infection to *Endothia parasitica* in chestnut trees.[214]

Resistance to pathogens in plants is usually based on the presence of antimicrobial phenolic compounds in plant cells and cell walls. For example, Patil et al.[215] showed that the level of chlorogenic acid in potato roots directly influenced the resistance of these plants to the Verticillium wilt pathogen. Also, they reported that resistant potato cultivars contained greater amounts of this compound than did susceptible ones. Similarly, substantial levels of chlorogenic acid was recorded in tobacco cultivars resistant to *Thielaviopsis basicola*.[216] Other investigators have shown that potato tubers resistance in potato tubers to *Actinomyces scabies* was related to the levels of chlorogenic acid in the cells close to the periderm of the tubers.[24] According to other data obtained by these authors, the following compounds prevented the development of *Fusarium nivale* in carrot root extracts: benzoic acid, caffeic acid, chlorogenic acid, gallic acid, and p-oxabenzoic acid. They stated that chlorogenic acid played an important role in protecting the plant against infection, and that the primary activity of this compound was not due to its natural presence in plant tissues, but to its formation in reaction to the presence of a pathogen.

Other phenolic and related compounds shown to influence the resistance of plants to pathogens was the increase of catechin and proanthocyanidin in cotton seedling hypocotyls infected with *Rhizoctonia solani*.[217-219]

Bound phenols. Some phenol compounds become bound to sugars to form glucosides or sugar esters in plant cells. These compounds accumulate in plant tissues and are generally considered stable compounds. The pigments, anthocyanins and flavonoids, are members of this group of compounds and are found throughout the plant. By binding in the form of glucosides, these phenolic compounds loose their direct antimicrobial action. However, when these glucosides are released because of injury or disease, these compounds, known as aglucones, manifest antimicrobial activity and may represent a factor in the defense of the plant. Release of aglucone is a result of the activity of glucosidase enzymes which are produced by various pathogenic microorganisms, the necrotrophic fungi in particular, into infected tissues (Figure 94). For example, anthocyanin and flavonoid aglucones probably take part in the development of necrotic tissues in hypersensitive reactions. Anthocyanins are highly toxic to a variety of fungal spores. Thus, the role of anthocyanins in resistant plant tissues is that their degradation releases toxic aglucones.[24]

Resistance in apple cultivars to *Venturia inaequalis* was attributed to phloretin aglucone, a polyphenol constituent of phloridzin glucosides.[202] Both compounds are synthesized in resistant and hypersensitive cultivars, preventing pathogen development. It was found that phloretin was more than phloridzin active against *Erwinia amvlovora*.[45]

Other compounds reported in necrotic plant tissues were scopoletin aglucone (6-methoxy-7-hydroxy-coumarin) and scopolin glucoside in resistant potato tissues infected by *Phvtophthora infestans*.[220] Scopoletin was shown to be fungistatic to *Ceratocystis fimbriata* at a concentration of 0.025% and fungitoxic at 0.1%.[221] In tobacco plants, scopoletin aglucone, scopolin, and chlorogenic acid accumulated in response to infection by *Pseudomonas solanacearum*.[222] Thus, scopoletin plays an important role in the protection of plant tissues against various fungi,[223] and viruses,[224] and forms in response to various mechanical injuries.[225]

Oxidized phenols. Oxidized phenols, also known as quinones, are the most reactive phenolic products found in plant cells and tissues. Quinones are formed during the oxidation of phenolic compounds. The oxidation of phydroxylic groups leads to the formation of quinone and water. Quinones are common in plants and play an active role in the hydrogen transport during respiration of amino acids, ascorbic acid and other compounds. In the oxidation-reduction processes, phenols are reversibly formed from quinones.

In infected or damaged cells, the metabolism of phenolic compounds results in an increase in the quantity of quinones, which disrupts the natural balance between phenolic compounds. The accumulation of oxidized phenols have been found particularly in tissues that have developed necrosis in response to virus infections, to infection by *Phytophthora infestans* on potato plants, to *Puccinia graminis* on wheat plants, to *Venturia inaequalis* and *Podosphaera leucotricha* on apple tissues, to *Verticillium albo-atrum* on resistance cotton tissues, and to *Botrytis cinerea* on cabbage tissues.[190]

Quinones are chemically very active and bind or polymerize with other organic compounds to form dark-colored resinous substances in plant cells and tissues. These substances usually permeate throughout the area and provide a chemical barrier around points of infection to prevent further spread of the pathogen involved. These dark-colored materials are byproducts

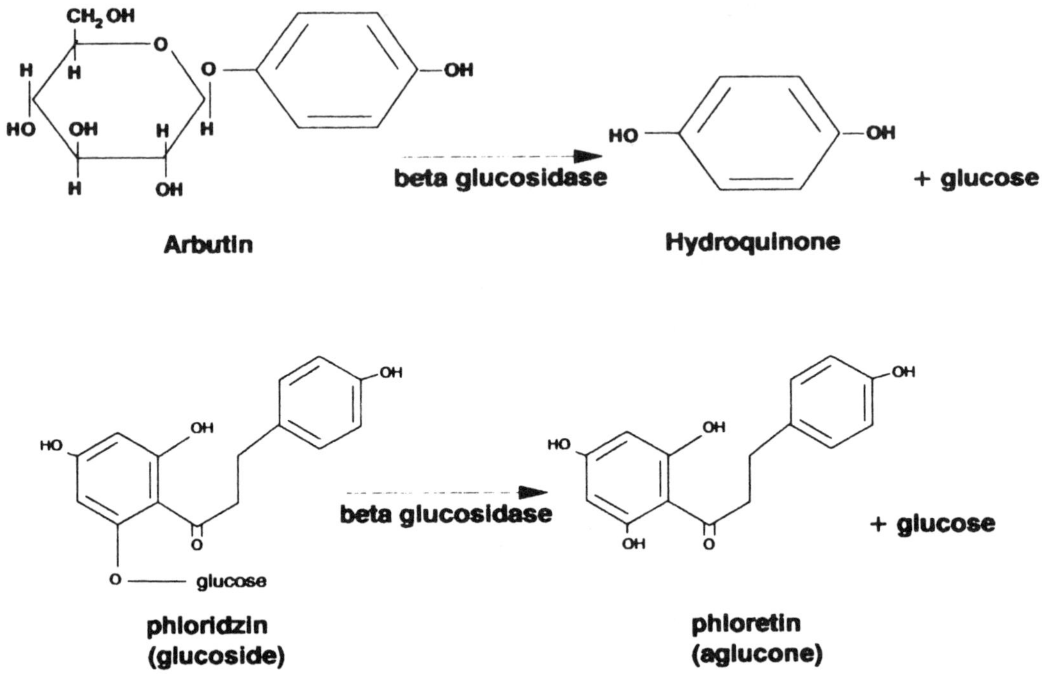

FIGURE 94. Aglucone release from glucosides due to glucosidase from plant pathogenic organisms, particularly by necrotrophic fungi in infected tissues.

of the reaction of phenolic compounds with various cell constituents, such as carbohydrates and aromatic acids.

Quinones with antiparasitic action have been found and studied particularly in tissues responding with hypersensitive reactions to infection. In these areas, quinones react with proteins, peptides, and amino acid groups to form polymers with corresponding C-N and C-S bonds.[198] Matthews[42] stated that if quinones have a second reactive group, the cross binding with protein molecules may also take place. Such reactions do occur with virus isolation and purification of viruses from infected plants. With the formation of the dark-colored polymers between quinones and cell constituents, melanoid pigments are formed. These pigments are insoluble amorphous polymers frequently linked with proteins and formed through the catalytic action of polyphenoloxidase in reaction with tyrosine.

Necrosis occurs as a consequence of a hypersensitive reaction. Kiraly[226] pointed out that such reactions are effective protective reactions against biotrophic fungal pathogens, as well as facultative and necrotrophic pathogens.

Certain phytopathogenic bacteria cause hypersensitive reactions in sensitive plants, such as *Pseudomonas syringae* pv. *tabaci* on tobacco plants and *Xanthomonas campestris* pv. *vesicatoria* on tomato plants.

In sensitive plants infected with certain viruses, the formation of necrotic spots prevents the further duplication and spread of virus particles, such as those of the tobacco mosaic virus in *Nicotiana glutinosa* and some pepper cultivars, and the alfalfa mosaic virus on bean leaves.

The activity of phenolic compounds and particularly their oxidized products also affects enzymes formed by various pathogens. These compounds probably react with the protein constituents of the enzymes and thus restricts or prevents their activity. The inactivation of enzymes by phenols is a general phenomenon and not specific in nature. It has been shown that various phenols and quinones inhibit the activity of amylase, protease, as well as pectolytic and respiratory enzymes.[24]

Oxidized phenols also inactivate extracellular enzymes produced by fungi such as *Fusarium, Sclerotium* and *Verticillium*. For example, species of *Verticillium* excrete pectolytic enzymes in advance of its growth to destroy plant tissues.[227] Also, it was shown that oxidized phenols inactivate polygalacturonase in apple fruits infected with *Sclerotinia fructigena*.[228] Stahmann[229] showed that caffeic acid and pyrocatechin inhibited pectinmethylesterase and polygalacturonase only upon transforming them into their oxidized forms. Other authors have established that leucoanthocyanins of high molecular weight and tannins inhibited pectolytic enzymes.[230,232] Reduced dehydrogenase activity was recorded in hypersensitive potato plants infected with *Phytophthora infestans* probably through the activity of oxidized polyphenols.[107] From studies in culture, Murkherjee and Kundu[212] showed that some phenolic compounds inhibited the activity of laccase, tryosinase and other macerating enzymes of *Helminthosporium oryzae* more effectively than the same enzymes produced by *Alternaria solani*.

Thus, these and other data have shown that phenolic compounds, particularly in their oxidized form, can partially or wholly inactivate extracellular enzymes, depriving the pathogens of one of their primary means of tissue destruction colonization. Also, the formation and activity of phenolic compounds as enzyme inhibitors was correlated with resistance in plants to various pathogens.[233,234]

Phytoalexins. Phytoalexins are metabolic phenolic compounds formed in injured plant cells and perform an antimicrobial compounds against the causal agent involved. Phytoalexins are formed as a general reaction to any biotic or abiotic factor and have been defined as "stress" metabolites and sometimes referred to as "plant antibiotics".

Phytoalexin synthesis in damaged or infected plants has not been fully characterized. In a study of the host-parasite relationship between potato tissue and *Phytophthora infestans*, Metlitskii[235] presumed that these metabolites produced by the host plant reacted with specific receptors on the pathogen membranes, and therefore pathogens excrete substances that induce phytoalexin synthesis. This excretory product penetrates host cells and stimulates them to produce the phytoalexins. This presumption was supported further by data defining the genetic basis for resistance between hosts and pathogens.[246]

Several authors emphasized a number of features of phytoalexins that distinguished them from other natural compounds.[236-238] Phytoalexins form as a result of mutual influence between the host and causal agent, and only cells nearest the points of damage take part in the synthesis, but movement of phytoalexin from cell to cell is possible. Phytoalexins may also be detrimental to the host tissues in which they are formed. They are not pathogen or causal agent specific, since various stimuli will induce their production. Phytoalexins are produced more rapidly and in greater quantities in resistant than in susceptible plants. The extent of pathogen colonization may depend on the speed in which phytoalexin production begins.

Phytoalexins are made up of a variety of compounds which are chemically different and some are characteristic of certain plant families.[238,239] For example, acetylenes are characteristic for Asteracea, diterpenes for Poaceae, furanoterpenes for Convolvulaceae, isoflavonoids for Fabaceae, naphthaldehydes for Malvaceae, and sesquiterpenoids for Solanaceae.

Hoffmann et al.[190] classified phytoalexins into three groups based on the plant family and host of origin and chemical structure: Fabaceae: *Phaseolus vulgaris* and *Glycine max* - phaseollin, and *Pisum sativum* - pisatin; Orchidaceae: *Lorolossum hirsutum* - hircinol, and *Orchis militaris* and *Orchis* sp. - orchinol; and Solanaceae: *Lycopersicon esculentum* and *Solanum tuberosum* - rishitin (Figure 95). Phaseollin is formed in bean endocarp and leaves following infection with several plant pathogens including *Colletotrichum lindemuthianum* and *Rhizoctonia solani*; pisatin is formed in pea pods by a variety of fungal pathogens and nonparasitic causes; orchinol is formed in orchid plants following infection by *Rhizoctonia repens*; and rishitin is formed in

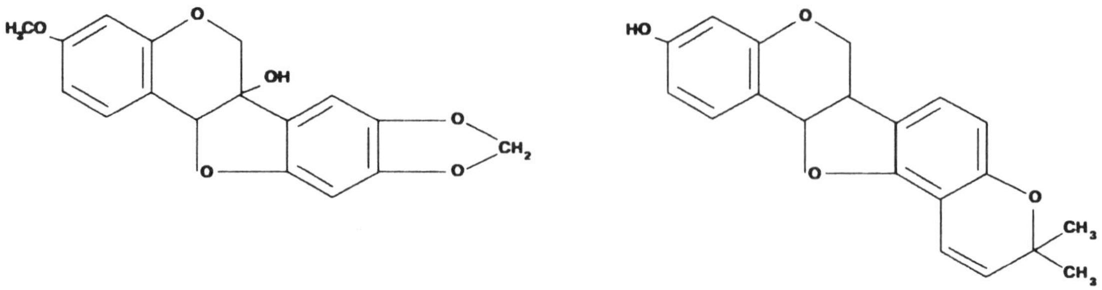

Pisatin **Phaseolin**

FIGURE 95. Structure of two phytoalexins, pisatin formed in *Pisum sativum* (common pea) and phaseolin formed in *Glycine max* (soybean).

potato tubers following infection with *Phytophthora infestans* as well as inoculation with nonpathogenic fungi.[190] According to the same authors members of the flavonoid, stilben, terpenoid, and tropolon phytoalexins inhibit a variety of wood rot fungal agents.[190]

Finally, research on phytoalexins continues to better understand their role in the defense system in plants and the possibilities of manipulating their production for use in practical disease control.

Additional lignins. Lignin is one of the most important natural compounds related to disease resistance in plants. As such, it has been classified into two types, structural and reactive, by plant pathologists. Structural lignin is formed during normal cytogenesis and histogenesis and is a normal part of the components of the structure of plant cells and tissues. In this form it acts as a passive agent for the control of certain plant pathogens (see section on passive parasitic action of phenolic compounds).

Reactive lignin is formed as a reaction to injuries and infections. Reactive lignin is usually formed in cells or tissues immediately adjacent to points of injury or infection and belong, together with phenols, flavonoids, phytoalexins, and various metabolites to the common protective system against infection.

Lignification of developing seedlings plays an important role in the protection of plants against infection by various fungi, including *Botrytis, Pythium* and *Rhizoctonia*, which cause pre- and postemergence damping-off of seedlings. Seedlings with nonlignified or partially lignified cells are highly susceptible to damping-off. As secondary tissues develop there is a progression of lignification which evenutally provides the older seedlings with resistance to these causal agents. According to McClure and Robbins,[206] resistance in seedlings to *Pythium* spp. is accompanied by the deposition of lignin in all cells and tissues surrounding points of infection by this fungus.

Lignification is one of several structural changes that take place in plants infected with systemic plant pathogens such as those causing downy mildew and rust.[26] The quantity of lignin increased by 20% in Sudan grass plants infected with *Colletotrichum raminicola*.[237] Because of this increase in lignin, the digestibility and therefore its value as animal feed decreased. Lignification was also recorded in leaves of fruit trees with shot-hole disease.[240] Also, increased lignification and suberization of parenchyma cells and tissues were observed in papaya leaves infected with the anthracnose fungus.[241]

Finally, it should be emphasized that injuries and death of infected cells stimulate lignin formation and polyphenol oxidation in healthy and infected cells surrounding the points of infection. Thus, this additional lignin formation represents an important factor in direct protection by plants from pathogenic microorganisms and viruses.

In conclusion that phenolic compounds are widely distributed in plants and play an irreplaceable role in plant survival. Some phenolic compounds, for example, directly perform certain functions to aid in the continuous maintenance of the living cell system. Phenolic compounds are part of the cell structure of a plant and are constituents of plant pigments, where they can play an indirect role in pollination and reproduction. For plant pathologists, phenols are of particular importance because of their role in protecting plants from damage by biotic and abiotic factors.

XXV. PLANT DEFENSE STRATEGIES

Plant life has developed and continues to develop in the constant presence of a multitude of unfavorable factors endangering their preservation and survival.

Vascular plants, after leaving their water environment, developed root systems to maintain their link with the substrate, thus became immobile organisms. Due to this immobility, plants developed conductive systems between the soil and the atmosphere in which they live. This internal linkage of living cells, tissues and organs has been established and interconnected with plasmodesmata. It is through this linkage with soil and atmosphere that vascular plants become dependent on the ecological conditions of their environment determining their anatomical and morphological structure and development.

Being immobile, plants can only defend themselves against the forces of nature. This type of defense has been compared to the defensive nature provided by forts and walled castles in the Middle Ages.[242]

Plants live under a constant threat of attacks by various plant pathogenic organisms. Many different types of plant pathogens with high reproduction and replication potential are capable of infecting various plant species. Therefore, as a general rule, there is no plant which in its course of development has not been attacked by one or more pathogens.

External abiotic factors also cause a number of injuries and damage to plants endangering their life. In crop production, injuries caused by various cultural practices also represent a permanent danger.

Plants have developed defensive "weapons" to protect themselves against unfavorable external factors, thus enabling them to survive. Since plants are only defensive in nature, they can be considered the most peaceful inhabitants of earth.

Having developed in an ecosystem made up of pathogens and nonpathogenic agents that cause them harm, plants have developed their own defense systems that enable them to oppose the action of all but the most unfavorable environments. Plant defense systems have developed during the course of evolution and continue to do so. Due to these defensive mechanisms, plants have successfully preserved themselves over the milleniums and constantly improve their organization for survival.

XXVI. PLANT CELLS — CARRIERS OF DEFENSIVE ACTIVITIES

Higher plants in contrast to higher animals have specialized immune systems for active defense against damaging agents. However, the living cells of plants carry on the defense activities within a living plant, particularly those of parenchyma tissue. These plant cells react to all injuries regardless of the nature of the injury, whether by a pathogen, mechanical or chemical. Approximately identical biochemical, physiological and structural changes take place which leads to the isolation of the causal agent or to healing of the wound in a damaged area.

Defensive reactions in plants occur not only in the immediate area of infected or damaged cells, but also in contiguous cells, nonaffected cells. The "message" for such reactions is transferred through intercellular bonds, such as plamodesmata, and by metabolites of either or both the infected or damaged cells or of the pathogen.

Infected and damaged cells begin their defense by intensifying or redirecting their biochemical processes to counteract the damage. This altered biochemical activity is accompanied by corresponding changes in the internal cellular structure, such as increases in the number of mitochondria or ribosomes. The increase in the number of ribosomes is necessary to meet the need for the synthesis of new structural proteins and proteins for the formation of enzymes, the number and activity of which increases with infection or damage. An increase in mitochondria is one factor contributin to increased respiration, which in addition to oxidation or degradation of compounds, releases chemical energy necessary for the biosynthesis of substances in infected and damaged cells.

The movement of nuclei toward damaged or infected point, called positive nuclear traumotoxis, also represents a defensive reaction. Following pathogen penetration, liposomes in the cytoplasm are degraded releasing hydrolytic enzymes capable of degrading pathogen cell constituents. Also the increase in the movement of cytoplasm and its constituents in the initial stages of infection or damage results in an increase in the metabolic activity of the cell and thus, contributes to the ability of the cell to defend itself.

The ability of meristematic tissues to divide more or less continuously is a particular function in plants that also contributes to the mechanisms of defense against disease and damage. This type of cell division provides for the substitution of new cells for damage or dead ones, or for the renewal of whole plant tissues or organs for damaged or killed ones.

An exceptional defense mechanism in plants is the ability of plants to transform the structure and function of cells through redifferentiation processes. Individually transformed cells, or new cells may be formed during these activities. Induced meristematic tissues provides for the healing of damaged plant parts. Induced peridermal tissue formation around points of infection or injury will limit or localize the spread of pathogens (see section on cytotransformative changes and induced tissue formation).

Another general defense of plant cells is their ability to absorb or resorb damaged cell parts and some foreign substances introduced into the plant. In ontogenetic plant development cell death may occur naturally or because of external factors (see section on cytolytic changes). Through absorption or resorption mechanisms, degraded substances are eliminated, thus providing for the normal maintenance of the structure and function of the plant tissues and organs.

In conclusion, plant cells are capable of numerous and a variety of defense mechanisms, which consist of various biochemical activities, structural, divisional, transformative and other changes, which can occur simultaneously and complimentary to influence protection of the plant.

Biochemical and physioloical defense mechanisms. Defense mechanisms in plant cells are complex biochemical and physiological processes occurring at the time of cell penetration and infection by pathogens or abiotic agents. The role of phenols and proteins in the defense reactions have received particular attention because of the importance of their role in the processes. Proteins play an important role in the "recognition"of pathogens after they enter the host and are part of the composition of enzymes, whose activity intensified at the beginning of infection influences cell structure and function. Protein-containing compounds have an inhibitory influence on enzyme activity of bacteria and fungi, thus depriving them of one of the mechanisms for degradation of plant cells. Acting directly on various pathogens, such proteins may arrest or prevent pathogen development in plant cells and tissues. Antiviral proteins have been isolated from tobacco plants resistant to the tobacco mosaic virus (see section on proteins and plant cell parasitization).

Phenolic compounds play an important role in the defense of plants against pathogens. Free phenols, which a caffeic acid, catechol, chlorogenic acid and pyro-catechuic acid, as well as phenols released from polyphenolic compounds play an active role in this regard. Other phenolic compounds, such as lignin, plays a passive role in the defense of the plant, since it acts as a mechanical barrier. Quinones also play a passive role in the defense of plants by combining with proteins and causing the death of cells, which in turn arrests the development of the pathogen involved (see section on antimicrobial phenolic compound action).

Phytoalexins belong to a group of phenolic compounds formed in response to injured or infected plant cells. They show antimicrobial activity against many species of pathogenic microorganisms and thus contribute to general plant resistance (see section on phytoalexins).

Phytoncides are antimicrobial compounds that are found consistently in plant cells and tissues of both higher and lower plants. Their antimicrobial action takes place either directly in the cells where they are synthesized or outside the cell in a volatile form within the biocenosis. They are known to be active against plant pathogens and the action is considered a biocontrol mechanism. For example, phytoncides from garlic and onion plants have an antibiotic influence on bacterial pathogens of tomato and can arrest the growth of *Phytophthora infestans* in potato tubers. Also, phytoncidal compounds from horseradish, radish, and hemp plants inhibit spore development of *Botrytis alii*.[24] Because of their wide-spread occurrence in plants and broad spectrum of activity against plant pathogens, they must be considered as having an active role in nonspecific resistance to disease agents.

Some compounds that are relatively resistant to degradation by pathogen enzymes are also found in plant cells. These compounds usually are made from complexes involving pectin, proteins and polyvalent cations, such as calcium and magnesium.[243] The presence or accumulation of these cations close to the points of infection leads to the formation of these complexes, which are resistant to the enzymatic degradation by plant pathogens. Thus, these compounds limits the spread of the pathogen, restraining them around the points of infection.

Biosynthetic products of cells, such as cutin, suberin and wax have a significant role to play in the protection of plants against foreign materials. Cutin, suberin and wax are cell products taking part in the structure of the cell. They are complex lipids which are highly resistant to both mechnical and enzymatic action and thus, very protective plant cell constituents. Suberin occurs in the structure of induced peridermal tissues surrounding infected or damaged cells (see section on induced tissues). Owing to these same traits, wax and cutin are also considered as special defense structures.

Other natural metabolic products of plants have been shown to be capable of inactivating various toxic compounds from pathogens resulting in the detoxification of such compounds in plants. For example, fusaric acid, the toxin produced by *Fusarium* spp., and piricularin, the toxin produced by *Piricularia oryzae* are inactivated in this manner.[244] Although the biochemical processes by which this is accomplished is not precisely known, they do contribute to plant resistance to pathogens.

Certain metabolic products excreted on the surface of plant organs may perform antimicrobial action. The action of these compounds may be either direct, i.e. toxic or inhibitory to pathogen enzymes, or indirect, i.e. altering the complex relationship between the biological entities. Among toxic compounds found on leaves, for example, are phenolic compounds, soluble organic culticle constituents, and organic acids. Compounds excreted from the surface of roots include amino acids, organic acids, proteins, polysaccharides, and hormonal substances.[245] Such excreted substance play an important role in preventing or delaying pathogen penetration of the host.

A protective role is plant by resinous or gummy substances excreted by cells into intercellular spaces surrounding infected or damaged plant parts. These materials also accumulate on the surface of plant parts. These substances are complex polysaccharides comprised of phenolic compounds. They are terminal compounds, resistant to pathogens, and serve to isolate damaged or infected cells from healthy ones. Such secretions are found primarily in stone fruit trees.

Finally, it should be stressed that biochemical and physiological processes are the basis for the defense of plants against disease and damage. These processes are numerous, varied, and can be mutually linked and coordinated in the plant's phylogenetic development. These defensive activities are genetically controlled by the living cells. Thus, changing in the cell genome represent the safest means to obtain new plant genotypes, including resistance or immunity to unfavorable environmental factors.

REFERENCES

1. **Wheeler, H. E.**, Permeability alterations in diseased plants, in *Physiological Plant Pathology*, Horsfall, J. G., and Williams, P. H., Eds., Springer-Verlag, Berlin, 1976, 413.
2. **Thatcher, F. S.**, Osmotic and permeability relations in the nutrition of fungal parasites, *American Journal of Botany*, 26, 449, 1939.
3. **Thatcher, F. S.**, Further studies of osmotic and permeability relations in parasitism, *Canadian Journal of Research*, Section C, 20, 283, 1942.
4. **Thatcher, F. S.**, Cellular changes in relation to rust resistance, *Canadian Journal of Research*, Section C, 21, 151, 1943.
5. **Dorohova, N. A.**, Fiziologiya kariofelynago rasteniya, porozhennogo fitoftoroy. in Tr. *In.-ta fiziolim.*, K. A. Timirjazeva, Ed. 3, 189, 1940.
6. **Kuprevich, F. V.**, *Fizioloiva bolnvnogo rasteniya*, Izd. Akad. nauk SSSR, Moscow, 1947.
7. **Sen, H., and Joshi, H. C.**, Relation between permeability of the plasma membrane and resistance to parasites, in *Modern Developments in Plant Pathology*, Maheshwari, P., Ed., University of Delhi, Delhi, 1957, 150.
8. **Wheeler, H. E., and Black, H. S.**, Effects of *Helminthosporium victoriae* and victorin upon permeability, *American Journal of Botany* 50, 683, 1963.
9. **Kokin, A. J.**, *Issledovaniya bolynago rasteniva*, Petrozavodsk, Gos. izdvo Karelo-Finskoj SSR, 1948.
10. **Suhorukov, K. T.**, *Fiziologiya imuniteta rasteniya* M. izd-vo AN SSSR, 1952.
11. **Wheeler, H. E. and Hanchey, P.**, Permeability phenomena in plant disease. *Ann. Rev. Phytopathol.* 6, 331, 1968.
12. **Wheeler, H. E.**, Disease alterations in permeability and membranes, in *Plant Disease An Advanced Treatise* m , Horsfall, J. G. and Cowling, B. E., Eds., Academic Press, Inc., 1978, 327.
13. **Glišić, Lj.**, *Opšta Citologija*, Unija Biol. Naućnih Drustava Jugoslavije, Blegrade-Zenum, 1980.
14. **Lundegardh, H.**, *Plant Phvsiologv* (English translation by F. M. Irvine and W. 0. James), American Elsevier Publishing Co., Inc., New York, NY, 1966.
15. **Gäuman, E.**, *Principles of Plant Infection* (English edition by W. B. Brierley), Crosby Lockwood, London, 1950.
16. **Rubin, A. B. and Artsikhovskaya, E. V.**, *Biohimiya i Fiziologiva Imuniteta Rasteniy*, Iudat. Akademij Nauk SSSR, Moscow, 1960.
17. **Kozlowski, T. T.**, How healthy plants grow, in *Plant Disease An Advanced Treatise III*, Horsfall, J. F., and Cowling, E. B., Eds., Academic Press, Inc., New York, NY, 1978, 19.
18. **Popović, Ź.**, *Fiziologija Bilja*, Naućna Knjiga, Belgrade, 1976.
19. **Schramm, R. J., Jr., and Wolf, F. T.**, The transpiration of black shank in-fected tobacco, *Journal of the Elisha Mitchell Science Society*, 70, 225, 1954.
20. **Threlfall, R. J.**, Physiological studies on the Verticillium wilt disease of tomato. *Annals of Applied Biology* 47, 57, 1959.
21. **Talboys, P. W.**, Water deficits in vascular disease, in *Water Deficits and Plant Growth II*, Kozlowski, T T., et al., Eds., Academic Press, Inc., New York, 1968, 255.
22. **Dimond, A. E.**, Biophysics and biochemistry of the vascular wilt syndrome, *Annual Review of Phytopathology*, 8, 301, 1970.
23. **Duniway, J. M.**, Pathogen-induced changes in host water relations, *Phytopathology* 63, 458, 1973.
24. **Rubin, A. B., and Artsikhovskaya, E. V.**, *Biohimiya i Fiziologiya Imuniteta Rasteniy*, Izdat.-Vissaya Shkola, Moscow, 1968.

25. **Dimond, A. E. and Waggoner, P. E.,** The water economy of Fusarium wilted tomato plants, *Phytopathology*, 43, 619, 1953.
26. **Talboys, P. W.**, Dysfunction of the water system, in *Plant Disease An Advanced Treatise III*, Horsfall, J. G., and Cowling, E. B., Eds., Academic Press, Inc., New York, 1978, 141.
27. **Boyer, J. S.**, Resistances to water transport in soybean, bean, and sunflower, *Crop Science*, 11, 403, 1971.
28. **Duniway, J. M.,** Resistance to water movement in tomato plants infected with *Fusarium, Nature*, 230, 252, 1971.
29. **Duniway, J. M.,** Water status and imbalance, in *Physiological Plant Pathology,* Heitefuss, R., and Williams, P. H., Eds., Springer-Verlag, Berlin, 1976, 430.
30. **Farella, G. M., Preece, T. F., and Wren, M. J.**, Effects of infection by *Phytophthora infestans* (Mont.) deBary on the stomata of potato leaves, *Annals of Applied Biology*, 63, 265, 1969.
31. **Priehradny, S.,** Water uptake of barley infection by powdery mildew (*Ervsiphe graminis* D.C.), *Biol. Bratislava*, 26, 507, 1971.
32. **Epstein, E.**, *Mineral Nutrition of Plants: Principles and Perspectives*, John Wiley & Sons, New York, NY, 1972.
33. **Huber, D. M.**, Disturbed mineral nutrition, in *Plant Diseases An Advanced Treatise III*, Horsfall, J. G. and Cowling, E. B., Eds., Academic Press, Inc., New York, 1976, 163.
34. **Steinberg, R. A., Bowling, J. D., and McMurtrey, J. E., Jr.**, Accumulation of free amino acids as a chemical basis for morphological symptoms in tobacco manifesting frenching and mineral deficiency symptoms, *Plant Physiology*, 25, 279, 1950.
35. **Hale, M. G., Frey, C. L., and Shay, F. J.**, Factors affecting root exudation, *Advances in Agronomy,* 23, 89, 1971.
36. **Butler, J. E., and Jones, G. S.**, *Plant Pathology*, MacMillan and Co., Ltd., London, 1949.
37. **Aiyar, S. P.**, The effects of potassium deficiency on rice, *Proceedings of the Indian Academy of Science*, Section B 28, 202, 1948.
38. **Martin, C.**, Etude comparee de l'activite de la polyphenoloxidase chez le tabac sain et inoculé par une maladie á virus, *Comptes. Rendua. Academic des Seances*, 246, 2026, 1958.
39. **Fric, F.,** Translocation of ^{14}C-labeled assimilates in barley plants infected with powdery mildew (*Erysihe graminis* f. sp. *hordei* Marchal), *Phytopathologische Zeitschrift,* 84, 88, 1975.
40. **Mandryk, M.**, Suppression of rooting of tobacco cuttings by yellow dwarf virus, *Journal of the Australian Institute of Agricultural Science*, 20, 50, 1954.
41. **Bergman, E. L. and Boyle, J. S.**, Effect of tobacco mosaic virus on mineral content of tomato leaves, *Phytopathology* 52, 956, 1962.
42. **Matthews, F. E. R.**, *Plant Virology,* 2nd ed., Academic Press, Inc., New York, 1981.
43. **Džamić, M.**, *Osnovi Biohemije II*, Belgrade, 1978.
44. **Jolivet, E., and Moyse, A.**, Les types métaboliques végétaux du point de vue photosynthetique, *Comptes Rendus des seances de l' Academie d'Agriculture de France,* 11, 839, 1982.
45. **Goodman, G., Kiraly, Z., and Zaitlin, M.**, *The Biochemistry and Physiology of Infectious Plant Disease*, D. van Nostrand Co., Inc.,London, 1967.
46. **Heber, U.,** Metabolic exchange between chloroplasts and cytoplasm, *Annual Review of Plant Physiology*, 25, 393, 1974.
47. **Wood, R. K. S.**, *Physiological Plant Pathology,* Blackwell Scientific Publications, Oxford & Edinburgh, 1967.
48. **Allen, P. J.,** Changes in the metabolism of wheat leaves induced by infection with powdery mildew, *American Journal of Botany*, 29, 425, 1942.
49. **Yarkina, A. M.,** *Sots. Zernovove Khozyaistvo Saratov*, 4, 137, 1940
50. **Montalbini, P., and Buchanan, B. B.**, Effect of a rust infection on photophosphorylation of isolated chloroplasts, *Physiological Plant Pathology*, 4, 191, 1974.
51. **Owen, P. C.**, The effect of infection with tobacco mosaic virus on the photosynthesis of tobacco leaves, *Annals of Applied Bioloy* 45, 456, 1957.
52. **Zaitlin, M., and Hesketh, J. B.**, The short term effects of infection by tobacco mosaic virus on apparent photosynthesis of tobacco leaves, *Annals of Applied Biology,* 55, 239, 1965.
53. **Šutić, D.**, *Biljni Virusi*, Nolit, Belgrade, 1980.
54. **Lovisolo, 0., Pennazio, S., and Redolfi, P.**, *Le infezioni da Virus Nelli Piante*, Piccin Editore, Padova, 1978.
55. **Ivanović, M.**, *Poylymyxa betae* Kesk., parazit šecérne repe i vektor virusa nekrotićnog žutila nerave repe, Doktorska Disertacija, Poljoprivredni Fakultet, Zemun, Yugoslavia.
56. **Gáborjányi, R., Balazs, E., Kiraly, Z.**, Ethylene production, tissue senescence and local virus infection, *Acta Phytopatholoical, Academy of Sciences*, Hungary, 6, 51, 1971.
57. **Marco, S., and Levy, D.,** Involvement of ethylene in the development of cucumber mosaic virus-induced chlorotic lesions in cucumber cotyledons, *Physiological Plant Pathology*, 14, 235, 1979.
58. **Wynn, W. K.**, Photosynthetic phosphorylation by chloroplasts isolated from rust-infected oats, *Phytopathology,* 53, 1376, 1963.

59. **Magyarosy, A. C., Shurmann, P., and Buchanan, B. B.**, Effect of powdery mildew infection on photosynthesis by leaves and chloroplasts of sugar beets, *Plant Physiology*, 57, 486, 1976.
60. **Magyarosy, A. C., Schurmann, P., Montalbini, P., and Buchannan, B. B.,** Effect of infection by obligate parasites on photosynthesis. In: *Current Topics in Plant Pathology*. Z. Kiraly (ed.). Akadémiai Kiadó, Budapest, 89, 1977.
61. **Mathre, D. E.**, Photosynthetic activities of cotton plants infected with *verticillium albo-atrum*. Phytopathology 58, 137, 1968.
62. **Goffeau, A., and Bové, J. M.**, Virus infection and photoysnthesis 1. Increased photophosphorylation by chloroplasts from Chinese cabbage infected with the turnip yellow mosaic virus, *Virology* 27, 243, 1965.
63. **Zaitlin, M., and Jagendorf, A. T.**, Photosynthetic phosphorylation and Hill reaction activities of chloroplasts isolated from plants infected with tobacco mosaic virus, *Virology,* 477, 1960.
64. **Spikes, J. D., and Stout, M.**, Photochemical activity of chloroplasts isolated from sugar beet infected with virus yellows, *Science* 122, 375, 1955.
65. **Last, F. T.**, Metabolism of barley leaves inoculated with *Erysiphe graminis* Marchal., *Annals of Botany* 27, 685, 1963.
66. **Duniway, J. M., and Slathyer, 0. R.**, Gas exchange studies on the transpiration and photosynthesis of tomato leaves affected by *Fusarium oxysporum* f. sp. *lycopersici*. Phytopathology 61, 1377, 1971.
67. **Kosuge, T.**, The capture and use of energy by diseased plants, in *Plant Disease An Advanced Treatise III*, Horsfall, J. G., and Cowling, E. B., Eds., Academic Press, Inc., New York, 1978, 85.
68. **Hall, A. E., and Loomis, R. S.**, An explanation for the difference in photosynthetic capabilities of healthy and beet yellows virus-infected sugar beets (*Beta vulgaris* L.), *Plant Physiology* 50, 576, 1972.
69. **Hunter, C. S., and Peat, W. E.**, The effect of tomato aspermy virus on photosynthesis in the young tomato plant, *Physioloical Plant Patholoy*, 3, 517, 1973.
70. **Goodman, N. R., Kiraly, Z., and Wood, R. K.,** The Biochemistry and Physiology of Plant Disease. University of Missouri Press, Columbia, 1986.
71. **Magyarosy, A. C., and Buchanan, B. B.**, Effect of bacterial infiltration of photosynthesis of bean leaves. Phytopathology 65, 777, 1975.
72. **Kawashima, N., and Wildman, S. G.**, Fraction I protein, *Annual Review of Plant Physioloy*, 21, 326, 1970.
73. **Hirai, A., and Uildman, S. G.**, Effect of TMV multiplication on RNA and protein synthesis in tobacco chloroplasts, *Virology* 38, 73, 1969.
74. **Crosbie, E. S., and Matthews, R. E. F.**, Effects of TYMV infection on growth of *Brassica pekinensis*, Rupr., *Physiological Plant Pathology*, 4, 389, 1974.
75. **Daly, M. J.**, The carbon balance of diseased plants: Changes in respiration, photosynthesis and translocation, in *Physioloical Flant Pathology*, Heitefuss, R., and Williams, P. H., Eds., Springer-Verlag, Berlin, 1976. 450.
76. **Kozlowski, T. T., and Keller, T.**, Food relations of woody plants, respiratory study of powdery mildew of wheat. *Botanical Review*, 32, 293, 1966.
77. **Allen, P. J., and Goddard, D. R.**, A respiratory study of powdery mildew of wheat, *American Journal of Botany*, 25, 613, 1938.
78. **Millerd, A., and Scott, K. S.**, Host-parasite relations in powdery mildew of barley, III: Utilization of respiratory energy, *Australian Journal of Bioloical Science*, 16, 775, 1963.
79. **Samborski, D. J., and Shaw, M.**, The physiology of host-parasite relations, II. The effect of *Puccinia graminis* Eriks. & Henn. on the respiration of resistant and susceptible species of wheat, *Canadian Journal of Botany*, 34, 301, 1956.
80. **Heitefuss, R., and Fuchs, W. H.**, Unterschungen zur pathologischen Physiologie von *Peronospora brassicae* (Gäum.) auf *Brassica oleracea, Phytopathologische Zeitschrift*, 37, 348, 1960.
81. **Grimm, R. G., and Wheeler, H.**, Respiratory and enzymatic changes in Victoria blight of oats, *Phytopathology*, 53, 436, 1963.
82. **Collins, R. P., and Scheffer, R. P.**, Respiratory responses and systemic effects in *Fusarium*-infected tomato plants, *Phytopatholoy*, 48, 349, 1958.
83. **Johnson, M. W., and Markham, R.**, Nature of the polyamine in plant viruses, *Virology*, 17, 276, 1962.
84. **Takahashi, T., and Hirai, T.**, Respiratory increase in tobacco leaf epidermis in the early stage of tobacco mosaic virus infection, *Plant Physiology*, 17, 63, 1964.
85. **Takahashi, T., and Hirai, T.**, Oxidative and phosphorylation activities of mitochondria from tobacco leaf epidermis infected with tobacco mosaic virus, *Plant Physiology*, 18, 219, 1965.
86. **Dwurazna, M. M., and Weintraub, M.,** Respiration of tobacco leaves infected wlth different strains of potato vlrus X, *Canadian Journal of Botany*, 47, 723, 1969.
87. **Owen, P.C.**, The effect of infection with tobacco etch virus on the rates of respiration and photosynthesis in tobacco leaves, *Annuals of Applied Biology*, 45, 327, 1957.
88. **Yamaguchi, A., and Hirai, T.**, The effect of local infection with tobacco mosaic virus on respiration in leaves of *Nicotiana glutinosa, Phytopathology*, 447, 1959 .

89. **Owen, P. C.**, Photosynthesis and respiration rates of leaves of *Nicotiana* glutinosa infected with tobacco mosaic virus and of *N. glutinosa* infected with potato virus X, *Annals of Applied Biology* 46, 198, 1958.
90. **Roberts, D. A., and Corbett, R. M.**, Reduced photosynthesis in tobacco plants infected with tobacco ringspot virus, *Phytoathology*, 55, 370, 1965.
91. **Jensen, S. G.**, Photosynthesis, respiration and other physiological relationships in barley infected with barley yellow dwarf virus. *Phytopathology*, 58, 205, 1968.
92. **Tu, J. C., and Ford, R. E.**, Effect of maize dwarf mosaic virus infection on respiration and photosynthesis of corn, *Phytopathology*, 58, 282, 1968.
93. **Nicolas, G.**, De l'influence qu'exercent les fumigines sur l'assimilation chlorophylienne et la respiration, *Rev. Gen. Bot.*, 25, 385, 1913.
94. **Dorohov, L. M.**, Izmenenie fiziologicheskih procesov tomatov pod vliyaniem *Cladosporiumfulvum, SAN SSSR* 21, 85, 1938.
95. **Daly, M. J., Bell, A. A., and Krupka, R. L.**, Respiratory changes during development of rust diseases, *Phytoatholoy* 51, 461, 1961.
96. **Shaw, M., and Samborski, D. J.**, The physiology of host-parasite relations III. The pattern of respiration in rusted and mildew cereal leaves, *Canadian Journal of Botany* 35, 389, 1957.
97. **Bateman, D. F., and Daly, J. M.**, The respiratory pattern of *Rhizoctonia* infected bean hypocotyls in relation to lesion maturation. *Phytopathology*, 57, 127, 1967.
98. **Beckman, H. C.**, Defenses triggered by the invader: Physical defenses, in *Plant Disease An Advanced Treatise V*, Horsfall, J. G., and Cowling, B. E., Eds., Academic Press, Inc., New York, 1980, 225.
99. **Scott, K. J., Craigie, J. S., and Smillie, M. R.**, Pathways of respiration in plant tumors, *Plant Physiology* 39, 323, 1964.
100. **Farkaš, G. L., Lovrekovich, L., and Klement, Z.**, Increased activity of glucose-6-phosphate-dehydrogenase in tobacco leaves affected by *Pseudomonas tabaci, Naturwissenschaften*, 50, 22, 1963.
101. **Baur, J. R., Richardson, B., Halliwell, S. R., and Langston, R**. Effect of tobacco mosaic virus infection on glucose metabolism in *Nicotiana tabacum* var. Samsun, III. Investigation of hexose-monophosphate shunt enzymes and steroid concentration and biosynthesis, *Virology*, 32, 580, 1967.
102. **Bell, A. A.**, Respiratory metabolism of *Phaseolus vulgaris* infected with alfalfa mosaic and southern bean mosaic virus, *Phytoathology*, 54, 914, 1964.
103. **Merrett, M. J. and Sunderland, D. W.**, The metabolism of carbon-14 specifically labelled glucose in leaves showing tobacco mosaic virus induced local necrotic lesions, *Physiological Plant*, 20, 593, 1967.
104. **Taylorson, R. B., and Hendricks, S. G.**, Aspects of dormancy in vascular plants, *BioScience*, 26, 95, 1976.
105. **Kahl, G., Lange, H., and Rosenstock, G.**, Substratspiegel, Enzymaktivaten, und genetische regulation nach depression in pflanzlichen speihergeweben, *Zeitschrift Naturforsch*, 24, 911, 1969.
106. **Kahl, G.**, Metabolism in plant storage tissue slices, *Botanical Review*, 40, 263, 1974.
107. **Rubin, B. A. and Akseenova, V. A.**, Uchastie polifenolosidaznoy sistemi v zashchitnih reakciyah kartofelya protiv *Phytophthora infestans, Biohemiya*, 22, 202, 1957.
108. **Kiraly, Z. and Farkaš, G. L.**, Über die parasitogenen induzierte Atmungssteigerung beim Ueizen, *Naturwiss*, 42, 213, 1955.
109. **Kiraly, Z. and Farkaš, G. L.**, On the role of ascorbic oxidase in the parasitically increased respiration of wheat, *Archives Biochemistry and Biophysics*, 66, 474, 1957.
110. **Frić, F., and Majernik, O.**, A study of terminal oxidases in barley leaves infected with the obligate parasite *Erysiphe graminis* f. sp. *hordei* Marchal., *Phytopathologische Zeitschriefte*, 50, 17, 1964.
111. **Heitefuss, R., Stahmann, M. A., and Walker, J. C.**, Oxidative enzymes in cabbage infected by *Fusarium oxysporum* f. *conglutinans*, *Phytopathology*, 50, 370, 1960.
112. **Sanwal, B. D.**, Investigations on the metabolism of *Fusarium lycopersici* Sacc. with the aid of radioactive carbon, *Phytoathologische Zeitschrift*, 25, 333, 1956.
113. **Herzmann, H.**, Biochemische Untersuchungen über den Stoffwechsel gesunder und brennfleckenkranker Bohner. I. Untersuchungen über Kohlenhydrate, Stickstoffverbindungen, Carbosäuren und einige Oxydationsenzyme. *Phytopathologische Zeitschrift* 33, 349, 1958.
114. **Nielsen, G.**, Weitere Untersuchungen zur pathologischen Physiologie von *Peronospora parasiticia* auf *Brassica oleracea, Phytopathologische Zeitschrift*, 40, 117, 1960.
115. **Maine, E. C.**, *Physiological Responses in the Tobacco Plant to Pathogenesis by Pseudomonas solanacearum, Causal Agent of Bacterial Wilt*. Ph.D. Thesis, North Carolina State College, Durham, 1960.
116. **Harel, E., Mayer, A. M., and Shain, Y.**, Purification and multiplicity of catechol oxidase from apple chloroplasts, *Phytochemistry*, 4, 783, 1965.
117. **Nye, T. G., and Hampton, R. E.**, Biochemical effects of tobacco leaf virus infection on tobacco leaf tissue. II. Polyphenoloxidase activity in subcellular fractions, *Phytochemistry*, 5, 1187, 1966.
118. **Drawert, F., and Gebbing, H.**, Bound phenoloxidase and its release from plant cell structures, *Naturwissenschaften*, 54, 226, 1967.

119. **Ruis, H.,** Particulate and soluble forms of o-diphenoloxidase from potato tumors, *Phytochemistry,* 11, 53, 1972.
120. **Farkaš, G. L. and Kiraly, Z.**, Role of phenolic compounds in the physiology of plant diseases and disease resistance, *Phytopatholoische Zeitschrift,* 44, 105, 1962.
121. **Nagy, R., Riker, A. J., and Peterson, H. W.**, Some physiological studies on crown gall and contiguous tissue, *Journal of Ariculture Research,* 57, 545, 1938.
122. **Tumaru, K., Shiroya, T., and Takanami, Y.**, Changes in polyphenoloxidase activity in cucumber mosaic virus infected tobacco leaves and its nonresponsibility for the changes in virus concentration in vitro, *Annals of Phytopathological Society of Japan,* 35, 41, 1969.
123. **Farkaš, G. L., Kiraly, Z., and Solymosy, F.,** Role of oxidative metabolism in the localization of plant viruses, *Virology* 12, 408, 1960.
124. **Raa, J.,** Cytochemical localization of peroxidase in plant cells, *Physiological Plant,* 28, 132, 1973.
125. **Ivanova, T. M., Davydova, A. M., and Rubin, B. A.**, On mitochrondrial peroxidase and its possible role in oxidative processes, *Biochimiya,* 31, 1167, 1966.
126. **Ivanova, T. M., Davydova, M. A., Rubin, B. A.**, On catalytic functions of peroxidase, *Biochimiya*, 32, 607, 1967.
127. **Parish, R. W.**, The intracellular location of phenoloxidases and peroxidase in stem of spinach beet (*Beta vulgaris* L.), *Zeitschrieft Pflanzenhvsiologie*, 66, 176, 1972.
128. **Darimont, E., and Bacter, R.**, Ribosomal and mitochrondrial peroxidase isoenzymes of the lentil (*Lens culinaris*) root, *Planta*, 110, 205, 1973.
129. **Penon, Pj., Chechin, R., Miassod, J., Pinna, M., Ricard, J., and Teissere, M.,** Peroxidase associated with lentil ribosomes, *Phytochemistry,* 9, 73, 1970.
130. **Stafford, H. A., and Bravinder-Bree, S.,** Peroxidase isozymes of first internodes of sorghum, *Plant Phvsiology*, 49, 950, 1972.
131. **DeJong, D. W.,** An investigation of the role of plant peroxidase in cell wall development by the histochemical method, *Journal of Histochemistry and Cytochemistry,* 15, 335, 1967.
132. **Rudolph, K., and Stahmann, A. M.**, Interaction of peroxidases and catalases between *Phaseolus vulgaris* and *Pseudomonas Phaseolicola* (halo blight of bean), *Nature*, 204, 474, 1964.
133. **Menke, G. H., and Walker, J. C.**, Metabolism of resistant and susceptible cucumber varieties infected with cucumber mosaic virus, *Phyopathology*, 52, 1149, 1963.
134. **Orlob, G. B. and Arny, D. C.,** Some metabolitic changes accompanying infection by barley yellow dwarf virus, *Phvtopathology* 51, 768, 1961.
135. **Tosić, M.**, Virus diseases of wheat in Serbia. II. Some changes in wheat plants infected with wheat streak mosaic virus (USMV) and brome mosaic virus (BMV), *Phytopathologishe Zeitschrift*, **71, 327,** 1971.
136. **Pešić, Z., and Djurišić, V.,** Akivnost nekih oksidativnih fermanata u sejancima kukuruza inficiranim virusom mozaika, *IV Simpozjum Savremeni Problemi Virusnih Infekcija,* Vrnjacum, Banja, 1977.
137. **Pešić, Z., and Jovanović, Z.,** Uticaj kompleksne infekcije virus mozaika kukuruza i virusa mozaika bromusa na aktivnost nekih fermentata u sejancima kukuruza šecérca. *III Jugoslovenski Simpozijum Zastiti Bilja,* Sarajevo, 1978.
138. **Serova, Z., Ges, D., and Rentskaya, L. M.**, Changes in peroxidase and catalase activity in virus-infected tomato and cucumber, *Vestsi Akademii Nauk BSSR, Biol Nauk*, 5, 48, 1972.
139. **Fric, F.**, Oxidative enzymes, in *Physiological Plant Pathology,* Heitefuss, R., and P. H. Williams, Eds., Springer-Verlag, Berlin, 1976, 617.
140. **Kiraly, Z., and Farkas, G. L.,** Decrease in glycolic acid oxidase activity of wheat leaves infected with *Puccinia graminis* var. *tritici*, *Phytopathology* 47, 277, 1967.
141. **Sanwal, B. D., and Waygood, E. R.**, Glycolic acid oxidase and fusariose wilt of tomatoes, *Canadian Journal of Botany*, 41, 55, 1963.
142. **Scott, K. J., and Smillie, R. M.**, Enzyme changes in barley leaves after infection with powdery mildew. Plant Physiology Supplement 37:1 VII.
143. **Lunderstadt, J.,** Die Aktivität einiger Enzyme des Kohlenhydrastoffwechsels in Weizenkeimpflanzen nach Infektion mit *Puccinia graminis tritici, Phytopathologische Zeitschrift*, 50, 197, 1964.
144. **Malca, I., and Zscheile, F. P., Jr.**, Dehydrogenases activity in relation to development of the Helminthosporium leaf spot disease of maize, *Phytoathology*, 54, 1281, 1964.
145. **Scott, K. J.,** Respiratory enzyme activities in the host and pathogen of barley leaves infected with *Erysiphe graminis, Phytoathology* 55, 438, 1965.
146. **Farkaš, G. L., and Solymosy, F.**, Simultaneous activiation of pentose phosphate shunt enzymes in a virus-infected local lesion host plant, *Nature,* 195, 835, 1962.
147. **Solymosy, F., and Farkaš, G. L.**, Metabolic characteristics and the enzymatic level of tobacco tissues exhibiting localized acquired resistance to viral infection, *Virology,* 21, 210, 1963.
148. **Kuć, J.,** Changes in intermediary metabolism caused by disease, in *Plant Disease An Advanced Treatise III,* Horsfall, J. G., and Cowling, B. E., Eds., Academic Press, Inc., New York, 1978, 349.
149. **Allen, P. J.**, Toxins and tissue respiration, *Phytopathology* 43, 221, 1942.

150. **Krupka, L. R.**, Metabolism of oats susceptible to *Helminthosporium victoriae* and victorin, *Phytopathology*, 49, 587, 1959.
151. **Verleur, J. D.**, Studies on the increased respiration of potato tuber tissue after infection with *Gibberella saubinetii* (Mont.) Sacc., *Acta Botanica Neetherlandia*, 9, 119, 1960.
152. **Swamy, R. N.**, Respiration in *Cercospora*-infected groundnut tissue: Part II. Effect of some enzyme inhibitors and 2,4-dinitrophenol on respiration, *Indian Journal of Experimental Biology*, 2, 193, 1964.
153. **Yarwood, C. E.**, Heat of respiration of injured and diseased leaves, *Phytopathology*, 43, 675, 1953.
154. **Shaw, M.**, The physiology of host-parasite relations of the rusts, *Annual Review of Phytopathology*, 1, 259, 1963.
155. **Weintraub, M., Ragetli, J. W. H., and Dwurazna, M. M.**, Studies on the metabolism of leaves with localized virus infections: Mitochondria activity in TMV-infected *Nicotiana glutinosa* L., *Canadian Journal of Botany*, 42, 541, 1964.
156. **Samborski, J. D., Rohringer, R., and Kim, K. W.**, Transcription and translation in diseased plants, in *Plant Disease An Advanced Treatise III*, Horsfall, J. G., and Cowling, B. E., Eds., Academic Press, Inc., New York, 375, 1978.
157. **Uritani, I.**, Protein metabolism, in *Physiological Plant Pathology*, Springer-Verlag, Berlin, 509, 1976.
158. **Van Dijkman, A., and Sijpesteijn, A. K.**, Leakage of preabsorbed ^{32}P from tomato leaf disks infiltrated with high molecular weight products of incompatible races of *Cladosporium fulvum*, *Physiological Plant Pathology*, 3, 57, 1973.
159. **Tomiyama, K., Lee, H. S., and Doke, N.**, Effect of hyphal homogenate of *Phytophthora infestans* on the-potato-tuber protoplasts, *Annals of the Phytopathology Society of Japan*, 40, 70, 1966.
160. **Kojima, M. and Uritani, I.**, Possible involvement of spore agglutinating factor(s) of host plant manifestation of host specificity by various strains of black rot fungus *Ceratocystis fimbriata*, *Plant and Cell Physiology* (Japan), 15, 733, 1974.
161. **Sequeira, L.**, Induction of host physical responses to bacterial infection: A recognition phenomenon, in *Plant Disease Control*, Staples, C. R., and Toenniessen, H. G., Eds., 143, 1981.
162. **Sölheim, B. and Paxton, J.**, Recognition in *Rhizobium*-legume systems, in *Plant Disease Control*, Staples, C. R., and Toenniessen, H. G., Eds., 71, 1981.
163. **Samborski, J. D., Kim, W. K., Rohringer, R., Hohes, N. K., and Baker, R. J.**, Histological studies on host-cell necrosis conditioned by the Sr6 gene for resistance in wheat to stem rust, *Journal of Botany*, 55, 1445, 1977.
164. **Klement, Z., and Goodman, R. N.**, The hypersensitive reaction to infection by bacterial plant pathogens, *Annual Review of Phytopathology*, 5, 17, 1968.
165. **Goodman, R. N., Huang, Yr., and White, A. J.**, Ultrastructural evidence for immobilization of an incompatible bacterium, *Pseudomonas pisi*, in tobacco leaf tissue, *Phytopathology*, 66, 754, 1976.
166. **Sequeira, L., Goard, G., and de Zoeten, A. G.**, Attachment of bacteria to host cell walls: Its relation to mechanisms of induced resistance, *Physiological Plant Pathology*, 10, 43, 1977.
167. **Akseenova, A., Nguen, D. G., Kozhanova, H. O., and Sultanov, J. U.**, Nekatorie osobennosti procesa sinteza belka v inficirovannih tkanjah rasteniy, *Patologisheskaya Fizilolgiya i Imuniteta Rasteniy Izd. Moskovskogo Univerziteta*, Moscow, 79, 1976.
168. **Loebenstein, G.**, Further evidence on systemic resistance induced by localized necrotic virus infections in plants, *Phytopathology*, 53, 306, 1963.
169. **Macovei, A., Šutić, D., and Savulescu, A.**, New development of the study on influence on *Prunus spinosa* on the appearance of sharka symptoms, *Annals of Phytopathology*, 3, 171, 1970.
170. **Ladigina, E., et al.**, Harakteristika trudnorastvorimoy frakcii belka mitohondriy pri virusnom patogeneze v tabaka, *Fiziologica Rast. T.* 19, 4, 1972.
171. **Rubin, B. A., Ladigina, M. E., and Taymla, E. A.**, Antivirusnie svoystva belkovoy frakcii mitohondriy tabaka, *Dokl. VASHNIL*, 3, 1973.
172. **Uritani, I., and Stahmann, A. M.**, Changes in nitrogen metabolism in sweet potato with black rot, *Plant Physiolgy*, 36, 770, 1961.
173. **Heitefuss, R. and Bauer, M.**, Der einfluss von 5-fluorouracil auf den Nucleinsaurestoffwechsel von *Phaseolus vulgaris* nach Infektion mit *Uromyces Phaseoli*, *Phytopathologische Zeitschrift*, 66, 1, 1969.
174. **McCarthy, D., Jarvis, B. C., and Thomas, B. J.**, Changes in the ribosomes extracted from mung bean infected with a strain of tobacco mosaic virus, *Journal of General Virology*, 9, 1, 1970.
175. **Albersheim, P. and Anderson, A. J.**, Proteins from plant cell walls inhibit polyglacturonases secreted by plant pathogens, *Proceedings of the National Academy of Sciences*, U.S.A., 68, 1971.
176. **Stewart, R. G. and Larher, F.**, Accumulation of amino acids and related compounds in relation to environmental stress, in *The Biochemistry of Plants V*, Academic Press, Inc., New York, 1980, 609.
177. **Shaw, M., and Colotelo, N.**, The physiology of host parasite relations: The effect of stem rust on the nitrogen and amino acids in wheat leaves, *Canadian Journal of Botany*, 39, 1351, 1961.
178. **Fuchs, U. H., and Rohringer, R.**, Biochemische Veranderungen in Weizenblatt durch Infektion mit *Puccinia graminis tritici*, *Naturwissenschaften*, 42, 20, 1955.
179. **Bowman, Margaret S. and Rohringer, R.**, Tomato metabolism and betaine formation in healthy and rust-infected wheat. *Can J. Bot.* 48(4), 803, 1970.

180. **Yamamoto, M., and Honda, X.,** Aminokisloti v listyah kartofelya na raznih stadiyah rasta i v klubnyah kartofelya porazhennih *Phytophthora infestans, Annals of the Phytopathological Society of Japan*, 25, 209, 1960 (in Russian).

181. **Lipsic, D. V.,** O vklyucenii metionina S^{35}v belki ustoychivih i vospriimchivih k raku sortov kariofelya, *DAN USSR*, 146, 947, 1962.

182. **Patel, P. N., and Walker, J. C.,** Changes in free amino acid and amide content of resistant and susceptible beans after infection with halo blight organism, *Phytopathology*, 53, 522, 1963.

183. **Neish, A. C., and Hibbert, H.,** Studies on plant tumors III. Nitrogen metabolism of normal and tumor tissues of the beet root, *Archives of Biochemistry*, 3, 159, 1943.

184. **Klein, R. M.,** Nitrogen and phosphorus fractions, respiration and structure of normal and crown gall tissues of tomato, *Plant Physiology*, 27, 335, 1952.

185. **Commoner, B., and Nehari, V.,** The effects of tobacco mosaic virus synthesis on the free amino acid and amide composition of the host, *Journal of General Physiology* 36, 791, 1953.

186. **Heitefuss, R., Buchanan-Davidson,- J. D., Stahmann, A. M., and Walker, J. C.,** Electrophoretic and immunochemical studies of proteins in cabbage infected with *Fusarium oxysporum* f. *conglutinans*, *Phytopathology*, 50, 198, 1959.

187. **Schuch, U.,** Effects of two protein defective strains of TMV on the stability of chloroplast ribosomal RNA, *Virology*, 60, 592, 1974.

188. **Reid, M. S., and Matthews, R. E. F.,** On the origin of the mosaic induced by turnip yellow mosaic virus, *Virology*, 28, 563, 1966.

189. **Farkaš, L. G.,** Senescence and plant disease, in *Plant Disease An Advanced Treatise III*, Horsfall, J. G., and Cowling, B. E., Eds., Academic Press, Inc., New York, 1978, 391.

190. **Hoffmann, M. G., Nienhaus, F., Schönbeck, F., Ueltzien, C. H., and Wilbert, H.,** *Lehrbuch der Phytomedizin*, Paul Parey, Berlin, 1976.

191. **Neish, A. C.,** Biosynthetic pathways of aromatic compounds, *Annual Review of Plant Physiology*, 11, 55, 1960.

192. **Kratzl, K.,** Lignin, its biochemistry and structure, in *Cellular Ultrastructure of Woody Plants*, Syracuse University Press, Syracuse, New York, 1965, 157.

193. **Butcher, N. D.,** Secondary products in tissue culture, in *Plant Cell, Tissue, and Organ Culture*, Reinert, J., and Bajaj, S. P. Y., Eds., Springer-Verlag, Berlin, 1977, 668.

194. **McIntyre, L. J.,** Defenses triggered by previous invaders: Nematodes and insects, in: *Plant Disease An Advanced Treatise V*, Horsfall, J. G. and Cowling, B. E., Eds., Academic Press, Inc., New York, 1980, 333.

195. **Gorz, H. J., Haskins, F. A., and Manglitz, G. R.,** Effect of coumarin and related compounds on blister beetle feeding in sweet clover, *Journal of Economic Entomoloy*, 65, 1632, 1972.

196. **Kiraly, Z., and Farkaš, G. L.,** Relation between phenol metabolism and system rust resistance in wheat, *Phytopathology*, 52, 657, 1962.

197. **Barnes, E. H., and Uilliams, B. E.,** A. Biochemical response of apple tissues to fungus infection, *Phytopathology*, 59, 844, 1960.

198. **Ladigina, E., and Edreeva, M. A.,** Biohimicheskie mehanizmi patogeneza pri Peronosporoze i tochechnom nekroze u tabaka, *Patologicheskava Fizioloiya i Imunitet Rasteniy*, Isdat. Moskovskogo Univerziteta, Moscow, 1976, 60.

199. **Handobina, M. L., Geraskina, V. G., and Valizneva, A. T.,** Izucenie nekatorih zashchitnih reakciy morkovi pri porazhenii yeyo *Phoma rostrupii, Patologischeskava Fiziologiya i Imunitet Rastiniv*, Isdat. Moslovskovo Universziteta, Moscow, 1976, 113.

200. **Sequeira, L., and Kelman, A.,** The accumulation of growth substances in plants infected by *Pseudomonas solanacearum*, *Phytopathology* 52, 439, 1962.

201. **Geissmann, T. A.,** The flavonoid constituents of normal and virus in-fected peach and cherry leaves, *Archives of Biochemistry and Biophysics*, 60, 21, 1956.

202. **Tarr, J. A. S.,** *The Principles of Plant Pathology* MacMillan Press, London, 1972.

203. **Cowling B. E.,** The engineering mechanics of pathogenesis, in *Plant Disease An Advance Treatise III*, Academic Press, Inc., New York, 1978, 299.

204. **Beuchat, R. L.,** *Food and Beverage Mycology*, AVI Publishing Co., Inc., New York, 1979.

205. **Keen, T. N.,** Evaluation of the role of phytoalexins, in *Plant Disease Control*, Staples, C. R., and Toenniessen, H. G., Eds., Wiley-Interscience Publications, Inc., New York, 1981, 155.

206. **McClure, T. T., and Robbins, W. R.,** Resistance of cucumber seedlings to damping-off as related to age, season of year, and level of nitrogen nutrition, *Botanical Gazette*, 103, 684, 1942.

207. **Moreau, M.,** Le perissement des oeilletes, *Encyclopedia of Mycology*, 30, 215, 1957.

208. **Kimmins, U. C.,** Wound induced resistance in plant virus infections, in *Cell Wall Biochemistry Related to Specificity in Host-Plant Pathogen Interactions*, Solheim, B., and Raa, J., Eds., Universitetsforlaget, Oslo, 1977, 271.

209. **Mount, S. M.,** Tissue disintegration, in *Plant Disease An Advanced Treatise III* . Horsfall, J. G., and Cowling, B. E., Eds., Academic Press, Inc., New York, 1978, 279.

210. **Kirk, T.**, Lignin-degrading enzyme system, in *Cellulose as a Chemical and Energy Resource,* Wilke, C. R., Ed., Wiley and Sons, Inc., New York, 1975, 139.
211. **Greathouse, G. A. and Rigler, N. E.,** The chemistry of resistance of plants to Phytomototrichum root rot, IV. Toxicity of phenolic and related compounds, *American Journal of Botany,* 27, 99, 1940.
212. **Mukherjee, N., and Kundu, N.,** Antifungal activities of some phenolics and related to compounds to three fungal plant pathogens, *Phytopathologische Zeitschrift,* 78, 89, 1973.
213. **Walker, J. C., and Stahmann, A. M.,** Chemical nature of disease resistance in plants, *Annual Review of Plant Physiology* 6, 351, 1955.
214. **Nienstaedt, H.,** Tannin as a factor in the resistance of chestnut, *Castanea* spp., to the chestnut blight fungus, *Endothia Parasitica* (Murr.) A. and A., *Phytopathology* 43, 32, 1953.
215. **Patil, S. S., Zucker, M., and Dimond, A. E.,** Biosynthesis of chlorogenic acid in potato roots resistant and susceptible to *Verticillium albo-atrum, Phytopathology,* 56, 971, 1966.
216. **Gayed, S. R., and Rosa, N.,** Levels of chlorogenic acid in tobacco cultivars, healthy and infected with *Thielaviopsis basicola, Phytopathology* 65, 1049, 1975.
217. **Hunter, R. E.,** Inactivation of pectic enzymes in polyphenols in cotton seedlings of different ages infected with *Rhizoctonia solani, Physiological Plant Pathology,* 4, 151, 1974.
218. **Hunter, R. E.,** Effects of catechin in culture and in cotton seedlings on the growth and polygalacturnosae activity of *Rhizoctonia solani, Phytopathology,* 68, 1032, 1978.
219. **Ramasami, R., and Shanmugam, N.,** Possible role of sugars, phenols, and gossypol in cotton seedlings and disease resistance, *Indian Journal of Mycology and Plant Pathology,* 1, 52, 1977.
220. **Sokolova, V. E.,** Prevrashchenie hlorogenovoy kisloti i ustoychivoshy klubney kartofelya k fitoftorozu. Sb. *Biohimiya Plodov i Ovoshchev* 7, 96, 1962.
221. **Uritani, I.,** The role of plant phenolics in disease resistance and immunity. *Symposium on Biochemical Plant Phenolic Substances,* Colorado, 1961, 99.
222. **Sequeira, L.,** Inhibition of indolacetic acid oxidase in tobacco plants infected with *Pseudomonas solanacearum, Phytopathology,* 54, 1078, 1964.
223. **Uritani, I., and Akazawa, T.,** Alternation of the respiratory pattern of plants, in *Plant Pathology An Advanced Treatise I,* Horsfall, J. G. and Dimond, A. E., Eds., Academic Press, Inc., New York, 1959. 349.
224. **Hampton, R. E., Suseno, R., and Brumagen, M. D.,** Accumulation of fluorescent compounds in tobacco associated with tobacco mosaic virus hypersensitivity, *Phytopathology* 54, 1062, 1964.
225. **Akazawa, T., Uritani, I., and Kubata, H.,** Isolate of ipomeamorone and two coumarin derivatives from sweet potato roots injured by the weevil, *Cyces formicarius elegantulus, Archives of Biochemistry and Biophysics,* 88, 150, 1960.
226. **Kiraly, Z.,** Defenses triggered by the invader: Hypersensitivity, in *Plant Disease An Advanced Treatise V,* Horsfall, J.G., and Cowling, B. E., Eds., Academic Press, Inc., New York, 1980, 201.
227. **Deese, D. C., and Stahmann, A. M.,** Formation of pectic enzymes by *Verticillium albo-atrum* on susceptible and resistant tomato stem tissues and on wheat bran, *Phytopathologische Zeitschrift,* 46, 53, 1963.
228. **Bateman, D. F., and Bashman, G. H.,** Degradation of plant cell walls and membranes by microbial enzymes, in *Physiological Plant Pathology,* Heitefuss, R. and Williams, P. H., Eds., Springer-Verlag, Berlin, 1976, 316.
229. **Stahmann, M. A.,** Biochemische Probleme der kranken Pflanzen, *Votrage eines Svmposiums i Institut fur Phytopathologie Aschersleben der Deutschen Akademie der Landwirtscharftswissenschafften zu Berlin,* 1965.
230. **Byrde, R. J. W.,** Natural inhibitors of fungal enzymes and toxins in disease resistance, in *Perspectives in Biochemical Plant Pathology,* Rich, S., Ed., Connecticut Agricultural Experiment Station 663, 1963, 4.
231. **Cole, M.,** Oxidation products of leuco-anthocyanins as inhibitors of fungal polygalacturonase in rotting apple fruits, *Nature,* 181, 1596, 1958.
232. **Williams, A. H.,** Enzyme inhibition by phenolic compunds, in *Enzyme Chemistry of Phenolic Compounds,* Pridham, J. B., Eds., Pergamon Press, New York, 1963, 87.
233. **Mahadevan, A., Kuć, J., and Williams, E. B.,** Biochemistry of resistance in cucumber against *Cladosporium cucumarinum* I. Presence of a pectinase inhibitor in resistant plants, *Phytopathology,* 55, 1000, 1965.
234. **Pridham, J. B.,** *Phenolics in Healthy and Diseased Plants,* Pergamon Press, New York, 1959.
235. **Metlitskii, L. V.,** Phyto-immunity molecular mechanisms, 31st Annual Bakh Symposium, Nauka, SSSR, Moscow, 1976, 1.
236. **Müller, K. 0. and Borger, H.,** Experimentelle Untersuchungen über die *Phytophthora* Resistenz der Kartoffel, *Arb. Biol. Reichsanst. Land-Forstwirtsch.* Berlin-Dahlem. 23 189 1940.
237. **Tarr, J. A. S,** Diseases of Sorghum, Sudan Grass and Broom Corn, *Commonwealth Mycological Institute,* Kew, Surrey, 1962.
238. **Cruickshank, M. A. I.,** Defenses triggered by the invader: Chemical defenses, in *Plant Disease An Advanced Treatise V,* Horsfall, J. G., and Cowling, B. E., Eds., Academic Press, Inc., New York, 1980, 347.
239. **Groos, von, D.,** Phytoalexine und verwandte Pflanzenstoffe, *Fortschr. Chem. Org. Naturst.,* 34, 188, 1977.
240. **Samuel, G.,** On the shot-hole disease caused by *Clasterosporium carpophyllum* and on the shot-hole effect, *Annals of Botany(*London), 41, 375, 1927.

INDEX

N

O

P

Q

R

S

T

U